CHEMICAL LECTURES OF
H. T. SCHEFFER

Chemists and Chemistry

VOLUME 14

A series of books devoted to the examination of the history and development of chemistry from its early emergence as a separate discipline to the present day. The series will describe the personalities, processes, theoretical and technical advances which have shaped our current understanding of chemical science.

The titles published in this series are listed at the end of this volume.

CHEMICAL LECTURES OF H. T. SCHEFFER

Edited by

Torbern Bergman (1775)

and

J. A. Schufle (translator)

New Mexico Highlands University,
Las Vegas, New Mexico, U.S.A.

SPRINGER SCIENCE+BUSINESS MEDIA, B.V.

Library of Congress Cataloging-in-Publication Data

Scheffer, H. T. (Henrik Teofilus), 1710-1759.
[Chemiske föreläsningar, rörande salter, jordarter, vatten, fetmor, metaller och färgning. English]
Chemical lectures of H.T. Scheffer / edited by T.O. Bergman and translated by J.A. Schufle.
p. cm. -- (Chemists and chemistry ; v. 14)
Translation of: Chemiske föreläsningar, rörande salter, jordarter, vatten, fetmor, metaller och färgning.
ISBN 978-94-010-5100-2 ISBN 978-94-011-2494-2 (eBook)
DOI 10.1007/978-94-011-2494-2
1. Chemistry. I. Bergman, Torbern, 1735-1784. II. Title. III. Series.
QD39.S2913 1992
540--dc20 92-8726

ISBN 978-94-010-5100-2

DEDICATION

To my father, Albert Bernard Schufle
(1876-1951),
who gave me a love for languages;
and my mother, Daisy Susanna (Frick)
Schufle (1881-1972),
who gave me a desire for learning.

Translator's Preface

I came upon this collection of lectures in chemistry by Henric Theophil Scheffer delivered in about 1750, edited by Torbern Bergman and published in Swedish in 1775, when I was spending a sabbatical leave in Uppsala, Sweden, in 1977, gathering material for my biography of Bergman (Coronado Press, 1985). I have written a chapter about "Scheffer's Lectures" in my Bergman biography. I refer the reader to this biography for further discussion. Bergman says in his "Foreword" to "Scheffer's Lectures" that he designed this collection for use by "those like me who teach chemistry". So this is one of the first, if not the first, textbooks designed to be used in classes in chemistry. That phrase, "one of the first, if not the first", is so typical of Bergman's work in many fields. Bergman is considered to be the father of quantitative chemical analysis, physical chemistry, modern chemical nomenclature, the concept of equivalent weights, and the electromotive series. And he is the discoverer, or made significant contributions to the discovery, of at least eight of the chemical elements, platinum, molybdenum, tungsten, manganese, cobalt, nickel, barium, and oxygen. As a crystallographer, he is considered to be the forerunner of Hauy in founding this science. He wrote the first great treatise in physical geography, his "Physical Description of the Earth". And finally he contributed greatly to our modern philosophy of science in his emphasis on "Ariadne's Thread", experiment, as the true essence of science. In this he was anticipating the view of Sir Karl Popper that criticism, based on experiment, is the essence of science.

In "Scheffer´s Lectures" Bergman presents the first of his three successively improved Tables of Elective Attractions. The second was published in his "Treatise on Elective Attractions" (1776), and the third in his "Opuscula Physica et Chemica", Volume III (1783). This table is one of the earliest attempts to present all the chemical elements and their properties in a single table, and preceded the modern periodic table by almost a century. It is on the basis of this table that Bergman is considered to be the father of physical chemistry.

And it is in "Scheffer´s Lectures" that Bergman shows that it was Scheffer, and not Lavoisier, who first did the "pelican experiment" to prove that one chemical element could not be changed into another.

J.A. Schufle
Las Vegas, New Mexico
February, 1991

TABLE OF CONTENTS

SECOND SECTION: CONCERNING EARTHS

FOURTH SECTION:
CONCERNING COMBUSTIBLE SUBSTANCES

FIFTH SECTION: CONCERNING METALS

SIXTH SECTION: CONCERNING DYEING

FOREWORD

(By Torbern Bergman)

Many years ago, the head of the Swedish Board of Commerce, and Knight of the Royal Order of Wasa, Mr. P. AHLSTROMER, left me all the notes which he had made during the years 1749-1751 of all the chemical lectures of the late Director SCHEFFER which the Director had given during that period. Since I wanted to review them thoroughly, and since after all this time they are still being used, but are little known abroad, I decided to publish them at this time. This year we have seen many clear evidences of the Commerce Counselor´s [Ahlstromer´s] extraordinary zeal for experimenting in science, in which we find much of Director SCHEFFER´S knowledge that has not appeared elsewhere.

Manuscripts were taken from complete notebooks concerning a great many experiments, together with much reasoning about them, concerning (1) salts, combustible materials, and metals, matters properly treated under the subject of chemistry; (2) the art of analysis; (3) the art of dyeing; all of these were reviewed by Dir. SCHEFFER and in many places were corrected in his own hand.

Careful reading of this material

shows, quite naturally, that it consists largely of well known material, but also in various places this material is accompanied by very ingenious observations, and after more than 20 years, we still find that it contains many new discoveries that have even yet not become widely known.

Mr. SCHEFFER had these notes copied down for his own use and desired them to be put into published form, along with some other transcripts made by his students. I have collected a large number of the latter, which appear annually, and some of them years later; Mr. Scheffer reviewed these some years before his death, but somehow foolishly neglected [to publish] them; some subjects are not found in some manuscripts but are discernible in all of them put together, and were lost in some of them during their abstracting.

To comply with Mr. AHLSTROMER´S praiseworthy desires in every possible way, I have thought it would be best to present here one brief abstract, in proper handbook form, especially designed for those like me who teach chemistry. Such a collection will be useful in that it will both save these materials from being forgotten, and preserve them for other editions in the future. Mr. SCHEFFER has discussed many of the same subjects which I myself usually discuss in public or private lectures, except for mineralogy which he seldom discussed unless it would be useful in chemistry. Besides, I have found it was difficult for my students to record my lectures; yes, it is impossible for students to always get the correct meaning, especially beginning students who have not yet gained a firm understanding. I have therefore followed the manuscripts and Mr. SCHEFFER´S later discoveries whenever the context required rigor in reference to the subject, but for the rest I did not feel bound to use the same words or order of topics [that Scheffer used] unless it was the same as the method which I usually used, and which I consider to be the clearest. For the sake of clarity the topics are taken up in sections, chapters, and paragraphs, and when the latter are lengthy in many instances the experiments are put in the form of problems. I have excluded the author´s medical observations, if he used more than were desirable in that material included in the experiment itself, if they did not properly belong to my subject. Throughout I have added various obesrvations of my own, which are in greater detail, partly to better illustrate certain topics, and partly to fill some gaps with respect to later discoveries, but in many places I have had to reduce them because they exceeded the limits I had set for my text, and I would have ample opportunity in my lectures to explain and augment whatever I feel needs it. In the text, instruments, furnaces, and operations are seldom fully described that are not commonly used, because such things are best learned from my own demonstrations. Chemistry is entirely a practical science, and can therefore never be taught correctly without much practice; yes, we can never properly evaluate certain experiments until after we have tried them ourselves. But although practice is of extreme necessity, it ought still always to lead to reflection and reasoning, while we search for the most accurate result; but if the result is not clear, or not the result of trustworthy experiments, we ought never to rush to conclusions or accept anything ambiguous as settled. Here, therefore, is required a mature and reasonable examination of the facts, so one side or the other is not neglected, which usually is clear enough if it is presented in the right manner, but nevertheless is often lost in a fog of confusion before it can appear in full luster. Established prejudice and preconceived opinion are not part of Chemistry, and are usually a very important obstacle as unfortunately the history of science shows; because the best discoveries are sometimes used for personal interest or gain, although this is often denied, and they are presented with much exaggeration, even pursued just to make a noise, even by those who have a love for science and know that they are thereby exposing their own great failings to posterity; for the glory of truth prevails sometimes, and at other times is repressed for a while, but eventually it comes through and shines upon its adversaries to their great shame.

To make the present work most useful to my readers, I have added two tables in copper-engraving, one of which presents the principal chemical symbols, and the other which presents the attractive power exerted between the materials included therein which are used in Chemistry. It is quite customary to give definitions for other types of symbols, and something similar is needed with the chemical symbols. Formerly science was shrouded in much ignorance. The most obscure phrases and the most absurd symbols were used to represent [chemical] substances, which one wanted to keep from general knowledge, and perhaps more often to hide our ignorance under the appearance of great learning. It would almost be better if we eliminated all scientific signs and characters. But a moderate position here, as elsewhere, is best. Such [chemical] symbols are especially useful when we want to note something down in a hurry, because with the help of symbols we can save both time and space. Besides they serve to represent the attractive power of bodies which are near to each other. I have therefore chosen the most useful symbols and devised some new ones, which are presented in the table. Those which can be read without explanation are undoubtedly the best, such as the symbols for acid of air, cinnabar, antimonium crudum, white gold, etc., but such is not the case for all substances. Under each of the 6 columns [in Table I] stands a scheme, or arrangement, which represents the behavior of attractions [of various chemical substances]. It is known that all natural bodies have an inclination towards union. That which holds the heavenly bodies in their orbits has been found by Mathematicians to be almost directly proportional to their masses and inversely proportional to the square of the distance between them; but between smaller bodies on our planet which are in proximity this is not the rule, but though masses and distances are equal, we still find considerable differences [in attraction] according to the different nature of the substances. That the same power in large bodies shows an entirely different behavior in small [bodies] I have had many occasions to observe; that the shape is not noticably influenced by the motion of the heavenly bodies; every diameter becomes infinitely small with distance, and thus [bodies] take on the quality of point masses, whereas when the distance [between bodies] is small, every shape and position of the particle con- tributes greater effect. Since no spy- glass now can help us to distinguish the size and shape of particle bodies, so it becomes impossible to make any calcula- tions; frequent trials are the only way which can teach us the relative strength [of attraction] which is found to be dif- ferent for every substance; this relation is usually called affinity, or elective attraction. GEOFFROY the elder in 1718 first thought to present these [attrac- tions] in the form of a table in which these strengths were represented by dif- ferent positions in a fixed order. This first attempt could naturally not have been complete, but this suggestion has undoubtedly advanced science to a great degree, even though other [attractions] are yet today undetermined. Meanwhile some improved tables have been produced, but altogether these various known works seem to deny the validity of all such collections of relative strengths [of attraction], and to make such tables worthless. The job is a large one, to treat such a touchy subject in this Foreword; recently tables have been given which show the results obtained by myself and others from known experiments in this field, and which I will publish elsewhere so that everyone will be able to judge them on this basis. I feel I should communicate this so that the exceptions presented hitherto may almost all be explained, depending on the circumstances, so that this discovery should not come to nothing. The principal job of Chemistry consists of analysis and synthesis, separate subjects, but yet united; knowledge of the laws of relation between substances is thus of the greatest importance, and its utility is apparent to our eyes.

The attached Table of Attractions [Table II] is double: one part for wet methods [via humida] and one for dry methods [via sicca], and consequently Table II is of entirely different construction than any produced up to now. At the head of every staff or column is the symbol for a substance [chemical element, as Bergman defined it], and underneath are listed in fixed order all the important elective attractions of that substance, which attractions show greater differences as they become weaker. Consequently, for the most part, a nearer approach or a further precipitation can be prevented by circumstances which usually but not always allow reaction, expecially in dry methods.

I have been careful to introduce in every case the most simple substances which would react readily with the principal substance, although in many instances its proper place was not known, nor had it been investigated hitherto, which then is shown by a vertical dashed line to the right near the edge of the column; a similar vertical dash to the left of the symbol signifies that the substance has almost the same attractive power as that substance preceding it in the column, and doubled over the edge indicates that, on the contrary, the rest of the series below is indefinite. [B. abandoned this dashed line in publishing his later Tables of Attraction].

When two substances are united and a third separates, either of necessity or because it takes hold of itself, I call this <u>simple attraction</u>, <u>attractio electiva simplex</u>. Four examples of this in symbols [B.´s reaction "schemes"] are found in the first Table, 2 for wet methods, which are marked by the symbol for water in the center, and 2 for dry methods, which display the fire symbol [inverted water symbol] in the center. The first [scheme] shows what happens when common salt and vegetable alkali react in water: the constituent elements of common salt are shown on one side set down inside the vertical bracket, while those of lye salt we do not yet know, and it does not decompose here either; the upper and lower brackets which point upward indicate that a new union is formed, and the mineral alkali given up remains in the water [in solution]. When no precipitation is observed we usually conclude that no decomposition has occurred, but if crystallization occurs, we conclude the opposite.

The scheme under the second column shows what happens when vitriolic acid is added to a solution of Kalklefver [liver of chalk; hepar calces; calcium sulfide]: the sulfur separates from calx [calcium], which forms gypsum with the acid; the first substance [sulfur] precipitates completely until all the calx is saturated, and the latter sunstance [gypsum] also precipitates for the most part, if water is not abundant. See §72 in the text.

The scheme under the fourth column gives the usual result when Salmiak [sal ammoniac; ammonium chloride] is distilled with the addition of vitriolic acid, giving the acid of salt first, and afterwards yielding the vitriolate of volatile alkali [ammonium sulfate], or the so-called "secret sal ammoniac".

Arranged under the fifth column is the result when lead glance [lead sulfide] and a certain amount of iron are melted together, lead falling to the bottom and lying above it is a sulfur pyrite (§283).

Double attraction, attractio electiva duplex, or as it is called, composite affinity, is when two substances, each of which can be separated into two constituent elements, exchange their respective elements during mixing.

Under the third column is found an example in the wet method. It is known that pure volatile alkali or caustic [ammonium hydroxide] does not precipitate calx nitrata but causes a mild kind of crystallization, although this latter result is quite small in all cases; the reason is that in the first case volatile alkali cannot react, since it possesses less attraction for saltpetersyra [acid of saltpeter; nitric acid] than calx, [and thus] cannot produce any decomposition; but in this case it is bound to acid of air [carbonic acid] which has a very strong attraction for calx, so that a connecting attraction between volatile alkali and [nitric] acid becomes stronger then the sum of the attractions between the alkali and acid of air, and between calx and the [nitric] acid, and consequently the constituent elements are exchanged; the regenerated calx precipitates and the volatile alkali remains dissolved, as the direction of the brackets indicates. In the calx solution saturated with [nitric] acid the pure volatile alkali clearly causes no precipitation, but if the solution is shaken for a long time, or one passes his breath through it, or otherwise adds acid of air, some precipitation soon begins in proportion to the amount [of acid of air] added, for the reason just mentioned.

The sixth scheme shows what happens when quicksilver vitriolate is distilled with common salt. Experience has taught me that such symbolic representations are very useful to convey to students a clear idea, but these schemes are also useful for making an arrangment what happens in the whole reaction, and takes little space, about half or less of that space which is required otherwise to present the same idea in writing.

In conclusion, the memory of Mr. SCHEFFER'S extraordinary services ought to be acknowledged with a few words. He was evidently one of the greatest chemists that Sweden has had, properly trained in mathematical and physical knowledge, endowed with acute natural intelligence, to discover the best paths for the examination of nature and to use these methods with skill and patience. He availed himself of every possible method, both on a large scale and small, as the circumstances required. To find the correct solution to a problem in hand he did not need to have twenty pounds or a barrel of material to examine, but could experiment carefully with tiny bits of material to make a sure analysis, as Mr. CRONSTEDT tells us. It is true that to measure something with more exactness, more attentiveness and more facility, we work with large amounts, but many experiments are still possible, for if we want to get information very much we sometimes can obtain a great store of information from very small amounts of material. Mr. SCHEFFER tried various experiments with 100 gran of platinum [1 Swedish gran = 0.062 gram] and discovered the principle properties of the white gold; Herr BergsR. von SWAB detected pure antimony with a blowpipe, not to mention other similar discoveries which are still buried in obscurity, and which no one has uncovered or understands how to examine properly in small amounts. When there is plenty of material, a wise chemist usually begins his experiments with small amounts, because he does not know in the beginning how an unknown substance ought to be handled properly; in this way not only are many needless expenses saved, but also considerable time, which can be used for large scale experiments. Mr. SCHEFFER taught himself first to work with quarterspanna [1 spanna = 2 bushels], so that he could perform his experiments in very large tubs; a small experiment in a crucible showed him how the ore ought to be treated in a large smelter furnace, etc. Besides we are easily exposed to error when large amounts are examined altogether. Everything has its fixed limit. Thus when a retort of two tums diameter [1 tum = 1 inch (approx.)] is half filled, then the fire requires more passages in distillation to separate the whole mass, than in one of 8 tums diameter, even if only half filled. Several mixtures, which Mr. POTT could not get to melt in his oven, were brought to flux in a moment with a blowpipe. Thus the circumstances determine the most suitable amount for a given experiment, and before someone cries out against small amounts, which on many occasions are not only useful but convenient, he may discover thereby his inability to carry out a reaction with proper accuracy [on a larger scale].

But something more than scale troubled Director SCHEFFER, for he made several useful discoveries himself, which failed to reward him but rather almost ruined him, and left him more stubborn and obstinate than before.

What the foregoing did accomplish, as it is well proven for once, was to show him a trustworthy way to distinguish what was important knowledge, which to some extent is mentioned in § 386, part 6, and which took him a long time to dis- cover, along with how to make his living. This was that in studying science he must above all seek the truth and its manifestations like an honest cosmopolitan. But he also realized that science would never give him a decent living, even if he worked harder than someone who worked only to make himself rich. But he worked at it just like a craftsman, and thus earned the right to be counted a craftsman, since he met every requirement of a craftsman. But Mr. SCHEFFER wanted only some compensation for his expenses and trouble, in order to improve his very straightened circumstances. And he was not unreasonable in his requests, nor was he demanding, but only after showing the direst need. He that knows great expenses, like those for work and travel which accompany the job itself, does not take these quietly in case of necessity, nor with bad grace; and I strongly maintain that to someone who can make his own living in some other way, mankind ought not deny compensation for searching out important information.

What his contradictory spirit accomplished was to make him careful when he had a difficult job to do, and to note thereby that he was not capable of observing all the facts himself. So consequently, though he believed himself to be competent, he always carefully consulted others; he was moreover of a very lively and passionate nature, so that his zeal was obvious. This was true nevertheless only when he was in a hurry, and he never allowed anything to appear in print in the way of polemical statements or remarks; he had more important things to do than to spread errors of others, although he knew full well that the best corrections sometimes occur that way, producing something better.

His friends knew quite well how this reticence would disappear. Director of Mines von SWAB has several times told me that when he wanted anything explained quickly, and had time for it, he went to Director SCHEFFER, and after deliberation he [Scheffer] would explain it to him without reserve, and with great authority. It was not necessary to awaken all our Disputator´s passion, nor did he stop at ineffectual words, but he would work both night and day in the attempt to investigate questions. When he finally found that his critics were right, he had no hesitation to acknowledge it; he preferred the truth in his heart no matter from whence it came.

Director SCHEFFER´S life story can be read in CRONSTEDT´S memorial discourse, given before the Royal Academy of Science. That his lectures are saved from oblivion, the public can thank the head of the Board of Commerce and Knight, Mr. AHLSTROMER, without whose praiseworthy zeal this would hardly have happened, and who has spared no pains to give me the necessary explanations of many passages. It is desirable that any other papers he left should come to light; they would probably give important information on many subjects.

Upsala 7 November 1774.

T. BERGMAN

H.T. Scheffer

Framledne Direct: och Kongl. Vet. Acad. Ledamots

HERR H. T. SCHEFFERS

CHEMISKE FÖRELÄSNINGAR,

Rörande SALTER, JORDARTER, VATTEN,
FETMOR, METALLER och FÄRGNING,
Samlade, i ordning ſtälde och
med Anmärkningar utgiſne.

UPSALA,
Tryckte på Bokhandlaren M. Swederi
bekoſtnad hos Joh. Edmañ, Kongl. Acad.
Boktr. 1775.

1. Title Page, "Herr H.T. Scheffers Chemiske Föreläsningar."

Herr H.T. Scheffers

CHEMISKE FÖRELÄSNINGAR

Rörande SALTER, JORDARTER, VATTEN, FETMOR,
METALLER,
och FÄRGNING, Samlade, i ordnung stälde
och med
Anmärkningar utgifne.

UPSALA
Tryckte på Bokhandlaren M. Swederi
bekostnad hos
Joh. Edman, Kong. Acad. Boktr. 1775.

SCHEFFER'S CHEMICAL LECTURES

Concerning Salts, Earths, Water,
Combustibles, Metals, and Dyes; edited
with addition of copious Notes by
Torbern Bergman, and published in
Uppsala, Sweden, in 1775.

Translated into English (1977-1988) by
Joseph A. Schufle,
professor of chemistry, New Mexico
Highlands University,
Las Vegas, New Mexico.

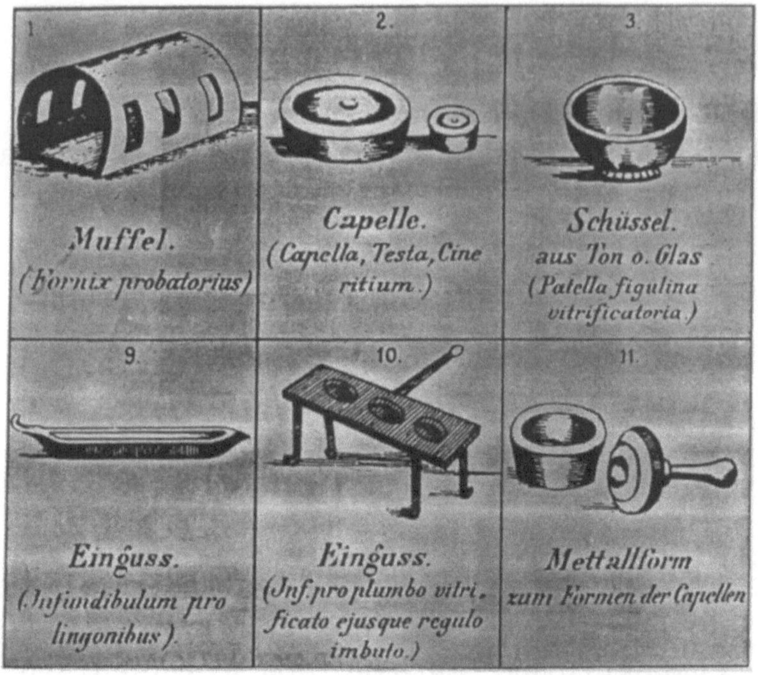

2. Metallurgical apparatus from the time of H.T.Scheffer and Torbern Bergman.

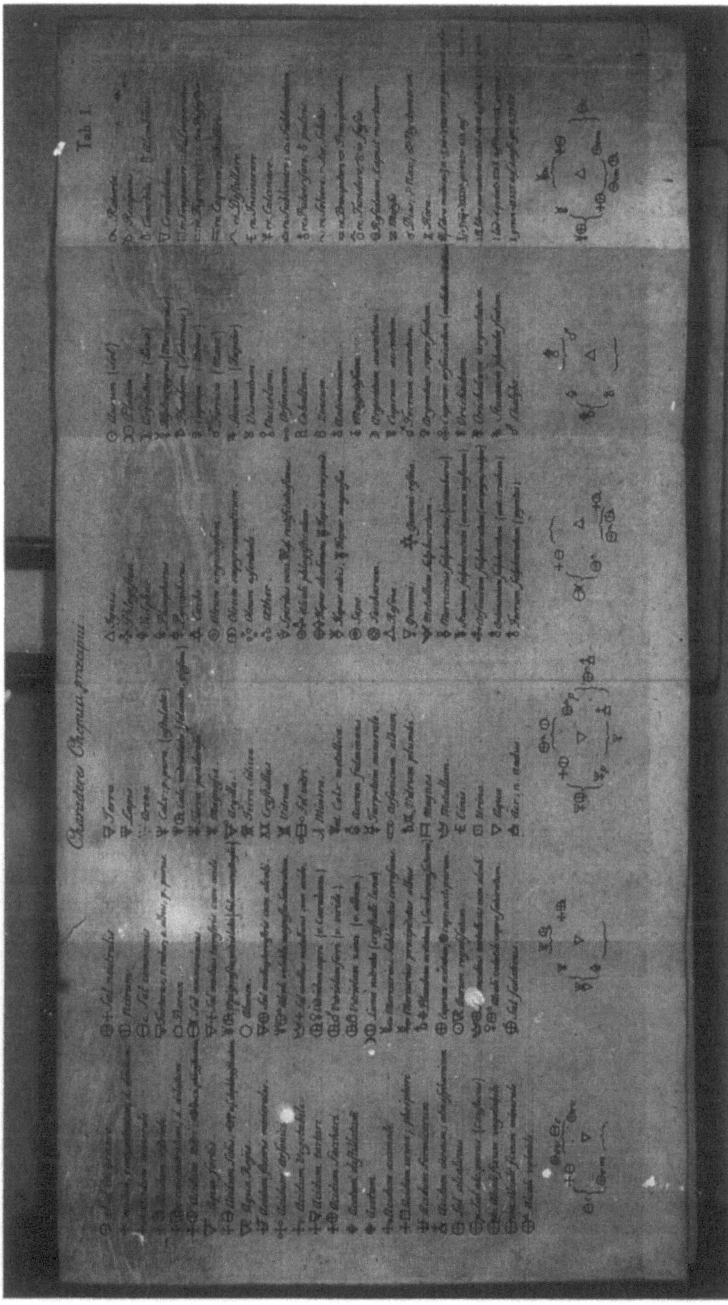

3. Characteres Chemia praecipui (Characters for the Chemical elements) (Tab. I).

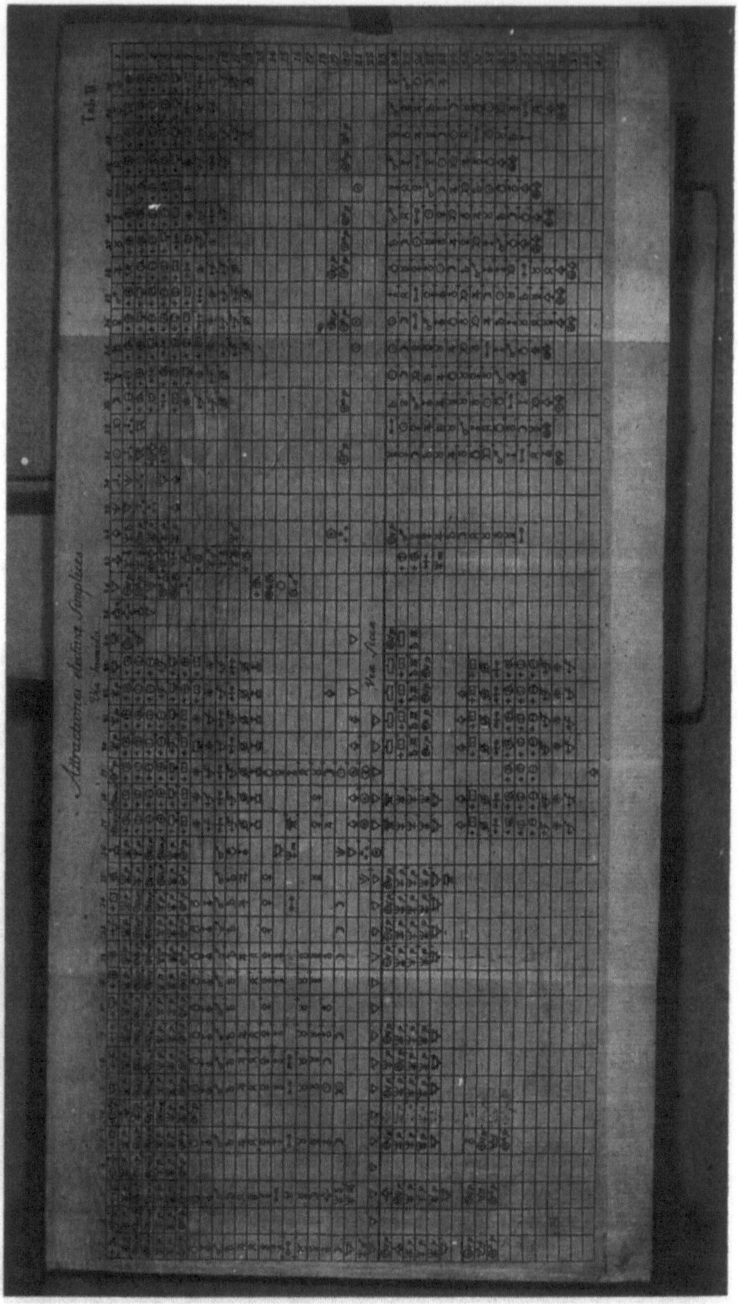

4. Attractiones electivae Simplices. 45 elements (Tab. II).

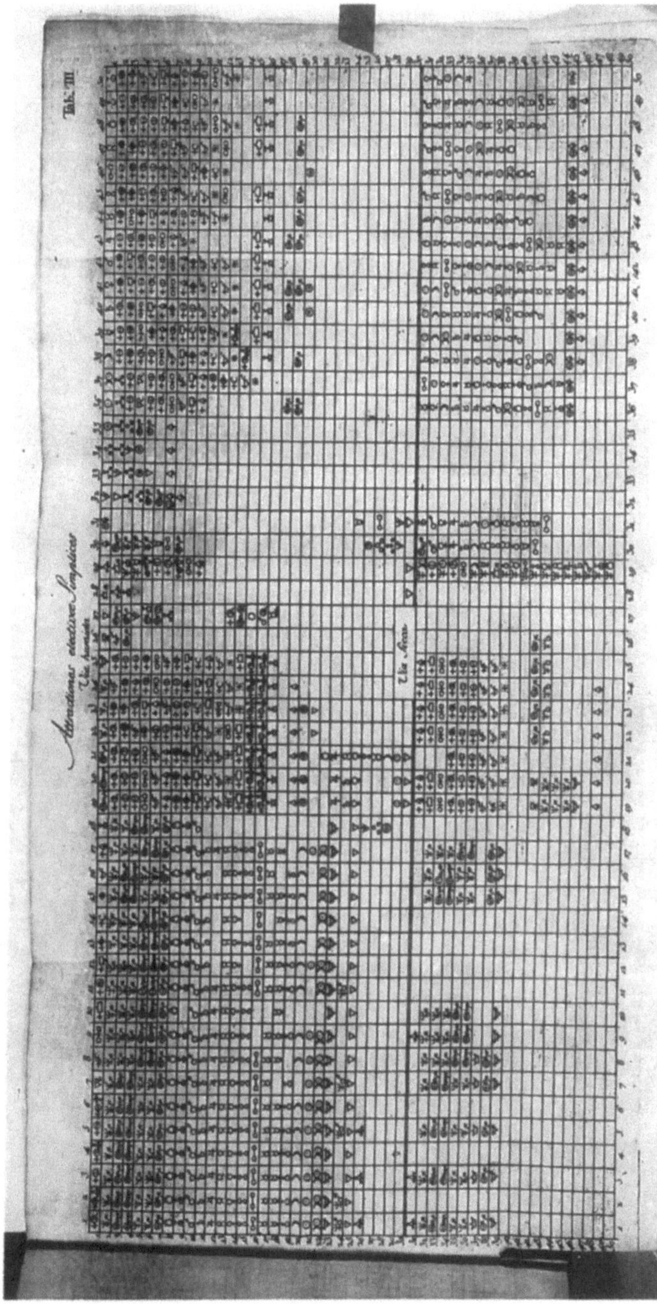

5. Attractiones electivae Simplices. 50 elements (Tab. III).

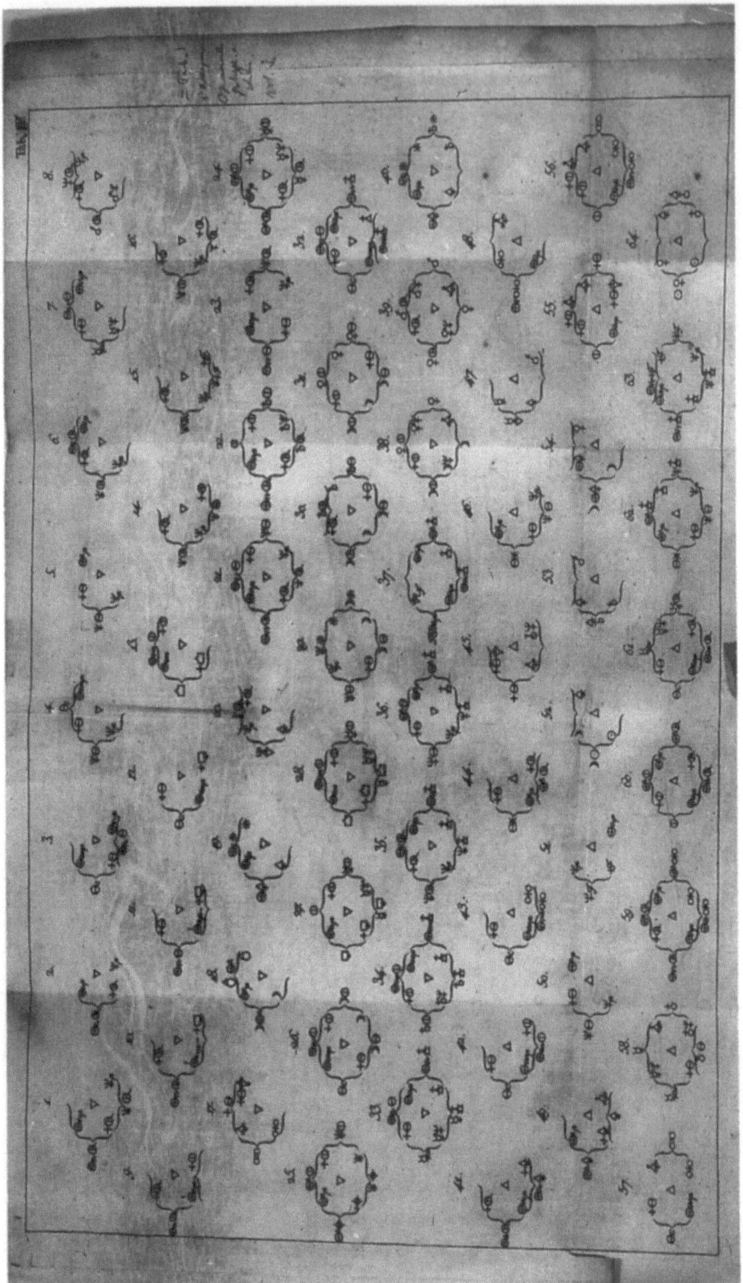

6. Bergman's Chemical "Schemes". (64 Schemes) (Tab. IV).

The Pelican Experiment of H. T. Scheffer

J. A. Schufle
New Mexico Highlands University, Las Vegas, NM 87701

The classic paper of Antoine Lavoisier,[1] "Sur la natur de l'eau", published first in 1770, was a strong refutation of the previous reports that had appeared to prove the possibility of changing water into earth. Johann Baptista van Helmont,[2] a Belgian doctor, had performed an experiment which was not published until 1648, after his death, in which he proved to himself that a tree grown in an earthen vessel over a period of four years gained "164 pounds of wood, bark and roots. . .out of nothing but water". Since wood was an "earthy" material, here was proof that water could be converted into "earth". Other well-known investigators, Robert Boyle among them, were puzzled by this and similar arguments. So, according to Case #5, "Plants and the Atmosphere", in "Harvard Case Histories in Experimental Science",[3] edited by James B. Conant, Lavoisier set out to prove once and for all that transmutation of water into earth was impossible.

Lavoisier described in great detail how he had boiled water in a closed vessel, causing it to be distilled over a period of 101 days continuously. Conant says Lavoisier

> hit upon the idea of using a pelican [Fig. 1], a special kind of distilling vessel used by the alchemists. When the lower bulb, containing the liquid to be distilled, is warmed gently, the vapors pass into the cooler upper chamber. Here they are recondensed, and the liquid so formed drains back into the lower bulb. Thus it is possible to obtain, in an hermetically sealed vessel, a semiautomatic and almost indefinitely repeated distillation of a liquid.

Lavoisier weighed the pelican before and after the experiment, and the pelican, water and earth found in the pelican, after the experiment, and found it had changed only 0.25 grains in weight. The earth was removed from the water and found to weigh 20.40 grains, whereas the pelican had lost weight equal to 17.38 grains. Lavoisier's evidence was therefore not exactly airtight, but it seemed to prove that the earth obtained was gained from the dissolution of the glass of the pelican. This is very interesting, but Lavoisier was not the first to "hit upon the idea" of using a pelican in this experiment. In "Scheffer's Lectures",[4] a book of lectures by Henrik Theophil Scheffer (1710–1759) (see Fig. 2 for title page), delivered in 1749–1751, and edited by Torbern Bergman (1735–1784) with the addition of copious notes by Bergman and published in 1775, there is described the identical experiment described by Lavoisier, carried out by Scheffer before 1750, at least 20 years earlier than Lavoisier's experiment.

Conant says, "The first part of Lavoisier's paper presents a lengthy historical review of both lines of evidence", for and against transmutation of water into earth. Later on Conant says, "But Lavoisier nowhere acknowledges that his experiments are precisely those whose importance had been suggested a century earlier by Boyle." Apparently both Lavoisier and Conant also overlooked, or did not know about, the work of Scheffer.

In his Foreword to "Scheffer's Lectures", Bergman says:

> Many years ago the head of the Swedish Board of Commerce, and Knight of the Royal Order of Wasa, Mr. P. Ahlstromer, left

Figure 1. A pelican.

Framledne Direct och Kongl. Vet. Acad. Ledamots
HERR H. T. SCHEFFERS
CHEMISKE
FÖRELÄSNINGAR,
Rörande SALTER, JORDARTER, VATTEN,
FETMOR, METALLER och FÄRGNING,
Samlade, i ordning ställde och
med Anmärkningar utgifne.

UPSALA,
Tryckte på Bokhandlaren M. Swederi
bekostnad hos Joh. Edman, Kongl. Acad.
Boktr. 1775.

Figure 2. Title page for "Scheffer's Lectures".

[1] Lavoisier, A. L. "Sur le natur de l'eau"; 1770. Republished in: Lavoisier, A. L. "Oeuvres"; Paris, 1862–1893; Vol 2, pp 1–28.
[2] Helmont, J. B. van. "Ortus Medicinae"; Leyden, 1648, p 109. English translation: Chandler, J. "Oriatrike"; London, 1662, p 109.
[3] Conant, J. B.; Nash, L. K. "Harvard Case Histories in Experimental Science"; Harvard University Press: Cambridge, MA, 1966, 2 Vols.; Case #5 is in Vol 2.
[4] Scheffer, H. T. "Chemiske Forelasningar"; Bergman, Torbern, Ed.; Swederi: Upsala, 1775. (See Fig. 2 for title page).

Reprinted from Journal of Chemical Education, Vol. 63, page 928, November 1986

7. The Pelican Experiment of H.T. Scheffer. Reprint from the Journal of Chemical education, Vol. 63, p. 928, Nov. 1986.

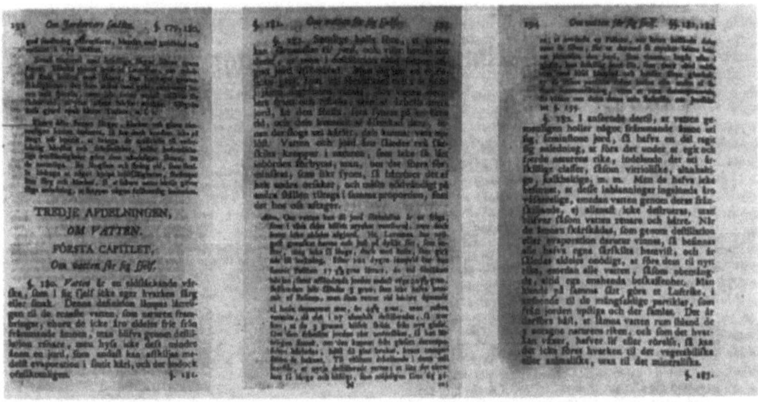

TREDJE AFDELNINGEN,
OM VATTEN.
FÖRSTA CAPITLET,
Om vatten för sig sjelf.

Figure 3. First three pages of Tredje Afdelningen (Third Section), "Scheffer's Lectures", containing reference to "Pelican-glas", p 193, 4th and 5th lines.

> me all the notes which he had made during the years 1749–1751, with the late Director Scheffer of all the chemical lectures which the Director gave [at the Mining College in Stockholm] during that time. Since I wanted to review them thoroughly, and since after all this time they are still being used, but are little known abroad, I decided to publish them at this time.

Bergman then went on to explain that he had used the notes of Scheffer's lectures as the basis for this book, designed "for those like me who teach chemistry", but did not hesitate to add notes of his own to bring them up to date. Bergman's notes eventually exceeded in length the original lecture notes from Scheffer, so that it became in effect Bergman's textbook.

It was in the Third Section, "Concerning Water", First Chapter, "Concerning Water by Itself", that Scheffer described his pelican experiment. The original text is reproduced in the original Swedish in Figure 3. Beginning with the second paragraph, Paragraph No. 181, we read in translation:

> Par. 181. Some people formerly maintained that water can change into earth, and wanted to demonstrate that in distillation they always found some earth left behind. But I have, in a glass Pelican, which was well sealed and was held for two years in continual digestive heat, allowed the water to circulate back and forth without obtaining any more earth than that which showed itself in a short time, and which came partly from dust [dam] contained in cracks in the vessel, and partly could be dissolved [from the Pelican]. The water and the earth are therefore two different bodies in nature, which are not readily interchanged, unless the former [water] is diminished, as it appears; so it must come from an entirely different source, and must require that the latter [earth] increase in the same proportion as it decreases somewhere else.

The note, by Bergman, which follows, reads:

> Whether water can be changed into earth is a question which has been much discussed in our time, but has still not been completely settled. Mr. Lavoisier has very recently examined this [problem] and in the very same manner as our author [Scheffer], but not for as long, but still with heat that was close to boiling. After 101 days [over] a lamp flame, he has found the Pelican 17-$^{4}/_{10}$ grains lighter than at the beginning of the experiment, and at the same time he separated earth of the weight of 20-$^{4}/_{10}$ grains, a difference of 3 grains which did not come from the Pelican; but since the water, on opening up the vessel, had not deposited more than 4-$^{9}/_{10}$ grains, other than the deposit gained when it was distilled in a new alembic, so he believes that the 3 grains were taken from the new glass. If the earth so obtained is investigated properly, it can be determined readily if it can be derived from decomposition of the glass, especially if the composition of the glass used is known. We need to know in this instance that distilled water was used; how long and how strongly it circulated to make reaction possible; and in respect to the change in the Pelican, whether the composition of its parts is known, in order to better estimate the amount of earth which is obtained. If ash makes up part of the glass, can we obtain a distinct earth?; and this must always be mixed with flint and perhaps even glass dust, since when more soluble particles dissolve, often others of firmer composition are dissolved without decomposition. See more on this subject in my book Physical Description of the Earth, Par. 155.

So Bergman was aware of Lavoisier's work with a Pelican, even if Lavoisier was not aware, apparently, of Scheffer's similar work. Scheffer did not give his results on weighing of the Pelican, water and "earth" he used, but only stated that the weight of earth obtained after two years digestion of water in the Pelican did not exceed that obtained "in a short time". Bergman points out that Scheffer's test was carried out for a much longer period than Lavoisier's.

FIRST SECTION: CONCERNING SALTS.

FIRST CHAPTER: Salts In General.

§ 1. A salt is a solid substance which dissolves in water and has a noticable taste. [Material from Scheffer's Lectures proper is set in large (10 characters per inch) type, such as that in this first paragraph.] [Notes by Schufle are enclosed in square brackets.]

Note 1. [Material from Bergman's Notes is indented 5 spaces and set in small (13 characters per inch) type, such as shown here.] To better understand this, and that which follows, it should be noted that any liquid substance is said to dissolve another substance, or to loosen or melt it, when it brings it to a state of complete clarity, thus laying hold of it so that it seems to disappear by degrees and finally to become entirely invisible. The former substance is called the solvent, the menstruum, or the solution-medium; the latter substance is called the solute; and the process itself is called solution or dissolving. When something produces a deposit after the solution which dissolves the solute is formed, and something is precipitated, deposited or thrown down, the former is called the precipitant, to distinguish it from the precipitate or deposit, and the process itself is called precipitation or deposition. The whole process is based principally upon attraction and its varying degrees.

Note 2. In general, salts are defined in a related way, unless they are not solids, in which case a definition seems superfluous; although where particles of pure acids are already solid themselves, then they still have a strong attraction for water, which they do not have (at least not all of them) for the other, the more liquid substance. It is almost the same with entirely pure alkalies. The rest are soluble in water and able to excite the taste of the tongue to an undeterminate amount, in which case no definition can be other than tentative and uncertain. [Here B. is wrestling with the difficulty raised by his inclusion of acids and bases with the salts.] Often a substance is really soluble, but not in the way which is usually employed. Breaking up [the substance] and heating it especially facilitates the formation of a solution. A substance which cannot be noticably attacked in a larger piece may be brought into solution if pulverized, and when mechanical division is insufficient the desired solution may be obtained by chemical reaction. All this is based upon the fact that the greater the surface [of the substance] is the more the solvent can attack it, because of the work connected with attraction which is increased as much as the surface is increased. In general, a substance is considered to be insoluble if it needs much more water than that required to encompass it about one time. Heating is another means by which solution is brought about, partly by expansion of the substance and [consequent] enlarging of the pores, so that water penetrates better and reaches more surface, and partly by producing greater fluidity in the solvent, the performance of which is increased by different degrees of heat. Consequently hot water dissolves better than cold; e.g., 100 degrees is the highest temperature that we expect water to reach, and generally is used for insoluble substances; and of those which are not dissolved by this, still others can be further affected by heat, if the vessel is completely sealed. Occasionally, however, it is clear that a substance is only slightly soluble, but is dissolved by the usual processes; I have also in V. Ac. Handl., 1773, and Jordklotets Physiska [Physics of the Terrestrial Globe], given an account of how chalk, along with magnesia, flint, etc., are actually salts; but these nevertheless can be grouped under the name Earths, if only to show the limits marked out for solubility. Of all the substances which are classified as true salts, there are none which require over 200 times their weight of water [for solution]; one can therefore classify all difficultly soluble substances as Earths. The name itself means nothing but is used only to avoid ambiguity, and although the progression from more to less soluble gives a consistent series of all substances, by their very nature these substances are properly classified among the salts, and artificial separation is permitted and is even often useful as an aid to understanding.

§ 2. Salts can be divided into two principal classes, of which the first includes within it those which consist of mere salts, while in the latter are placed those which, in addition, contain any other intermixture.

Note. This is an entirely natural classification. The former class could be called proper salts [Sales proprii], and the latter analogous [salts]. It should still be noted here that with the latter are included substances which are truly soluble and are not merely mechanical mixtures, which latter class is quite commonly denoted by the name suspension.

§ 3. To the first class belong Sales alkalini and [Sales] neutrales or medii. Because pure acids are not found in the solid state, they cannot be placed here (§§ 1, 2).

Note. It is shown immediately from the above that solidity is not an essential property of salts, and thus acids should be classified here first of all since they are seen to be more saline than the others, that is they have solubility and taste to a higher degree. As to the others, the writer [Scheffer?] uses Neutral-[salts] and Medium-salts [Medelsalter] as synonyms, as they are commonly used, but these terms can be used to denote two different combinations; the first [combination] to distinguish salts of acids combined with alkalis, which are well suited to reveal the taste or reactions of either of the ingredients and therefore shall receive the name Neutral-[salts]; whereas if earths are dissolved to saturation, there is usually found still clear evidence of solvent or base, and such composition is therefore not improperly called Medium-salts [Medelsalter], because they have a sort of medium quality, but can still be brought to full neutrality.

§ 4. In the second class are placed acid-salts, Medium-salts [Medelsalter], and alkaline-salts.

Note. The following classification appears to be natural. Earths and metals are by themselves usually not soluble in water, but through combination with natural salts they are brought completely to a soluble state. As they are now generally of three kinds (§ 3, Note), so there arise three separate combinations, (a) one of which can be called Acidic Medium-salts [Syrlige Medelsalter], which contains earths or metals dissolved to a certain degree in acids; (b) Alkaline [Medium-salts] [Alkalinischke Medelsalter], which have alkalis for the solvent; and (c) Neutral Medium-salts [Neutrale-Medelsalter], which have somewhat neutral salts intimately bound to an earth or metal. Examples of each kind will be given in the following [tables].

§ 5. Under acids salts are classified: (1) those which consists of acids with earths, either metallic, as happens with vitriolic [acid], or earthy, such as are produced by alum; (2) those which in addition to acids contain phlogiston, such as sal volatilis succini, and the so-called Essential salts, which are compounds of acids, earths, and phlogiston.

<u>Note.</u> The author [Scheffer] places here all compounds of saturated acids, such as [combinations of] metals to vitriolic acid and earths similar to alum, although this name is properly used most reasonably when the acid is vitriolic.

The Essential salts, which we know somewhat better such as tartarus and Sal acetocellae, consist of vegetable fixed alkali [potassium] with more acid than is required for saturation, and contains but few earths that are completely pure.

§ 6. To medium salts of both classes belong those which consist of acids saturated with alkaline earths.

§ 7. Alkaline [salts] of the same class include those which contain earths dissolved in alkalis.

<u>Note 1.</u> Alkalis attack not only earths but even several metals, and permit us thereby to distinguish between crystals.

<u>Note 2.</u> Here at the end of the salts may be placed most conveniently those classes which I

consider to be most suitable from their nature. [Hereafter follows B.´s tables of Proper Salts and Analogous Salts.]

Table of Proper and Analogous Salts

Proper Salts

- Acids (Acid-Salts
  - Mineral Acids
    - Vitriolic acid
    - Acid of Saltpeter
    - Acid of Salt
    - Acid of Fluorspar
    - Acid of Arsenic
  - Vegetaable Acids
    - Acid of Winestone
    - Acidum Aceto-cellae
    - Acid of Vinegar
    - etc.
  - Animal Acids
    - Acid of Urine
    - Acid of Ants
  - Common in all Kingdoms
    - Acid of Air
- Alkalis (Alk. Salts)
  - Non-vVolatile Alkali (Fixed)
    - Vegetable Alkali
    - Mineral Alkali
  - Volaatile Alk.
- Neutral Salts
  - Mineral Acids with
    - Vegetable Alkali
    - Mineral Alkali
    - Volatile Alkali
  - Vegetable Acids with
    - Vegetable Alkali
    - Mineral Alkali
    - Volatile Alkali
  - Animal Acids with
    - Vegetable Alkali
    - Mineral Alkali
    - Volaatile Alkali

Analaogous Salts

- Acidi Salts
  - Mineral Acids with
    - Earths
    - Metals
  - Vegetable Acids with
    - Earths
    - Metals
  - Animal Acids with
    - Earths
    - Metals

Acidic Salts (Cont.)	Acid of Air with	Earths
		Metals
	Aqua Regia or others of the compound acid type with	Earths
		Metals
Alkaline Salts	Veg. Alk. with	Earths
		Metals
	Min. Alk. with	Earths
		Metals
	Vol. Alk. with	Earths
		Metals
Neutral Salts with	Earths	
	Metals	
	Middle Salts	

Note 3. The examination or study of salts is undoubtedly among the most important parts of Chemistry. When all of the compounds in the foregoing Table have been described, certainly many of Nature's secrets will have been uncovered, and they who consider such work useless, or devised merely to confuse, only show their ignorance of the methods which are continually employed in Nature's workshop. The Neutral-salt and Middle-salt classifications can be better classified as shown later in its place. (See Notes in §§ 51, 71, and 80.)

SECOND CHAPTER

About Acids

§ 8. The ordinary properties of acids are the following:

I. To effervesce with alkaline salts.

Note 1. This is true with both alkaline salts and salts of earths, but only when they are more or less saturated with acid of air; for those entirely pure, or those which are otherwise called caustic, they show not the least reaction.

II. To make syrup of violets and juice of Heliotrope red.

Note 2. These have about the bluest color in the vegetable kingdom, yet not the bluest of all, since Indigo can dissolve in acid of vitriol without change, and afterwards this solution is used to give wool, silk, and linen a beautiful but rather changeable blue color.

Blue vegetable juices have different sensitivities to acids. Thus juice-paper reddens in nitric acid but is not affected by vinegar; litmus tincture reddens in acid of air but not syrup of violets, etc. When therefore all blue juices are tested for their sensitivity, a convenient progression is obtained to measure the strength of acids in regular progression, one after the other.

III. To precipitate those alkaline salts which are soluble.

Note 3. Sometimes this will take place with neutral salts, [a reaction] which is caused by addition of acids which do not dissolve that substance which they have precipitated. For the rest the product forms more quickly, the more completely saturated the alkali. For many acids can prevent precipitation by redissolving the precipitate.

Note 4. To these related properties we should also add the property that acids, when properly diluted, arouse a feeling in the tongue which we call taste, which cannot be clearly described but is very easily recognized, when we experience it.

Concerning Acid of Vitriol

§ 9. The name of this acid distinguishes it from those substances which differ from it, such as acid of sulfur and acid of alum, which KUNCKEL first found to be similar to it and which were made from vitriol. Since they have different strengths they should have different names. Oil of vitriol denotes the strongest acid known up to now, and certainly its oily consistency gives reason to so name it; but on the other hand, it quenches fire and does not have other properties which oils have. If the acid is diluted so that it is no longer viscous, it is called spirit of vitriol [spiritus vitrioli]; whereas, if it smokes, it is called fuming [fumans].

Note. Rationally the name should be known for its usefulness or it should be given up, and by degrees exchanged for another name which better suits the substance; so I use the name concentrated acid of vitriol for oil of vitriol, and dilute acid of vitriol for spirit of vitriol, names which immediately make good sense.

§ 10. This acid exhibits the following properties:

(a) It precipitates Sugar of Saturn [lead acetate].

Note. This precipitate consists of finely divided grains or crystals which consist of lead-vitriolate [lead sulfate], and which precipitates because it is so little soluble.

(b) It removes the blackness from ink and makes it [colorless] like water, a property which also belongs to acid of saltpeter [nitric acid] and acid of salt [HCl], lemon juice, and salt of aceto-sella.

(c) We do not obtain sweetness with minium, litharge, or white lead, which should still be more closely examined.

(d) By union with phlogiston, fuming spirit of vitriol is produced.

(e) It is not made volatile except by the most intense heat, and is the most stable of all acids.

Note. Water-free acid of urine is also completely stable to heat.

(f) It is the strongest of all acids.

Note. This statement has several exceptions. Vinstens-syra [tartaric acid] decompses Tartarus vitriolatus [K_2SO_4], which also occurs to some extent with acid of saltpeter. See § 52, Note.

(g) Also, to obtain its densest form, it must be heated strongly to boiling, and it holds this heat long afterwards.

Note. The specific gravity of the strongest acid of vitriol which can be obtained is 2.125; that which reads 1.8475 is considered quite excellent. Mr. SAGE puts the usual specific gravity of acid of urine at 2.6875.

(h) It has strong attraction for water, so that vigorous spitting and heat are produced when they are mixed. If this is carried out in a glass vessel we must be careful or it will break.

Note. In open air it can draw moisture to itself until it doubles in weight.

(i) Oil of vitriol dissolves silver, quicksilver, tin, bismuth, cobalt, and regulus of antimony; and spirit of vitriol dissolves copper, iron, nickel, and zinc.

(k) It devours hide, and both wool and linen.

It can be used as a caustic to eat away dead flesh from wounds. Internally it might be suitable for cooling and refreshing in a manner similar to lemon juice, but it damages by its strong astringent power.

Concerning Distilled Acid of Vitriol

§ 11. Distillation is a process in which anything is carried over in the form of a vapor, and at the same time is condensed back to liquid form.

For the present purposes an easier preparation is used: (a) Iron vitriolate is calcinated (burned to a powder) in a pan or pot over a low fire, and with diligent stirring we should see a smoke rising from it and we should observe melting until all reaction ceases, at which time the vitriolate is removed from the fire and is said to be calcinated, although it appears to have only lost water, but since the color is changed ever so little to orange or red, an undetermined amount of acid may also be lost. In order to prevent the last mentioned loss the vessel must not be burned, and since this last is prevented with pots which have thicker bottoms, these are used in preference to pans. This is the ordinary calcinated [vitriol], but if finely powdered vitriol is spread on paper and slowly brought to whiteness in the sun or on warm sand, it is called philosophically [mildly?] calcinated vitriol [calcinatio vitrioli philosophica]. Usually one pound [of vitriol] loses 14 or 15 lod [1 lod = 1/2 oz.] of water through calcination, but the more redness it loses the more acid is driven off besides.

(b) The calcinated vitriol is broken into small pieces so that it can be filled in through the neck of a stone retort, which is filled until half or two-thirds full, fitted with or cemented tightly with a receiver, and is placed in a blast furnace and heated very carefully. Here, when all the acid is decomposed, the egg white [material] is tapped and sifted like unslaked lime, until it becomes like thick gruel, which is spread out to the space of two hands' breadth on rough paper or linen and there allowed to harden. Mixed with some calx like coal ash or white calcinated vitriol we get the so-called lutum sapientiae. One part French clay and 2 parts milk-white tile [powder], worked into a dough with water, is also used with similar results.

(c) If the retort remains hot, the receiver becomes filled with white smoke and some phlegm goes over; but if the fire gradually begins to increase to a higher temperature some additional acid passes over, which should be diluted quite a bit with water and is called spiritus vitrioli. Last comes oil of vitriol if the receiver remains hot and is not once allowed to cool, even though the passing over ceases. If the vessel is opened after cooling, we find in the receiver a thick dark-brown liquid, consisting of water, acid and phlogiston. The residue in the retort is red and is called residue of vitriol [colcothar vitrioli]; if this is extracted with water and afterwards the water is then evaporated, a fine salt is produced which is used as an emetic. The earth remaining produced by fire contains 45, 50 and even over 60 percent iron.

(d) Green vitriol consists of acid, water and iron. By calcination a large amount of water is driven off, which otherwise would dilute the acid; later, by means of heat, the acid is separated from the metal, which because of its incombustible nature stays behind in the bottom of the retort. All the water cannot be driven off without a large loss of acid in calcination, therefore much phlegm goes over in the beginning, and although it diminishes more and more at the end of the operation, yet a small part of the acid seems to still be discharged. A gold-brown color is produced by the phlogiston which the metal possesses, before that of the acid is transferred to the vitriol and then mixed in by the distillation. Combustion causes some white smoke to be given off during the distillation, but still water-clear oil of vitriol is produced.

Note 1. The proportion of water, iron and acid in the vitriol crystals is approximately 17, 10 and 5. BERNHARDT has produced in 10 days, in 24 retorts, from 600 pounds of yellow calcinated vitriol, 52 pounds of crystallized acid.

Note 2. Nowadays in England acid of vitriol is prepared with the least difficulty and cost from sulfur which, blended in certain proportions with saltpeter and blue [vitriol?], is burned in a closed vessel, the acid fumes collected in water, and the dilute acid which is obtained is finally concentrated by distillation in which the watery part goes over first.

Note 3. That phlogiston alone could make acid of vitriol dark is not likely, because carbon powder can be mixed with it without changing the color. But if some oil is mixed with it, the concentrated acid turns yellow or darkens immediately.

Concerning Concentrated Acid of Vitriol

§ 12. The acid obtained by the foregoing process (§ 11), or otherwise diluted, is put into a retort to which an alembic is cemented on with egg-white and unslaked lime. The retort is placed in a sand-bath, a small flask is placed under the alembic tube, and when all is in place slow heating is begun. As the water is more volatile than the acid, it is driven off first in the phlegm, used in certain complaints to wash the eyes with; and when most of it has gone over the stronger acid starts to rise, which condenses on the cement and gives a black stain, which is a sign that the heating should stop and the oven be allowed to cool.

From beginning to end the degree of heat ought to be very low, since otherwise the acid rises with the water and immediately corrodes holes in the cement, so that the entire installation must be made over again. A thin cloudiness, which slowly increases, is observed with too much heat, but as soon as a haze arises quickly in the retort, the heat should be decreased. According to the amount required, 3, 6 or more days may be required for this operation.

To Prevent the Dark or Yellowish Color.

§ 13. A dephlegmed, but still colored acid of vitriol is poured into the clean retort, which is placed in a sand bath with a flask attached, after which it is carefully heated. When the retort reaches moderate heat some thick dark-brown drops come over, after which [something] entirely clear follows, at which point the acid of vitriol also lying in the retort begins to boil and comes over quite colorless. This is usually called oleum vitrioli rectificatissimum or concentratissimum, and should be kept in a flask with a ground-in stopper, since if air gets in it brings moisture and the strength of the acid is reduced. On the bottom of the retort a salt is sometimes found after the boiling off, which can be nothing else than mineral alkali now changed to Glauber's Salt probably containing water, which is used for vitriol-cement, in case no flintstone, which is usually used, is to be had.

Note. If the above-mentioned salt in the residue, Glauber's [Salt], which consists of mineral alkali combined with vitriol, is produced, it should not be volatile in fire because it cannot go over like the acid in the first distillation because it does not have such volatility; and more than likely it comes from decomposition of the glass and probably is mixed with flintstone, since we know that heavy boiling of strong acid in alkaline glass sometimes attacks the glass and the flint dissolved therefrom is deposited as a fine powder. Prof. GAUBIUS says that he has found both white and green vitriol in the residue, in addition to Glauber's Salt and earthy salt.

Preparation of Fuming Spirit of Vitriol

§ 14. Anise of juniper berry or some other essential oil is mixed with acid of vitriol and distilled in a glass retort to produce a white, very choking, noxious smoke, which slowly deposits and is condensed into a fluid. It is remarkable that such volatility can be given to such a heavy acid by phlogiston.

Note. Pure concentrated acid of vitriol does not fume until it meets moisture.

Acid of Saltpeter

§ 15. This acid gets different names according to circumstances: Spiritus nitri [Spirit of nitre] when it is very concentrated; Glauber's spirit of nitre when it is slightly diluted, but not excessively with water; and aqua fortis when it is much diluted with water.

Note. These variations are designated more properly as follows: acidum nitri concentratum, concentratissimum, dilutum, for strong, strongest, and dilute acid of saltpeter. It is not yet settled whether this acid is really mineral, but until this fact is sufficiently established it will be classified as such.

§ 16. It exhibits the following properties:

(a) It detonates with phlogiston in fire.

Note. The acid must still be combined with alkali or other base, which holds it reasonably fast, so that all excess water can be expelled, since otherwise the necessary combination with phlogiston does not occur. The acid itself does not acquire enough heat for dephlegming without being volatilized, and consequently cannot be detonated without a base, and this must retain it quite strongly if it is to be suitable.

(b) It bleaches black ink.

(c) It dissolves silver, quicksilver, copper (with a blue color), iron, bismuth, regulus of cobalt, zinc, and gold under some conditions, but tin is only corroded, and regulus of antimony is precipitated.

(d) With redlead it produces litharge and sweet white lead.

(e) It devours hard leather, wool and linen, but only when air is available, for if linen is immersed completely in a flask or glass filled with nitric acid it is found entirely unharmed after several days, but if it is brought to the surface and washed quickly in cold water only a minute passes before the linen decomposes entirely. In air the acid decomposes more strongly, but what causes it to do this is unknown. What happens with Glauber's spirit of nitre has not been tested.

Note. It has been tried successfully with fuming concentrated nitric acid; but wool decomposes and more quickly than silk.

Distilled Nitric Acid

§ 17. (a) This acid is obtained from purified saltpeter, of which eight parts are mixed with seven parts of calcinated vitriol, both finely sieved.

(b) Afterwards a glass retort is filled with this mixture and the retort is sealed to a receiver containing a little water, so that the elastic fumes are slowly absorbed, which would otherwise break the vessel asunder. Sometimes the seal is not tight and brown smoke drives the air from the vessel; at other times the receiver has an opening with a ground stopper which prevents the escape of the fumes.

(c) The retort is placed in sand 2 fingers deep in the bottom of a bath, for that level can be properly heated and afterwards filled with sand almost entirely, and when the cement dries and it is heated cautiously at the beginning, the vessel will not blow up. When the retort becomes warm, the acid should rise more and more until finally it fills the receiver with dark brown smoke. If it drips before the brown smoke appears it is a sign that the vitriol has left some phlegm behind which rises sooner than the nitric acid, taking a little vitriolic acid with it, because that has not become strong enough to decompose the nitre before it loses most of its water. If the vitriol is properly calcinated it should drip one hour after the smoke appears and has not yet diminished.

(d) When the dripping stops the heat is increased little by little and the sealing of all joints is opened, until the sand anneals, after which the oven is cooled since no more acid can be expected.

(e) The cement is broken up and the acid taken out, from which you should avoid the noxious fumes which have not yet had time to condense.

Note 1. The noxious fumes are difficult to avoid during removal, so you should have a receiver prepared which has a tube at the bottom which can deliver into a flask through a sealed opening. With such an arrangement the acid flows out immediately and is quickly collected, and nothing else is needed but to remove the flask in due time and close it with a stopper ground to fit the opening.

(f) The preceding procedure is evidence that vitriolic acid is a stronger and more expulsive acid than nitric in uniting itself with fixed alkali.

Note 2. Here there is a force of attraction really difficult to evaluate, the attraction in both salts for other substances; vitriolic acid is united with alkali, and nitric acid with iron, but heat causes decomposition of the lighter last named compound so quickly that the acid, with the phlogiston in iron, becomes completely volatile, fuming and becoming more or less brown colored.

(g) Some believe, with BOERHAAVE, that some vitriolic acid is easily distilled, especially in an iron retort, but alkali undoubtedly retains the acid when it is used with a small amount of vitriol or only a little calcinated [vitriol]. See V. Ac. Handl., 1753.

(h) A large iron retort is used with a small lid on the top, and the residue removed and put into [this] new set-up, with a hole drilled in the bottom of the receiver equipped with a stopper, for drawing out the acid, which is always contaminated with acid of salt, because the hardened saltpeter is never entirely free from it. Unhardened [saltpeter] cannot be used because of the oiliness associated with it.

Note 3. With saltpeter and highly calcinated vitriol much difficulty is encountered with the entire operation after passing over 5/8 of the brown-red acid. For aqua fortis, usually uncalcinated vitriol is used, and only until the operation is 1/2 complete; for although more than half is commonly obtained, this acid is colored more and its separation less convenient.

Testing the Strength of Aqua Fortis

§ 18. The more the water mixed with the acid, the weaker it [the acid] is. To find out how much acid is contained in a given amount [of aqua fortis], fixed alkali is used, dissolved in 3 times as much water with boiling; to this is added little by little the aqua fortis until no more effervescence is observed, after which the weights of the added alkali and the saturated aqua fortis are taken equal. The ordinary aqua fortis, which is purchased commercially, uses 1/2 of the [weight of] alkali for saturation, and in the strongest, 11/16; but when we do the distillation ourselves it can be brought to use 7/8.

Note. Reaction paper is far more certain for finding the point of saturation, than is the [point of] ceasation of effervescence, which is hard is observe sharply.

To Precipitate Aqua Fortis

§ 19. Aqua fortis is always contaminated with acid of salt, and sometimes even with vitriolic acid. The removal of this contamination is called ´precipitating aqua fortis´, and occurs this [following] way. A small amount of fine silver is put into a distilling retort and in with it is put some aqua fortis, and this is put over a low fire or candle flame until all the silver is dissolved. The solution becomes very cloudy, because the acid of salt precipitates the silver as a white slime; this is precipitated both finely and heavily in the aqua fortis which is to be purified, which after much stirring is allowed to sit quietly until all the white sediment settles to the bottom. After this the clear liquid is transferred to a retort again with a little silver. If the new solution gives a precipitate again the process is repeated until none appears, because a clear solution indicates that the contaminating acid has now been eliminated.

The precipitated aqua fortis usually contains silver, since the precipitation cannot be carried out without taking up some. So, after this separation, the clear liquid is poured into a glass retort, and with very cautious heat so that it may not boil too violently, it is distilled over into a sealed receiving vessel, while the silver remains in the retort.

The reason for this operation is this, that acid of salt and vitriolic acid both have a stronger attraction for the silver than aqua fortis, but cannot keep it dissolved in liquid form.

Note. It often happens that the distilled aqua fortis is still somewhat milky with dissolved silver, although more often it is free from excess silver. Probably there is some dephlogisticated acid of salt at work here, which does not precipitate silver, before it is restored to its usual condition by contact with some combustible substance. See § 23, Note 2.

Glauber's Distilled Spirit of Nitre

§ 20. (a) Clean saltpeter is put in an iron pan, and to this is added enough water to dissolve it with heating over a smoking fire, but it must be continually stirred with a knife and at the same time it must be scraped away from the bottom and edges of the pan so that it does not become stuck or burned (it may explode), and part of the acid be lost. Lumps are broken up so that it is well heated throughout and it is finally turned into a white powder; and when no further sizzling is observed it is done. This is called cracking of saltpeter and serves to drive off all the water and also phlogiston, in case there is any present.

(b) Put into a retort 12 parts of this cracked saltpeter, and to this add 5 parts of concentrated oil of vitriol, and after this cement on a receiving vessel immediately. The retort is put in a sand-bath and heated very cautiously, because otherwise the oil of vitriol boils over. The receiver need not be as large as that for the nitric acid distillation, because the fumes are driven over with less heat and consequently less expansion.

Note 1. For distillation of an acid, which gives acrid and corrosive fumes, it is best to have an opening in the upper part of the receiving sphere, to which a bent glass tube is cemented fast and the other end of which is immersed in distilled water. In this way we are not only freed from noxious vapors, but is assured that the vessel will not be burst open, and at the same time the sourest acid is collected in the water, which should be changed when it begins to smoke.

(c) Even before the fire begins to act, the usual fumes appear in the vessel, which are darker in this case than when nitric acid is distilled. When so much has distilled that what remains in the retort becomes dry, all the fire must be removed except for 1 or 2 coals, because it now should melt to a mass, which boils up so vigorously that it boils over if the heat is too strong; but after the sizzling stops, it is heated by a strong fire until all which is in the retort becomes white and dry, so no more acid is to be expected.

Note 2. When nitric acid is distilled in the usual way, we are much troubled by noxious fumes which penetrate the sealing cement; but burned lime with addition of almost 1/6 common salt, and a little water, mixed well together 1-2 times, makes an excellent cement which does not leak fumes.

Clay washed free of lime, dried, sifted, and well mixed with linseed-oil, which by boiling dissolves 1/18 litharge, makes a good cement [also].

(d) When the vessel has been cooled, the acid is poured into a flask with a ground stopper, well secured. The color is completely dark and continuously fuming.

Note 3. If the gold-brown acid is blended with about 1/4 its amount with water, it produces much heat and a beautiful bright green color, but the fumes are still gold-brown; if approximately 2 to 3 parts water are taken with 2 parts acid, the color is still heavenly blue; more water gives a colorless acid. We sometimes get entirely different colors merely through phlogiston spreading or thinning out, and this was quite mysterious with green or blue nitric acid, and the Alchemists made much to do about this. The brown color undoubtedly comes from phlogiston, and can be distilled off on a very slow fire, so that what remains is like clear water but growing yellow again if ever so little phlogiston gets in; indeed Mr. SCHEELE has told me that he has had white concentrated nitric acid in a flask, which gets colored in sunshine and produces golden fumes. It is not true that much distillation can pass the retained phlogiston through the flask and cause blood red fumes, when both the saltpeter and the acid of vitriol are most carefully purified.

Note 4. Usually half as much concentrated vitriolic acid by weight is used as niter, and when 5/8 acid is allowed to react fumes occur later. With such an arrangement as those just described [in Note 3], hardly any acid is lost. With 10 pounds of niter and a like amount of heavily calcinated vitriol, BERNHARD has obtained 20 pounds of good aqua fortis (10 pounds of water being put into the receiver), and in the kiln-glass about half an ounce of transparent crystallized colorless acid, which gives red fumes. With clay, in some way which is mentioned later (§ 26), about half as much acid as saltpeter is obtained. Talc can also be used.

(e) Proof that this is perfectly true can be obtained thus: that as soon as it is put into distilled oil (or pressed fat), e.g., oil of cloves, the acid contains so very little water that it does not kindle the oil slowly, but rather sets fire to it suddenly, like gunpowder, so that the oil is burned with a clear, large and vigorous flame. When the trial is made it ought not to be held in the hand, and even the clothing should be protected, because whatever it touches is eaten up.

Note 5. How oils can be kindled with nitric acid will be discussed in the Third Section. Until its strength is more closely measured it can be partly neutralized with alkali (§ 18), and partly with stirring. The most highly concentrated nitric acid has a specific gravity of 1.583.

Note 6. Mr. POTT maintains that saltpeter alone goes over with strong heating, especially in an iron retort, setting free its acid. With white arsenic, sal sedativus [boric acid], and acidum urinae [phosphoric acid], it is expelled, and indeed Mr. MARGGRAF has found that 10-12 parts of acid of salt can, with the aid of heat, decompose 1 part of saltpeter, which is much more strange than that 2 parts nitric acid, or 8 parts of weaker acid, would expel the acid from 1 part common salt.

§ 21. Some maintain that nitric acid mixed with oil is changed into vitriolic acid, and that this takes place best with turpentine, but when this substance is collected it rises with a very strong effervescence; and a golden substance, more than doubled in mass, is obtained which is like sulfur in color, yet effervesces by itself without heating when it is poured on heated charcoal, giving neither the smell of sulfur nor that of saltpeter. If it is distilled with sand or brick flour the receiver is filled with white smoke, which has strangely been set free, even passing over water, and a black oil is set free separately, and not the least trace of sulfur is found. Nitric acid loses here all its properties and leaves no sign of itself, like a vegetable acid which can be robbed of phlogiston merely by heat. If a little of this swelling mass is ground with half its amount of lye, put into a crucible, and heated over a fire, it fumes heavily, with the same liberation as before, but will not flame particularly nor give signs of detonation, but at the end leaves a red ash.

Concerning Acid of Salt

§ 22. This is called acidum salis or spiritum salis, and since it is driven out of common salt by vitriolic acid, spiritum salis Glauberi after its discoverer.

Note. Dilute acidum salis, concentrated [acidum salis], and very concentrated [acidum salis], for weak, strong, and strongest acid of salt, are the denominations used most often.

§ 23. In respect to strength this acid is third in the order, and it exhibits the following properties:

(I) It precipitates silver from its solution.

Note 1. Acid of salt cannot take hold of silver directly, but when this metal is dissolved in aqua fortis sufficiently, acid of salt takes it [silver] to itself immediately and comes down with it in the form of a dirty white precipitate which is nothing else than a difficultly soluble salt, and which comes down in want of a solvent. This having happened, acid of salt may be free or bound, depending on whether or not vegetable alkali is present. Quicksilver and lead come over with silver under similar circumstances, and these three metals, because of their similarity in several instances, were once classified under the common name of metalla alba [white metal].

(II) It dissolves in a mixture with golden aqua fortis, and likewise quicksilver, lead and copper (with a green color), when they are calcinated (in metallic form copper is attacked little), but iron, tin, cobalt, antimony, and zinc, in both metallic and calcinated form. Fulminating gold is dissolved by the attack of nitric acid.

Note 2. If acid of salt is passed over brunsten [manganese dioxide] it is changed into a red vapor [chlorine gas] of a particular kind, and which can hardly be condensed into a liquid, but which recovers its former state entirely if it is allowed to unite with phlogiston. In this manner both analysis and synthesis are combined to show that phlogiston is a constituent part of acid of salt. If it is dephlogisticated it dissolves gold, and probably the acidum nitri in aqua regia serves no other purpose than to take this combustibility from acidum salis. See Mr. SCHEELE'S treatise on brunsten in V. Ac. H., 1774. If fulminating gold has some acidum nitri it will be discussed later.

(III) It takes away the blackness from ink.

(IV) It attracts moisture strongly, so that solutions with it deliquesce readily.

Note 3. Even though vitriolic acid has an attraction for water, it produces no deliquescent salts.

(V) It is the most volatile of all acids, the most penetrating, and the most difficult to confine.

Note 4. Thus this acid smokes when it is very concentrated, but smoke is observed only when there is moisture in the air.

(VI) It attacks the skin a little, but preserves it and makes it tough, while removing corns if it is daubed on frequently, and even a swollen cold sore can be healed with it, but the clothing should be protected carefully because they are badly damaged by it, especially linen. On the other hand this acid is used in cold drinks to slake the thirst, to cause urine and [kidney] stones to pass; it is useful against putrefaction, and even against hot fevers and colds.

To Drive the Acid Out of Salt With Vitriolic Acid

§ 24. (a) Common salt [NaCl] is put into a crucible with a crucible cover and put over gentle heat so that the salt begins to crackle and pop, which continues as long as water is present. This is called decrepitation. When this stops remove the crucible, and care should be taken that none of the salt is melted through excessive heat.

(b) The decrepitated salt is moistened a little, and without pulverizing more finely is put into a glass retort, to which is added 5/8 its amount of oil of vitriol, which if it is not already diluted considerably with water, is diluted with a like amount of water, but a little at a time and cautiously, so that the vessel in which it is mixed does not break. The diluted acid is then poured into the salt [in the retort] and a receiver is cemented on, because strong effervescence begins immediately, which irritates the nose, mouth and eyes.

Note 1. This operation can be carried out more conveniently in a retort equipped with an outlet tube. For larger quantities a stone retort is used, which requires greater heat, or much is left behind in the residue.

(c) The heat is not increased beyond that which causes gentle boiling in the retort. The cement seal should be very heavy and strong, because acid of salt corrodes linen very badly.

(d) When so much has distilled over that the salt is observed to stick to the edges of the retort and the liquid begins to disappear, all heat is removed except 1 or 2 coals, since if it is not cooled now the retort will burst from the salt which attaches itself to the glass very firmly and tightly. When still well dried the oven is heated up again and the rest of the acid of salt is distilled over. The salt must not be allowed to reach the temperature of melting in the retort because it hardens with much hissing. When the vessel cools it is opened and the acid is transferred to a servicable flask, giving off fumes constantly, which if they are weak are not unpleasant.

> Note 2. Usually dilute vitriolic acid is taken up by the salt, and 3/8 part of acid [of salt] is usually obtained for each part of salt. At the end an equal amount of vitriolic acid should have been used; see § 53. The vapors are collected in water (§ 20, Note 1) to get more than double the amount which is obtained in the receiving vessel, that is if the vitriolic acid is well concentrated, but if it is dilute, most of it is collected in the receiving vessel, and especially if the heat is applied slowly.

(e) If we wish to have the strongest possible acidum salis, then the most concentrated oil of vitriol is used. The acid so obtained is properly called Spiritus Salis Glauberi, is very strong and fumes constantly. The operation can hardly be carried out except with a retort equipped with a tube through which the infusion can be added and the fumes permitted to pass over at the same time.

(f) This does not kindle oil like Spiritus Nitri Glauberi (§ 20 (e)).

Concerning Rectified Acid of Salt

§ 25. The acid, which is given off when a little more than 5/8 part of oil of vitriol is used with [one part of] salt, is never free from contamination by vitriolic acid, or by phlogiston in case it is dark colored when it is used. Purification is therefore used. Purification is employed in this way, by distilling the impure acid of salt again over a like amount of decrepitated common salt which drives out or else takes up with the acid, which by this means becomes very strong and pure.

Note. A well concentrated acid of salt is in weight compared to a similar volume of pure water, as 1.140 is to 1.000; the strongest which has been obtained up to now is 1.190.

Driving Out Acid of Salt With Clay

§ 26 (a). One part decrepitated clay and 3 parts Colnisk pipe clay are ground to pieces, sifted, mixed, and moistened with a little water and kneaded first with a wooden spatula, and last with the hands, then forced into small balls which can pass through the neck of a retort. These are dried well in a warm place and thereafter put into a Waldenburg retort well rubbed with clay and sanded (glass will not tolerate the heat to be used here) until the retort is 2/3 full, the retort is placed on a grate in a blast furnace, and a receiving vessel cemented on with the help of a so-called Vorstoss, which leads out of the neck of the retort and into the receiver.

(b) The furnace is heated up cautiously and when the retort begins to get hot white fumes are emitted with much phlegm, which all collects in the receiver. The heat is increased after this to its highest point and, when nothing further will go over, the furnace is allowed to cool.

(c) In the receiver is found a quantity of acid of salt mixed with water, which is freed from the phlegm in the same way as was done with vitriolic acid (§ 12); in a retort with an alembic [the mixture] is boiled off very gently and cautiously with heat until the sealing is attacked, and although a lot of water is still left this cannot be avoided.

Note 1. To obtain somewhat stronger acid it is best not to seal on the receiver before most of the watery matter has distilled off and the receiver is still empty. If we have a re- ceiver which has a pipe at the bottom, it can then be sealed on and a lot of the phlegm is quickly separated, as desired.

(d) In the neck of the retort after the first distillation there is usually found some sal ammoniac, which shows its volatile alkali when it is ground with a little burned calx. This alkali comes from the clay.

(e) When the [clay] balls are poured out of the retort, some of them are found to be burned and to have a reddish color all the way through, but part of them have not reached this stage and are grayish inside. The clay requires so much of the alkali of salt to attach to itself that they have melted together, while some of the salt remains unchanged. POTT used 10 parts of clay to drive all the acid from one part of salt. We can determine if any salt is left undecomposed when the residue in the retort is extracted, and the extract filtered and evaporated to dryness (ad pelliculum). When only three times as much clay [as salt] is taken, some salt is always found in the residue.

Note 2. When 8 parts of clay are taken with 1 part of common salt, and the retort is kept heated for several hours, almost 1/4 of the salt is found to be undecomposed in the residue.

(f) The acid of salt which is found on the market is usually produced with bole [bolus], is contaminated from mishandling, sometimes even with nitric acid, which latter substance can be detected if a little bit of gold is heated with the acid in a glass container over a candle flame, and in the case where some [nitric acid] is present, some gold dissolves, but otherwise not.

Note 3. The reason why clay is able to drive out both of these acids from saltpeter and common salt has not yet been fully explained. Most maintain that it gets its strength from vitriolic acid which is more or less attached to the clay; and I will not yet deny that it would seem to be admissible, but it is still difficult to understand why this hidden acid can act in the dry way with the salts named when it is not separated by boiling in lye, yes not even by burning with porcelain, since genuine porcelain decomposes just like clay. Could it be that the clay acts like a solid substance which prevents the mass from melting, and thus allows greater heat? Pure sand, pulverized quartz, and even high melting glass decompose similarly, without being suspected of containing vitriolic acid. There is obtained from the residue Tartarus vitriolatus and sal Glauberi, so the matter is indisputable, but they are not obtained without addition of fixed alkali which often contains some of these [substances] and they are obtained in much less quantity than we would expect from what was decomposed.

Preparing Acid of Salt by the Use of Clay and Alum Together

§ 27 (a). With 1 part of decrepitated salt take one part of ground up alum and double this amount of burned clay, which are mixed together and distilled in a clay retort in the manner before described (§ 26), and thus is obtained a true acid of salt.

(b) When the residue is extracted with water we obtain, in addition to clay, Sal Glauberi, along with some common salt and even alum, which indicates too low a heat.

(c) Arch. HIERNE has obtained acid of salt by distilling 1 part of dry talc with 1 part of common salt. Other neutral salts, of which this acid makes up a part, similarly release the acid in distillation, although some are joined with a more strongly attracted base, which holds it strongly. This so-called sal commune regeneratum, distilled with twice its amount of saltpeter, releases acid of salt, but there also passes into the receiver some acidum nitri because of its volatility, whereby aqua regia is produced. In the residue we find perfect saltpeter.

Note 1. Common salt, when 1 part is distilled with 2 parts of concentrated nitric acid, leaves in the residue nitrum quandrangulare [sodium nitrate], and hardly shows any trace of common salt, if the heat was sufficient. On the other hand, Mr. MARGGRAF has found that 1 part saltpeter, or nitrum quandrangulare, with 8 parts of strong acid of salt produces in the residue after distillation little else but common salt. This reciprocal decomposition has not been satisfactorily explained up to now, in accordance with acknowledged laws of attraction.

Note 2. The previously described distillations, which used to be carried out in a sand bath, can also conveniently be carried out in the following manner. The retort is supended over a little portable heating furnace made of sheet iron made into a cylindrical shape, in which is placed a grating of heavy iron wire. With this we have a great advantage, in that we can observe the entire operation from beginning to end, and at the same time we can better regulate the heat than we can with a sandbath, and obtain a higher heat in a shorter time with a smaller amount of charcoal.

To Distill Royal Water or Aqua Regia

§ 28 (a). Since this [aqua regia] is a mixture of the acids obtained from common salt and saltpeter, therefore 1 part decrepitated common salt is weighed out with 5 parts of acidum nitri and distilled in the manner previously described (§ 24). Hardly any common salt is found in the residue.

(b) Aqua regia is found in the receiver, which dissolves gold, copper, iron, tin, bismuth, cobalt, zinc, and regulus of antimony, but lead and silver slowly. This aqua regia is better than that which is prepared extemporaneously by simple addition of common salt.

Note 1. Aqua regia cannot easily dissolve lead or quicksilver, since acidum nitri takes some part to itself so quickly that most of the rest is immediately left to the acidum salis, which for the most part precipitates with these metals as difficultly soluble salts.

(c) Nitric acid and acid of salt when mixed are the best solvent for gold. Aqua regia can be prepared quickly in the following manner. Common salt or sal ammoniac are put into aqua fortis, which also dissolves gold without precipitation, but if sal ammoniac is used some of the gold is dissolved at the same time as volatile alkali is distilled over.

Note 2. Nitric acid dissolves gold conveniently not only in the presence of common salt, but also sal ammoniac, or any other neutral or middle salt which is available, provided it contains acid of salt. Likewise acid of salt can be the liquid medium if saltpeter or some other salt, which contains nitric acid, is present.

Note 3. In addition to the acids just described, there are some others in the mineral kingdom which appear to differ in their nature from other known acids. In particular are the acids which can be obtained from fluorspar and white arsenic. The former resembles, in lightness and volatility, acidum salis, but different circumstances prevent us from regarding it, as Mr. BOULANGER does, as acidum salis, when it remains tightly combined with an earth. No better method is known up to now to obtain this acid than through the distillation of fluorspar with vitriolic acid. See V. Ac. H., 1771. That white arsenic contains an acid, we have many reasons to conclude from its salts; it [arsenic acid] is obtained in the residue when saltpeter, nitrum quadrangular, and [nitrum] flammans [ammonium nitrate] are distilled with it [white arsenic], according to Mr. MACQUER´S description; but Mr. SCHEELE has discovered how this acid can be obtained pure and flowing, and that is by means of the dephlogisticated acid, which takes the combustibility [phlogiston] to itself which had previously been a property of sulfur. See V. Ac. H., 1774. The so-called Sal seditivus [boric acid] is a peculiar substance which combines with alkalis like an acid, but we are not yet sufficiently familiar with it to decide on its proper nature.

Vegetable Acids

§ 29. Of all acids these are the weakest, and they always contain an abundace of phlogiston, but otherwise they have properties similar to acidum nitri, except that they do not dissolve silver, but gold very well when the heavily concentrated acid is mixed with acidum salis. With alkali they burn up in flame.

Note. The author [Scheffer] here is usually referring to vinegar, which he refers to under the name of acidum vegetabile. There is no doubt that this acid is vegetable, but recent observations and experiments show us a large number of different acids from the vegetable kingdom which might seem to be chance modifications of the same basic substance, but yet have different appearances, until their preparation is worked out and shows clearly otherwise. Those few, about which we know something, give us no reason to confuse them with vinegar. Tartaric acid produces, in equal amount with alkali vegetabile, Tartarus Tartarisatus [K_2Tart.$(1/2)H_2O$], and with the acid in excess it produces complete tartar [potassiuym acid tartrate, KHTart], whereas vinegar gives with the same alkali the so-called Terra foliata tartari [potassium acetate], entirely different from the above named salts. Tartaric acid is obtained in this manner: chalk is heated with a solution of crystalli tartari, by which its liberated acid reacts with the chalk and produces a little soluble salt which precipitates, and can be called calx tartarisata [calcium tartrate]; this is separated and mixed with dilute vitriolic acid which gradually takes hold of the chalk and leaves the tartaric acid free, which is retained. See Mr. REZI'S treatise in V. Ac. H., 1770. If instead of chalk burned lime is used, more calx tartarisata is obtained, because when tartar is decomposed completely, but in a liquid, which in the foregoing produced Tartarus Tartarisatus, in this case it is merely contaminated with the alkaline part. Tartaric acid is sour and allows itself to be crystallized easily. Mr. MONNET maintains that it is disguised acid of salt; he has distilled Cremor Tartari, immediately separated the acid passed over before the oil set free takes it up, saturated it with alkali and driven it off with vitriolic acid, which process was repeated several times, by which means he finally obtained an acid which precipitated quicksilver, and when the precipitate was sublimed it gave a salt very much like Mercurius dulcis [mercurous chloride]. Mr. SCHEELE at my request has repeated this experiment and found the same thing, but has discovered that here alkali tartari almost always contains some acid of salt, and consequently the conclusion is completely otherwise [from that of Mr. Monnet]. When the distilled acid is saturated several times with fresh alkali, and in between set free with acidum vitrioli, the amount of acid of salt gradually becomes greater and greater, until it finally shows itself plainly. Here again is proof of how cautious we must be in order not to be deceived and confuse one substance with another.

Sal acetosellae also contains a different acid of its own, about which we as yet know no way to precipitate it from all bases, as a result of which it has not yet been examined. In addition several berry and fruit acids are found which are believed to have different properties, but are usually of such a nature that they are destroyed by the heat of distillation; sometimes a vegetable substance is obtained, but at other times something which resembles vinegar. The purification and investigation of these acids remains a vast field for future discoveries.

To Rectify Vinegar

§ 30 (a). Here a glass retort with an alembic is used, which is sealed with flour paste on strips of heavy paper. A small flask is fixed under the alembic pipe, to collect the distillate. The distillation itself is carried out on a sand-bath.

Note 1. Metallic containers are to be avoided because it is possible, we know especially, that the compounds of vinegar with tin, copper and lead are dangerous. With large quantities glass [containers] cannot be used, but rather a large stone retort to which a glass alembic is fitted, but these are not available everywhere.

(b) That which first distills over smells like burnt sealing-wax, and burns even when it is brought to boiling over a low fire, which is sufficient to distill spiritus vini. After the phlegm distills over and the acid begins to come over, the oven is allowed to cool.

(c) The residue in the flask is thick and turbid from phlogiston and impurities; it is transferred to a retort which is fitted with a receiver sealed om with flour paste on linen patches. Then it is distilled over a low fire until it all goes over, and it is watched closely near the end for burning, because otherwise when it begins to become dry in the receiver a lightly smouldering phlogiston goes over. In this way completely clear vinegar is obtained in the receiver, which still contains a lot of combustible material and water, and only the caput mortuum (as all that remains after distillation is called) remains as a dry brown dough clinging tightly to the rim of the retort.

Note 2. The first half of the distillate, although not as sour as what follows, is almost free of oiliness and thus especially suitable for terra foliata tartari [potassium acetate], which is always dark colored even when evaporation takes place in a balneum mariae [water bath?], if that part of the vinegar is taken which is distilled over at the end. The reason is that the longer it distills the more concentrated the acid becomes, and so it gets a greater ability, in proportion, to react with the oil present. If we stop the distillation when the residue begins to become very dry, and this residue is mixed together with dry clay, formed into small balls, dried well, and finally heated with strong heat in a properly mounted retort, a very acrid acetum radicatum [radical vinegar] distills over.

(d) How acetum acerrimum is obtained, which dissolves tin, and mixed with acidum salis dissolves gold, is mentioned later on in the description of Spanish green.

Note 3. Vinegar is an acid which is obtained by fresh treatment (fermentatio) of wine or beer. This is always mixed with much water, and mixed with oil to make extracts. To obtain it pure and strong various methods are used, such as by freezing, wherein the water changes to ice; and after that is separated, by distillation, because water is somewhat more volatile than the acid, but the separation is so little that they cannot be separated completely by this method [distillation]; and by saturation with fixed alakli or some metal, from which compound afterwards it is driven out by vitriolic acid in distillation to form a very strong acid. This in its greatest purity has a very penetrating volatility, is acidic and corrosive, is coagulated by 12 to 15 degrees of cold, is set on fire and burns easily when it is heated to boiling beforehand. Vinegar appears therefore to be an acid which is undergoing subtle fermentation and is combined intimately with spirituosum vini. STAHL has obtained vinegar under digestion, in a manner to distinguish it from acid of citron, among others, saturated with crab's eyes and diluted with spiritum vini; this is successful also with vegetable slime, with gum arabic, with tartaric acid, etc., when to these is added spiritum vini and they are allowed to stand for some time; and Mr. SCHEELE has also had success when gelatin of stag horn is used here. Substances which do not contain spirituosum, or cannot themselves first undergo fermentation to a degree, must necessarily be mixed with spiritus vini, if vinegar is to be made from them.

Note 4. The author [Scheffer] does not mention any of the acids which are found in the animal kingdom. We do not know more than two, namely, one which is contained in human urine [acid of urine; phosphoric acid], and another which is found in ants [acid of ants; formic acid]. The former is distinguished by its property of producing a kind of sulfur with phlogiston, which is usually called phosphorus urinae. STAHL believed that this was acidum salis altered in some way, but the more we learn about this acid the less believable is this conjecture. Perhaps this acid comes from vegetables within the animal, and it is still really part of the vegetable kingdom, contrary to common belief; but meanwhile it does carry the name acidum phosphori or acidum urinae. Acid of ants resembles vinegar very greatly, but Mr. MARGGRAF has found their reactions to be different with iron and zinc, in addition to other differences.

Note 5. Acid of air is found in all the natural kingdoms, resembles air very greatly in lightness and elasticity, which is the reason I gave it this name; moreover it is always found in the atmosphere. I have described this acid in V. Ac. Handl., 1773, and still more fully in a treatise presented to the Royal Scientific Society of Upsala. It is of all known acids the weakest, is found in great abundance with alkaline substances, and is found in mineral waters. It reddens tincture of blue litmus completely, gives water a sharp acidic taste, reacts with alkaline salts and alkaline earths, and likewise with metals to form neutral and middle salts, precipitates lime water, liver of sulfur, and liquor silicum [water glass?], decomposes soap, etc., quite like acids in general.

THIRD CHAPTER

About Alkaline Salts

§ 31. Their common properties are the following:

(I) They effervesce with acids.

Note 1. Effervescence occurs when a stronger acid drives out acidum aereum, since this latter [substance] becomes elastic at the same instant and is separated and consequently, because of its lightness, it floats up in the form of bubbles. For the rest, see § 8, Note 1.

(II) They precipitate acids which are soluble.

Note 2. Next to phlogiston, alkali has the strongest attraction for acids, and the separation and precipitation depends on this, in case the substance which is separated can be dissolved itself, either in (1) water alone, in (2) the precipitating agent, or in (3) acid of air. An example of the first case is found when vegetable alkali is put into a solution of Glauber's salt, since although mineral alkali is separated from the acid, precipitation still does not occur, since it [the compound set free] is by itself alone soluble in water; the second case occurs when volatile alkali is put into a silver solution; the third case shows itself when magnesia alba is separated from any acid by crystallized alkali, which occurs if the water is sufficient, or there is little or no precipitate.

(III) They turn syrup of violets green, which is otherwise blue. The change to green does not result from separate alkaline salts but from a mixture; small additions give the color of celadon [grayish green], larger additions grass green, and still more a bright golden color, the latter of which colors does not show itself unless the mixing is allowed to continue for some time.

Note 3. Syrup of violets becomes green even in vitriol and other materials, so that this greenness alone does not demonstrate the presence of alkali. However, we cannot obtain either pure syrup of violets or anything like it, so that what we get is seldom or never uncontaminated. So I prefer to use paper painted sometimes with fernambuck [extract of brazilwood], sometimes with turmeric; the former becomes blue with alkaline substances, and the latter brown. Although most blue vegetable juices become green in alkalis, litmus does not become green, but only still more blue than it was before.

(IV) They produce a darker color with the juice of heliotrope.

Note 4. Tincture of litmus is usually red when it is concentrated, but can be diluted with water so that the color becomes a full blue. It should always be like this when its reaction is observed. Redness [in litmus] is also removed by alkalis, so this already blue litmus is not as useful for indicating a new reaction [as the red litmus], since alkalis can hardly make it still more blue; and so a little acid is [first] mixed in to saturate any alkali [originally present].

(V) They make infusion of gallapple darker in color.

Note 5. Here actually precipitation occurs.

FIRST SECTION, CHAPTER III, Alkaline salts

Concerning Incombustible Vegetable Alkali

§ 32. This alkali possesses the following properties in addition to those given above:

(I) It forms a red precipitate with a solution of corrosive sublimate of mercury.

Note 1. If vegetable alkali is completely saturated with acid of air, the precipitate is white. To obtain such saturation with confidence, it is not enough to let the alkali crystallize (Note 4.); such crystals still contain some caustic and therefore produce a partly white, partly brown precipitate, but if they are pulverized after lying for several days in the open air the desired precipitate is obtained; paper painted with fernambuck or litmus turns blue with this alkali, but golden turmeric slowly turns brown.

(II) It turns the color of copper vitriol from green to blue, as copper is precipitated.

Note 2. Alkalis corrode copper with a blue color. See § 84.

(III) It dissolves per deliquium [through deliquescence].

Note 3. That which is pure or caustic attracts moisture from the atmosphere as much as is needed to dissolve itself; but the crystals do not grow moist; I still have some which have remained unchanged 6 years in an unheated room.

(IV) It does not crystallize by itself.

Note 4. Caustic does not crystallize, but if allowed to stand a long time in the open air, or if it is saturated completely in solution with acid of air, it produces beautiful crystals in 4-

sided prisms, whose points consist of two planes rising in the shape of a roof from opposite corners; 100 parts contain by weight 20 [parts] of acid of air, 42 of alkali, and 38 of water of crystallization.

(V) It drives out a penetrating vapor of urine from those substances which contain volatile alkali.

(VI) With strong steady heat it becomes very corroded and tastes like strong unburned alkali.

Note 5. Heat expels acid of air, against which alkali is usually entirely stable; and to the same degree in which this is separated, the alkali regains its natural sharpness, which is called causticity, when it has been freed to such a degree that it does not effervesce further with acids. See § 8, Note 1.

(VII) It also obtains causticity in mixing with burned lime and water.

Note 6. What fire displaces from alkali by force is also displaced from lime, which also becomes caustic thereby, but the lime has a stronger attraction for acid of air than alkali, and therefore conveniently serves to make the latter more caustic.

(VIII) It dissolves all oils.

Note 7. Essential oils unite with alkali with difficulty.

(IX) It requires strong heat to melt it.

Note 8. It is heated to glowing a long time before it begins to flow.

(X) It turns to glass with the addition of earths.

(XI) It produces long radiating crystals with acid of saltpeter.

<u>Note 9.</u> The differences in the neutral salts which arise from this alkali and from min- eral [alkali] with the same acids, give the best method of distinguishing between them.

The foregoing properties distinguish vegetable alkali, although not all of them are clearly evident. Thus, when diluted, its effervescence with acids can become imperceptible, etc. The absorption of ac- ids, such as occurs with chalk, crab's eyes [kräftstenar], etc., also often confuses them [chalk, etc.] with alkalis; but since they are not soluble in water, nor are they converted to a flux in fire, they should not be placed among the salts, but among the alkaline earths.

<u>Note 10.</u> Since chalk and lime are actually salts, although their alkaline properties are weak, they are placed there [among the salts] in the following pages. See Second Section.

Because the above described fixed alkali is usually obtained from vegetables, it has received the name of vegetable [alkali vegetabile].

To Extract Alkali Fixum Vegetabile from Cooking Ashes

§ 33 (a). One kanna [1 kanna = 2.6 liters] of sifted ash from tree leaves is mixed with 1/4, 1/5, or 1/6 its amount of burned lime after having its volume reduc- ed by light calcination, is heated with fresh water in a pan for quite a while, is then removed from the fire and allowed to stand until it begins to thicken.

(b) After this it is filtered through doubled printing paper, (the gray kind is eaten up by the lye), which prevents the earth from coming through with the solution. What passes through is preserved in a clean pan. As this begins to thicken we should scrape down frequently with a knife that which adheres to the sides of the pan, stirring it on the bottom, until it is chopped up in small pieces, to avoid the formation of lumps. The heat must also be applied to it very lightly so that, with the stirring about, melting is prevented.

Note 1. Filtering does not separate all lime, since some dissoles and goes through the filter as lime water. Without this pretreatment it is found as abundantly as in the ash itself.

(c) This now impure grayish mass is called gray potash lye (to differentiate it from the raw or untreated calcinated gray Danzig potash), and is calcinated to white refined potash in this way: The gray potash is put into a crucible, which is placed on a low fire, and the mass is stirred frequently so that none of it sticks and burns. In this way phlogiston, which causes the grayness, is driven off, likewise the rest of the water; but if the heat is too strong and melting results, both these substances are retained. When the mass is observed to be bright white, the crucible is removed. This white kind deliquesces more slowly than the gray.

(d) We now have a good potash, but one contaminated with some earth, which stays behind on the bottom if it is boiled again with water. The solution is filtered and evaporated to dryness with heat, and what remains is an earth-free alkali fixum.

(e) It can still easily happen that some common salt is present, either because the wood used contained salt, or from some other source. This is separated if the earth-free alkali is boiled again with three times as much water by weight, is filtered into glass and put in a cold place. If there is some common salt present it crystallizes on the bottom.

Note 2. In this way the largest part of the common salt is removed, but not all of it.

(f) The white material which is not dissolved in paragraph (d), is claimed to be soluble in six times its amount of water, and cryztallizable, if the solution is evaporated until a scum forms [ad pelliculum], poured off and put in a cold place. The gray crystals which then separate appear much different from alkali, and are supposed to be Egyptian soda [natrum] or ancient niter. A similar material is obtained from potash itself by dissolving and recrystallizing it. This is sometimes nothing else than common salt, along with Tartarus Tartarisatus from vegetable acid which was not burned away, especially if calcination did not take place; but most often it is Tartarus Vitriolatus from vitriolic acid which intrudes here.

Note 3. Tartarus Tartarisatus seems seldom if ever to be found in potash; potash is so readily soluble that the moisture in the air is sufficient to make it liquid. That which in ashes is called common salt [koksalt] is often Sal digestivus Sylvii [potassium chloride]. Tartarus vitriolatus is sometimes so abundant that it constitutes one-fourth of the weight: the greater part can be separated if the lye is purged with cold water, a little at a time.

§ 34. In most cases potash is prepared in the following way:

(a) Wood is burned in a large pit to ashes, and then new wood is piled on those ashes (layer on layer) and burned. The logs are moistened a little beforehand so that the ash may not be lost. The ashes are piled with new wood again and burned, which process is repeated two or three times, until the ashes become quite hard and contain more salt, because the more is burned the more essential residue is separated and the more salt can be obtained from it.

(b) The ash is transferred to the refining house and is put into a kettle with a coal fire under it; warm water is added, drawn off afterwards, boiled to dryness leaving a brown mass, which is calcinated in a specially constructed oven.

Note 1. [Misprint in original calls this Note 2.] Boiling of the lye in an iron pot is believed to be the reason for the name potash.

(c) In Prussia a so-called Blue Crown is produced which is nothing else but a calcinated ash without the soaking beforehand. It looks bluish and contains much iron and other impurities.

(d) Soaked ash after pot-boiling is used with ordinary sand in wide window glass and green bottle glass, and is charged with saltpeter to fix acid of saltpeter.

(e) A test of good white potash is that it gives a red precipitate with Mercurius sublimatus corrosivus, whereas the unfit [potash] gives a white precipitate.

Note 2. See § 32, note. [§ 32, (I).]

(f) Alkali fixum can be obtained from all wood ash, but in differing amounts. Beechwood and birchwood give the most, but pine and fir very little. Oak contains a considerable amount, but it is very difficult to bring its abundant content to calcination because it fuses together so easily. Brackenwood produces more potash than some wood, and it can be used without calcination.

Note 3. The author [Scheffer] gives a detailed history of potash making in Vet. Ac. Handl., 1759, and the proper use of it.

Ash of brackenwood will contain 1/6 alkali fixum by weight, according to Mr. HOME'S experiments, which is almost as much as can be produced from any tree, and 1 part completely pure alkaline salt requires 600 to 700 parts of dry wood, from those that give the most. For the most part if we get greater weight, it comes from its iron content. Wood which has lain in water for a long time gives little and sometimes no alkali fixum, and Mr. BOERHAAVE has found that when Rosemarin is boiled with water no alkali is obtained from its ash afterwards, but is found again in the decoction; and this shows, in addition to many other circumstances, that alkali is found before burning in vegetables. Compare § 38, Notes 2 and 4. VAN HELMONT is thought to have been the first to come up with the idea that alkalis are the product of fire; all those who still agree with him have never reflected on the process. **HOMBERG** requires acid and iron [to make alkali?], **NEUMANN** requires acid and fatty material, of which the most subtle parts are combined with iron during combustion, etc.

To Extract Alkali Fixum From Tartar

§ 35. Up to now both white and red tartar have been used. Burning is carried out either in the open or, if we wish to lose nothing, first in a closed vessel.

(a) The residue after distillation of tartar (see § 66) is black from an abundance of oil, which is expelled if free air is kept away. To extract the salt, cold water is added in the retort (heat takes much of the fatty material with it), it is shaken, poured out, filtered, and evaporated to dryness, from which a white alkali fixum is obtained, which is called Sal tartarti in the apothecary's shop, and which must be immediately enclosed in a flask and well stoppered because it deliquesces and is then called, because of its viscosity, oleum tartari per deliquium.

(b) That which does not dissolve is calcinated in a crucible with constant stirring, where it burns with a strong glow, and is removed when it becomes entirely white.

(c) Then water is added, boiled, filtered, and boiled dry, when again alkali fixum is obtained.

(d) That which stays on the filter usually still contains some phlogiston and accompanying alkali; it is therefore calcinated again, extracted with water, etc., which process is continued as long as any alkali is obtained, finally leaving a completely white or reddish earth.

(e) Sal tartari is sometimes contaminated with acidum vitrioli which comes from smoked winecasks, since pure acid of saltpeter obtained from these often gives a precipitate with silver solution.

Note 1. For 1 part of so-called Sal tartari is required about 2-3/4 parts of cream of tartar or crystalli tartari. Probably an equal amount of alkali is obtained from Sal acetocellae. It should be noted, however, that the impure alkali always effervesces with acid, and consequently the presence of acid of air should be reckoned in its weight. Refer to § 66.

(f) French ash from wine lees [Fransosernes cendre gravelée] is prepared from the dregs or impure tartar which is left in the bottom of the wine cask, which is first dried, then burned to charcoal in an oven, and lastly calcinated until entirely white, which still takes a long time and leaves a completely impure product because all the earth remains which cannot be separated without redissolving and filtering.

Note 2. Burned alkali from the whitest cream of tartar sometimes seems to be contamin- ated with acid of sulfur [svafel-syra], which is so named for the moment, although the precipitate usually comes from acidum salis, of which we can convince ourselves because it behaves like Luna cornea [silver chloride], and which can be seen still more clearly with crystallized sugar of lead [lead acetate], the precipitate from which is soluble in vinegar. It is thus more difficult than we might believe to make completely pure vegetable alkali; saturation with acid of saltpeter, followed by precipitation with silver, detonation with charcoal, which is free from common salt, calcination, dissolving, and filtering, these are troublesome operations.

To Extract Fixed Alkali From Saltpeter

§ 36 (a). Saltpeter is ground and forced through a sieve, then mixed with 1/4 its amount of fine coal dust. After mixing, a little is put in an iron pan and set aflame with a glowing piece of coal or iron, when it immediately begins to burn vehemently and to hiss; when it is nearly finished some more of it is added, and this process continued until all of the mixture is burned. If the entire mixture is kindled in one mass, the reaction becomes so violent that some of the mixture is thrown out.

We can also melt niter in a crucible and then toss in the charcoal a little at a time, as it burns up. Charcoal is thought to sustain in this way a more powerful penetration of combustion than sulfur, of which more is required for decomposition, in addition to continual melting over fire.

(b) After the burning off, which is called detonation, the mass should either be immediately ground up and put in a closed container, or it should be allowed to deliquesce and after filtering is called liquor of fixed niter [liquor nitri fixi], and also Glauber's alkali [alkahest Glauberi], being very acrid and penetrating, and which was thought to be useful as a universal solvent. Pure alkali is produced also from the detonated mass, if water is added, the mixture boiled, filtered, and evaporated to dryness. If common salt is present it can be detected in the manner previously mentioned (§ 33 (e)).

(c) Self-detonation depends on acid of saltpeter, which is burned up by means of phlogiston.

<u>Note.</u> Here arises a Sulfur nitrosum which catches fire as soon as it is produced, and burns up, whereby the acid is entirely destroyed, which can [also] be accomplished with clyssus nitri, or the liquid which is collected when saltpeter is brought to detonation in a closed vessel.

To Extract Alkali Fixum From Tartar and Saltpeter Together

§ 37. Both materials are ground and mixed in different proportions for various reasons.

(a) Similar amounts of both substances, formed into a cone-shaped pile in an iron pan, are kindled by a pointed piece of glowing iron, when detonation occurs immediately and proceeds gradually to the bottom [of the pile], leaving a white mass which, in Analysis, is known by the name of <u>white flux</u>, alkali extemporaneum, or Dresdenite. The tartar is charred by heat and its phlogiston is burned up by the niter. The product should immediately be chopped into small pieces and placed in a well-stoppered bottle, or it will deliquesce.

Note 1. Hereafter a little more saltpeter is taken than that which is required for the burning of tartar, to produce a white color, and also some excess in weight of saltpeter, the acid of which becomes phlogisticated, so that it can be displaced by weaker acids and easily detected by its noxious odor. If the materials are not dried beforehand about half the weight will be lost.

(b) If 2 parts of red tartar are taken with 1 part of saltpeter, we then obtain the so-called black flux. A large crucible is filled with the mixture which is then set afire and covered with a stone of proportionate size. When detonation ends the black mass is quickly ground, so that it is not permitted to go out of itself.

If the crucible is not completely filled, then the flux has room to swell up and burns completely into ash. Placing the stone on it prevents loss during detonation, but if it is so heavy that the flux is not permitted to have some air get to it during burning, the crucible will crack. If the glowing mass is knocked out and is permitted to go out by itself it becomes entirely white.

Of these fluxes, which are used in analysis, the former [white flux] is not entirely reduced but is deprived of phlogiston; but the latter [black flux] contains some [phlogiston] which reduces and conserves metallic substances under it, whereas the alkaline part vitrifies minerals and earths; but still better is white flux mixed with a like amount of red tartar, because this contains more combustible matter [phlogiston]; this can be called Tartar flux.

Note 2. If the substance is not dried beforehand, over half its weight is lost in detonation. The author [Scheffer] uses the name Raw flux for that which in his text was just named Tartar flux, which latter name [Tartar flux] I believe to be more suitable, because the former name [Raw flux] is used with other meanings by some of the newer writers. Black flux contains alkali extemproaneum more abundantly than a similar amount of Tartar flux, but the latter on the other hand has more tartar than black flux, whose tartar has already lost much of its weight in fire.

§ 38 (a). Fixed alkalis from ash, saltpeter and tartar all have the same nature when purified, in that they all produce the same liquid oil of Tartar [oleum Tartari per deliquium].

(b) They always contain some iron residue, although often if they are dissolved, filtered, boiled, and calcinated, and the alkali saturated with acidum nitri, the iron is left on the bottom. The acid can be separated afterwards through detonation, but then earth is introduced from the coal dust.

Note 1. The earth which vegetable alkali releases is flint-like, and seems to be either in the vegetable itself and dissolved from it in the wet way, or becomes united with the alkali in burning. It is released gradually, especially if the alkali is dissolved in much water and is permitted to stand with a large surface exposed to free air; this is a precipitation by acid of air.

(c) The fatness in alkali is burned most readily by the addition of a little saltpeter or its acid, although it should be measured accurately so as not to use more than that amount which can be burned off. In glass-making accuracy is not so important because sand or stone meal drives away the excess acid.

(d) That vegetable fixed alkali can be changed to volatile [alkali] through putrefaction is observed from dry wormwood, which, when fresh after distillation, gives fixed alkali in the residue, but when rotted produces somewhat less of the incombustible [fixed alkali], and volatile alkali instead.

Note 2. This deserves to be repeated with care. We should gather wormwood in the same place, divide it into two similar amounts, burn and extract one part immediately, but leave the other to rot in a glass vessel so that the alkali fixum cannot be lost during the process, or in any other manner. If handled in this way, in both cases very similar amounts of [fixed] alkali are obtained. [B. is being kind to Scheffer in his critcism here, since what Scheffer suggests is nothing less than transmutation of elements.]

(e) If dissolved alkali fixum is evaporated, and as soon as salt begins to show itself it is removed from the fire and poured into a flask along with 1/4 or 1/2 its amount of spiritus vini, alkali settles out in the form of crystals.

<u>Note 3.</u> Acid of air is associated with crystallization. See § 32, <u>Note 4</u>. Here Spiritus attracts the water to itself.

(f) Alkali fixum is not injured or attacked by fire, since vegetables which contain acid give alkali after burning, but this is only transformation of form and not of essence, because phlogiston also seems to be transformed into alkali by burning, or perhaps the fire, when phlogiston is consumed, then alters some of the residue which was united with it. This occurs also without fire through putrefaction, when the material is exposed to acidum nitri with which the alkali can unite, whereby the added acid attains the state which it has after putrefaction.

<u>Note 4.</u> That alkali fixum is found in tartar in the same quantity and condition as is found after burning can be discovered readily by analysis, because it can be separated by means of mineral acids, as Mr. MARGGRAF showed, although much acid is required for complete decomposition. It is obtained most quickly of all through double attraction. Mr. MONNET uses quicksilver solution for this; the metal immediately unites with acidum tartari forming a completely insoluble salt which immediately precipitates, whereupon the alkali of the tartar joins with the acid of saltpeter to form perfect nitrum prismaticum [potassium nitrate]. When we know beforehand how much alkali is in the saltpeter, which causes full decomposition of tartar, it is easy to calculate the amount of alkali in a certain weight of tartar, which corresponds with what remains after burning. Although tartarus can also be regenerated from vegetable alkali and acid of tartar,

there is still nevertheless room for doubt as we leave this subject.

(g) Alkali fixum vegetabile is used in Medicine, the dyehouse, the glass factory, the soap boilery; and as a flux in Chemical Analysis, because it brings substances difficult to melt to the molten state.

Mineral Alkali

§ 39. This exhibits, in addition to the usual properties [of alkalis] (§ 31) the following properties:

(I) It is crystallized easily by itself.

Note 1. But not without acid of air. The crystals are 10-sided, or alum-shaped with two blunt points right opposite each other; only 16/100 of its weight is acid of air, which is not enough to precipitate white mercurius sublimatus.

(II) It weathers in air to a flour.

Note 2. This is called efflorescence [fatescera] and results from the loss of water of crystallization.

(III) In combination with acidum nitri it takes on cubic form.

Note 3. These [crystals] are not really cubic, but inclined cubic, like spar.

(IV) It is never obtained naturally free from acidum salis, except in saltbrick [?] [mursalt].

Note 4. This alkali, which is found in the orient and on the north coast of Africa, is in all cases nothing but common salt decomposed by extended exposure to changes in weather, since when we go deeply into it, [common salt] is found there, and the deeper we go, the more abundantly it is found.

(V) It causes corrosive sublimate to form with an orange color.

(VI) It dissolves oils, but not all of them.

(VII) It melts more readily than alkali vegetabile.

Note 5. When water of crystallization is absent, it burns before it melts.

(VIII) If it is still less fire-resistant, it is called [alkali] semi-fixum.

Note 6. This alakli should not be named minerale therefore, although we collect it from the mineral kingdom, since rather we obtain most of it from sea-vegetables [hafsväxter; sea-weed?], and so thus it is obtained mostly from the vegetable kingdom, partly free from any acid other than acid of air in that produced in the Orient and Africa, and partly combined with equal amounts of acid of salt in salt springs and rock-salt, not to mention other less important sources. This is the Natron or Nitrum of the ancients.

Extraction of Alkali Minerale From Soda

§ 40 (a). Soda is prepared in quantity in Spain from kali geniculatum [Arabic: qaliy - ashes of saltworth; Latin: geniculatum - knotted; therefore kali geniculatum - ash of knotted sea-weed. J.A.S.], which is cast into a pit, set afire and burned to ashes; this is mixed with fresh grass [sea-weed], which is burned, and this is repeated until the pit is full of ash, which hardens into clumps from burning, is taken out and sold. It is also manufactured from Salicornia [species of sea-weed]. Soda is used extensively in glass.

Note 1. In southern Europe the ash is burned from various sea-weeds, but all these do not contain the same amount of the pure alkali, so we find great diversity in soda [content]. The soda ash from Spain, which is generally found to be the best, is burned from Salsola Soda Linn., contains little common salt, is also free from liver of sulfur, tartarus vitriolatus, and alkali vegetabile; we seldom get this kind. Otherwise it is burned from Salsola kali, Salicornia europaea (which on the Baltic Sea is called herbacea, but on the Mediterranean Sea fruticola), etc. The worst kind is generally Soude de Varech [Soda from Varech], which is burned in Picardy, Normandy, and Flanders from Fucus viciculosus [a red sea-weed or rock lichen], containing much hepar, common salt, tartarus vitriolatus, and alkali vegetabile.

Pure soda-salt contains in 100 parts of crystals, 20 parts of alkali, 16 [parts] acid of air, and 64 water.

(b) Soda consists of salt, earth, and a little sulfur. Regions in which the ash is prepared are near the sea, and without doubt it contains both alkali and acid from sea water.

Note 2. Sal sodae is found in the juice of vegetables before they are burned, as Mr. MONTET'S research shows, and yet it is doubtful if this amount is increased in the burning by decomposition of common salt.

Mr. DU HAMEL'S research with Salsola kali is noteworthy; he has planted it far from the sea in salt-free ground, and afterwards, after it grows up, he found alkali vegetabile in its ash, which would not be the case if it grew near the sea. The seed from the first year's harvest he planted again in salt-free ground, and he thinks that alkali vegetabile appears more abundantly in the second year's harvest. Is it possible that alkali minerale may disappear completely in this manner? Or is it only diminished to a certain degree? This certainly needs to be investigated.

(c) Now, to extract the alkali from the soda, it is ground, forced through a sieve, put in a pan, a considerable amount of water is mixed in, and it is boiled down almost by half. After this it is taken from the pan, allowed to stand so that the cloudiness may settle out, it is filtered, evaporated again, held in a glass vessel and allowed to stand so that crystals may form, which are redissolved in water, filtered and boiled, at which point we have obtained an alkali which is called Sal sodae, and by the Italians Rochetta.

Note 3. A pure soda salt, like that previously mentioned [Note 1], when it stands for several years, loses a large part of its water. [If this is Rochelle Salt that B. is referring to here, it is potassium sodium tartrate, which crystallizes with 4 water molecules per molecule of salt. J.A.S.]. I have examined similar salt when it was 3-1/2 years old, and it then contained, in 100 parts [of the salt], 31 of alkali, 27 of acid of air, and only 32 of water; after 5 minutes of heating to glowing all the water is separated and similarly 6 parts of acid of air; good heat for 6 hours, but still not heated to glowing, causes 7 parts of acid of air to separate, and 30 parts of water. We see thus that after different ages the strength is changed, or the amount of alkali in 100 parts.

(d) This Sal sodae precipitates corrosive sublimate [mercuric chloride] black by virtue of its intrinsic hepar [sulfur], and copper vitriolate green, and gives the same reaction with iron and acid of Berlin blue alone, which [precipitate] color is nothing other than the sulfureous essence combined with iron.

To Extract Alkali Minerale From Common Salt

§ 41. This cannot be carried out in any other way except to decompose common salt with acid of saltpeter, by which we obtain nitrum cubicum (§ 56) [sodium nitrate], and from this in turn a pure alkali minerale, when it is detonated with coal dust, dissolved in water, filtered and boiled. This is given the proper name of alkali minerale, alkali salis communis, or [alkali salis] marinum.

Note. Mr. NARGGRAF tells us how this operation should be carried out. See § 27, Note 1. A shorter method is given in § 57, Note.

§ 42. Mursalt is the third kind of alkali minerale, which is also obtained entirely pure and free from foreign substances, is consequently purer and is crystallized more easily than sal sodae.

Note. Mursalt is not always pure alkali minerale, but is often mixed with various foreign substances.

Alkali Volatile

§ 43. Volatile alkali is found in all of nature's kingdoms. That it is found in clay has already been noted (§ 26 (d)).

Among the vegetables and their parts which yield this substance are particularly those which, when they are ground or pinched, sting the nose and burn the eyes, like onions, horse-radish, radishes, mustard, etc. Pepper might also be thought to belong here, but on the contrary is supposed to contain alkali fixum.

In the animal kingdom it is obtained from the blood, urine, and other animal parts.

§ 44. Its various properties are:

(I) A strong and penetrating odor, which is increased by adding caustic alkali fixum.

Note 1. The more caustic, the stronger the odor.

(II) It does not melt over fire, but fumes away completely.

(III) It turns corrosive sublimate into a white powder.

Note 2. Since when dissolved alkali is added to the brown-red precipitate obtained with alkali fixum, it quickly turns entirely white, it is consequently less strange that alkali fixum blended with 1/10 [alkali] volatile produces a white precipitate.

(IV) It effervesces with milk over heat, when it begins to boil.

(V) It dissolves copper, and its readily formed precipitate, both with the color of blue saphire.

Note 3. If the copper is dissolved in a closed and filled vessel, so that no air stands over the surface, which can draw phlogiston to itself, the solution becomes completely colorless; but if it is then opened [to the air] a blue color begins to appear at the surface, which spreads out like a lock of wool until it colors the whole mass. When the solution becomes very blue the color does not vanish if it is again closed up.

(VI) It dissolves zinc.

Note 4. Zinc dissolves better than copper, so that the former precipitates the latter.

(VII) It does not precipitate silver solution when the alkali is free from acidum salis, for although a little white powder can appear at the beginning, it immediately redissolves if more alakli is added.

(VIII) If the vapors of this alkali and those of acidum nitri react with each other they form a visible smoke.

(IX) It unites with all distilled fatty oils, but some with difficulty, as [for example] oleum terebinthinae aethereum.

(X) It crystallizes with water, but only in a closed vessel, since in an open vessel it vaporizes.

Note 5. The crystals seem to be alum- shaped with 4 cut-off corners, and cannot form without being saturated with acid of air, which makes up almost 53/100 of its weight.

To Drive Alkali Volatile Out of Mustard

§ 45 (a). To a retort almost filled with mustard seed, a receiver is cemented on with eggwhite and unslaked lime, the retort is placed on a sandbath, and when the cement is dry, the retort is heated.

(b) First to appear is thick and turbid water; after this an oily spirit; then when the heat becomes stronger, white clouds together with a salt of alkali volatile, which condenses in dry form in the retort and all around the neck of the receiver; finally, opening all joints, when the heat reaches its highest point, a thick oil comes over, like pitch, after which the oven is allowed to cool.

(c) The alkali obtained thus has all the above described properties (§ 44).

Note 1. Professor BOERHAAVE reports that he has found mustard seed to effervesce with acid.

(d) The phlegm and oil contain as much alkaline salt as they can absorb.

(e) The oil burns with a wick, like pressed oil, but does not catch fire all over, although it is one of the empyreumatic [oils], probably because some phlegm is still left in it.

(f) The caput mortuum [final residue] has enough phlogiston left in it to burn vigorously at first, and consequently needs a considerable time in which to be calcined white in an open crucible. If this latter is boiled in water, filtered, and evaporated to dryness, it is found to be a reddish earth, which in no manner changes the color of syrup of violets, which indicates that it does not contain any alkali fixum.

Note 2. That alkali fixum is readily found in vegetables is completely proven in our times, but the case is not yet so clear with alkali volatile. In Kalmuckiet [a species of seaweed?] native sal ammoniac is found in abundance; several extracts, or vegetable juices boiled over a low fire, give a pronounced odor of alk. vol. when they are ground with alkali fixum causticum or newly burned lime; and when fresh human blood from the veins is put in lime water, a similar odor is detected. These [facts], etc., give reason to believe that volatile alkali also on other occasions is separated alone.

To Drive Alkali Volatile Out Of Soot

§ 46. This takes place in the same way as with mustard or other plants, except that we proceed with this difference, that in the residue alkali fixum is found, which after calcination, is obtained by extraction and evaporation.

To Drive Alkali Volatile Out Of Hartshorn

§ 47 (a). The process is carried out in the manner described above, in a retort and sandbath, with well dried and ground hartshorn.

(b) First the so-called spiritus cornu cervi comes over which is nothing other than water containing some dissolved alkali volatile; then comes alkali volatile as a white smoke, which condenses in the neck of the retort, and to some extent in the receiver, in dry form, dripping somewhat out of the retort when it is molten, but solidifying immediately in cold. But this salt accompanies even the thick black oil, and the heat must be moderate if no oil along the way is to dissolve any of the salt and thus retain it for a net loss, since the oil is good for nothing.

(c) When nothing further comes over, though the valves are open, the oven is allowed to cool.

(d) Spiritus can afterwards be separated from the oil in a separatory funnel, and should immediately be put into a stoppered flask.

(e) The salt is a yellowish material, but it can be purified if it is ground together in a glass mortar with some burnt French clay or white bone ash; this mixture is put into a retort which is sealed to an alembic, and finally is sublimed (driven over in dry form) with careful heating. The clay or bone ash protects the material in case the heat gets too strong. If we wish to make spirits from the salt, it is put into a flask with a little pure water and shaken well, and we continue to add water a little at a time, over a period of half a day, until most of the material is dissolved, but some still should remain undissolved to indicate that the solution is as concentrated as it can be.

To Drive Alkali Volatile From Sal Ammoniac by Use of Fixed Alkali

§ 48 (a). Two parts of dry alkali fixum is mixed with 1 part Sal ammoniac, both finely ground, passed through a sieve, blended well, which must be done quickly because alkali volatile begins to be driven off, the mixture is put into a retort to which an alembic is sealed with eggwhite and unslaked lime, and this in turn with a tube to a small retort to receive the distillate. The operation is carried out in a sand bath.

Note 1. Potash is saturated with acid of air, but this is released before the acidum salis is taken from the Sal ammoniac, so that some alkali volatile and acid of air are united occasionally.

(b) When the heat begins to work, the first thing to distill over is a little of the so-called spiritus salis ammoniaci, water being an important part of it, while the rest is sublimed and takes on the appearance of ice. When it begins to distill more slowly the heat is increased, and this is continued until a blue-white film appears in the retort, at which point the oven is allowed to cool, because nothing further now is expected other than some acidum salis, which alkali fixum cannot retain, and which accompanies alkali volatile to the end in Sal ammoniac, condensing to a blue-white film.

Note 2. If 3 parts of aged and well dried potash is used with 1 part of Sal ammoniac, we usually obtain 5/6 [part?] alkali secretum, which still has the odor of strong caustic. With chalk I have obtained slightly more.

(c) The spirits and the salt are kept separately, in such a manner as their volatility requires. There is some acidum salis in this, which is detected with silver solution, (sometimes even a little alkali fixum is driven over, which also precipitates silver), and this can be removed by a new sublimation with alkali fixum.

(d) The feces (residue after sublimation) contain sal commune regeneratum, since a decomposition occurs here and alkali fixum takes acid of salt from alkali volatile, by virtue of greater attraction.

To Drive Alkali Volatile From Sal Ammoniac by Means of Burned Lime

§ 49 (a). For this, Sal ammoniac is used with one part of slaked lime, equal amounts of both to reduce the cost, although some salt in dry form is readily obtained. Fully slaked lime seems to contain more water than is needed to dissolve the salt, by which [water] the spirits [of ammonia] are weakened, but otherwise the more liquid entering into the distillation the less heat is required. If we use slaked lime containing a certain amount of water, and afterwards determine how much water is present with the dry salt which is produced, we discover the basis for determining how much ought to be present in the beginning in order to produce an entirely volatile form with the same amount of lime; and then we should use slightly less than this amount in order to obtain full strength spiritus [of ammonia]. The tested lime should not lie too long afterwards, since it takes up more water immediately from the air. Since alkali volatile is somewhat expandable, a large receiver must be used in the distillation, because if not, some air will penetrate the cement sealing, which occurs to some extent in any case.

Note 1. The alkali volatile produced here is caustic, and we also obtain some dry salt, if the lime is fully and sufficiently burned; and in the event that no acid of air is present, then no alkali volatile concretum results [alkali volatile concretum must be ammonium carbonate]. The pure or completely caustic alkali volatile seems to be to some degree weaker than the [alkali volatile] concretum, but this is not true; the precipitations which can be obtained with the latter, and not with the former, have their basis in double attraction. See § 78, Note, and the Preface.

(b) To purify spirits [of ammonia] it is redistilled in a retort with a hatt [alembic] (the neck of the retort cannot be made as free from the contents as the hatt, which is then put on), and over new lime.

Note 2. Distillation of alkali volatile causticum is carried out most conveniently by Mr. WOULF´S method, in a vessel such as that described above, § 20, Note 1, under separation of acids. The test for completely caustic [pure] alkali volatile is that it should not precipitate lime water. Its specific gravity is to that of distilled water about as 0.9908 is to 1.0000.

(c) Stinkspiritus [spirits of ammonia] is made in the apothecary's shop by the method just described, and is called spiritus salis ammoniaci aquosus, if it distilled with water; but [spiritus salis ammoniaci] vinosus if spiritus vini is added.

<u>Note 3.</u> In the apothecary's shop 1 kanna [2.6 liters] of water is usually added for each pound of Sal ammoniac, and this is then distilled as long as the distillate has an odor. Into the bottom of the retort is first put the Sal ammoniac, and over this the burned lime, so that nothing is lost in the mixing.

(d) These spirits are too expensive to be used in the dye-house.

A dash of it added to smelling water [smelling salts] improves its agreeableness, but this mixture should not touch clothing because alkali volatile alters many dyes.

<u>Note 4.</u> Acidic and alkaline salts are so simple that up to now their basic elements have not been determined. STAHL, and after him most others, believes acids are formed from water and earth. The properties which are found in neither of these ingredients, but sometimes belong to their combination, give reason to declare that two tasteless substances can, in certain combinations, acquire a taste. Others have also urged that combustible material [phlogiston?] and air are necessary elements here. We have believed that alkalis differ from acids principally in the proportion of these ingredients, and that all

salts in common were more or less compositions of a single principium salinum, most closely resembling acidum vitrioli.

Simple salts are so firmly combined that up to now they have not been decomposed. The only thing which can be stated with confidence is that which was said about acid of salt above in § 23 Note 2, [regarding decomposition of acid of salt when it is passed over manganese dioxide]. Alkali volatile digested with acid of saltpeter and brunsten is also decomposed, showing such constituent parts as Mr. SCHEELE'S dissertation on brunsten reveals. Alkali volatile has not yet been reconstituted from these two materials. As to the rest of what is written about this substance, it consists of more or less similar information. We observe and read much that is confusing about simple salts, not only that which can be repeated with accuracy and that has shown promising results, but also often that which cannot be repeated according to directions. Those who continue to have their secrets and will not give them up, do not gain particular confidence from the natural philosophers of our time, who are not in the habit of believing anything simply on someone's word; on the contrary, they are suspected of making groundless boasts and they seem to be contented with concealing their supposed knowledge rather than to be caught with erroneous experiments or conclusions, which seems to happen readily if they reveal their modos procedendi.

Alkali vegetabile is commonly believed to be able to change into [alkali] volatile, but it still would be enlightening if some sensible chemist would examine this substance more closely than has been done up to now. The least that is known is that upon decomposition of alkali vola-

tile no trace can be found of alkali fixum, which compound ought to be expected if it [alkali volatile] is found therein [in alkali vegetabile].

FOURTH CHAPTER

About Neutral Salts

§ 50. Middle salts [Medelsalter] are of two kinds, either those having alkaline salts for their base, or those having alkaline earths; neither of these causes change in the color of syrup of violets, nor effervescence with either alkalis or acids.

Note. The author [Scheffer] takes Sales neutrales and [Sales} medii as synonymous, as has already been noted; but since two names are not needed to designate the same substance, I believe that these names can be properly fixed for substances which are really of different nantures and compositions, the former name [Sales neutrales] to designate compounds between acids and alkaline salts, and the latter [Sales medii] to denote proper combinations of salts with earths or metals.

It was not known in the author's time that certain combinations of both essential ingredients require a surplus of either one or the other. Purified tartar and Tartar Tartarisatus have several different properties (§ § 63, 66), so that they are regarded as entirely different compounds in the Pharmacy, but if the compounds have no other differences, and both are quite pure, then the tartaric acid in the latter is just saturated with alkali vegetabile, whereas in the former there is a certain excess of acid. Borax also contains an excess of alkali, and has an entirly different nature than that saturated with sal sedativus. We therefore have good reason to separate neutral salts into two groups: complete, which are fully saturated, and incomplete, which contain an excess of one of the essential ingredients.

§ 51. When an alkaline salt is dissolved and dropped into acid, there occurs: (a) Strong effervescence.

Note 1. This reaction is not observed when complete caustic or pure alkali is used, but only when it is combined to a greater or lesser extent with acid of air which is driven off by all other acids known up to now, which are stronger; and because it [acid of air] at the moment of separation becomes free of combination, immediately exhibits its elasticity [gaseous properties], and must follow hydrostatic laws so that it floats up through the heavier acid in the form of bubbles of air. Then occurs a foaming action which can be called effervescence, and is wrongly called by the name of fermentation, which is quite another process.

(b) The nearer it gets to the point of saturation, the stronger is the internal reaction.

Note 2. A proper salt indicates its satiation or saturation when it cannot dissolve anymore of either ingredient. This saturation is either absolute, when it cannot react further with either ingredient, or relative, which only happens with certain materials. This distinction is illustrated in many cases. The author [Scheffer] refers to this here as the property of richness, but only when the alkali is to some extent caustic, since the pure caustic reacts first, and later that which contains acid of air which reacts to some extent, although resisting a little.

The best test for saturation is to use test paper dyed completely blue with litmus, which reddens at the least excess of acid; for example it becomes red with weak vinegar, and becomes blue with the least amount of alkali. The acid itself often requires no more than its regular amount of alkali, but when an excess of either ingedient is present crystals form as a consequence, and something like neutrality is produced; sometimes this excess can be eliminated by sufficient mixing, but not always. Acid of air is probably present in several, if not all neutral and middle salts, which makes them more solid. See § 55, Note 7, and § 61.

(c) Alkaline salts require for saturation more of the stronger and heavier acids than of weaker and lighter acids. If, for example, 16 lod of decrepitated common salt is distilled with 13 lod of strong vitriolic acid until no more distills over, and the residue is finally annealed, it is found to weigh 19-1/2 lod, or 3-1/2 lod more than the original common salt used. If now the amount of alkali minerale is the same, and excess vitriolic acid is driven out through good annealing, then most of the surplus consists of acid, of which more is required for saturation in the one case [Na_2SO_4?] than in the other [NaCl?]. Further, if 16 lod of common salt is distilled with only 8 lod of vitriolic acid, the residue is found to weigh only 18 lod, and has consequently lost 6 lod of acidum salis, but if 8 [lod of vitriolic acid] drives out 6 [lod of acidum salis], then 13 must drive out 9-1/2. Thus in 16 lod of common salt there are 9-1/2 [lod] of acid, and consequently 6-1/2 [lod] of alkali minerale, and since in 19-1/2 lod of Sal Glauberi there is no more alkali, there must be 13 lod of vitriolic acid.

Note 3. Out of all later transcripts [of notes of Scheffer's lectures] which I have had occasion to see, not one word is found in support of this statement except in the last, that of Ahlströmer, where the above quoted statement is found. It is strange enough that a weaker acid should be required in a lesser amount for saturation than a stronger acid, but it contrasts entirely with the experiment which HOMBERG describes in his investigation of strong acids. I have therefore thought that this case deserves further investigation. To finally settle whether the above statement is correct or not, requires no extensive preparation; of one and the same alkali, either [alkali] minerale or [alkali] vegetabile, similar amounts are weighed out, and these are each separately saturated with different mineral acids, evaporated to dryness, gently annealed to drive off all water and excess acid, collected carefully and weighed, when it appears evident that with acidum vitrioli we get the most neutral-salt, second most with acidum nitri, and least with acidum salis. Since now the amount of alkali in all cases is the same, then any difference [in weight] must necessarily depend on the acid used. If we know the absolute quantity of acid added, which in every case reacted with the alkali, we ought to know for the latter [alkali] its content of water and acid of air. The former [water] is driven out by gentle heating, but the latter [acid of air] in the following manner: a weighed portion of alkali is put into a large flask A, together with as much water as is requir-ed for solution, while in a smaller flask B is placed as much acid as is required for saturation. A and B, filled with the substances which we have just named, are corked and weighed very accurately; let their weights be [respectively] a and b. Now a small portion of B is poured into A and as soon as this occurs, [flask A] is closed with a cork. After a few minutes the process is repeated, and we proceed in this manner until the alkali is completely saturated, which is determined by test paper; if the acid is slightly in excess it doesn't matter. After this the flasks are weighed again, and let their weights now be $\underline{a}$ and $\underline{b}$. If (b - $\underline{b}$) is the amount poured into [flask] A, this difference should be equiv- alent to ($\underline{a}$ - a), but it happens otherwise if the alkali is not fully caustic because there is always found to be a loss in A, and this is the difference between (b - $\underline{b}$) and ($\underline{a}$ - a), which appears to be the weight of acid of air lost. If A is small, some moisture is always lost around the cork, along with the elastic acid [acid of air], which introduces error; but if the flask is large enough so that the solution fills it only to 1/8 or 1/16 of its volume, no noticable error occurs.

With pure alkali, I believe it to be free from water and acid of air. If we proceed in the manner described above, I have found that with 100 parts of pure alkali minerale the following amounts of various acids are required for saturation:

Acid of vitriol	177 [parts]
Acid of saltpeter	135-1/2
Acid of salt	125

Acid of air	80

With 100 parts of pure alkali vegetabile:

Acid of vitriol	78-1/2
Acid of saltpeter	64
Acid of salt	51-1/2
Acid of air	47

Consequently each alkali attracts more of the heavier acids than the lighter acids, and the stronger or vegetable [alkali] needs less [of each acid] than the mineral [alkali]. All determinations agree with this.

(d) When an acid becomes saturated, so that neither effervescence or change in syrup of violets shows any excess, it is filtered and evaporated ad pelliculum, and after that it is put into a glass container and allowed to crystallize.

Note 4. Neutral-salts can in general be prepared in three methods, namely: (1) through direct combination; (2) through decomposition in the wet method; or (3) through decomposition in the dry method. Consequently common salt can be manufactured if acidum salis is mixed with alkali minerale; if some earthy or metallic middle salt which contains acidum salis is precipitated equally with alkali minerale; or similarly, if the residue is extracted after sal ammoniac is distilled with the proper amount of alkali minerale.

Note 5. [Crystallization and Heat of Solution.] The amount of water and heat determines the quantity dissolved, if there is enough of the substance being dissolved. It follows from this that when salt is dissolved in water it can be forced to separate through diminution not only of the amount of solvent, but also of the amount of heat. In evaporation it is usually directed to continue until a film appears on the surface, which arises from fine crystallization, because the uppermost layer or stratum of solution nearest to the air at first must lose the amount of water necessary for solution. The foregoing rule has on many occasions proved sound, but it is not necessarily to be followed in all cases. The reason is that some salts are dissolved in many cases better by hot water than by cold, and consequently no film appears during boiling before so much [water] is evaporated that it [the solute] congeals in a lump. Saltpeter and Glauber's salt, for example, belong here; they crystallize on cooling and should not be boiled longer than to the point where one drop put on cold glass or iron crystallizes in a few moments. Other salts, like common salt, are dissolved only slightly more abundantly in hot than in cold water, so they require constant evaporation and cease crystallization almost as soon as strong evaporation stops. Crystallization is without doubt the work of attraction between substances, but the proper reason for its symmetry, and the usually different compositions for different materials, is still a puzzle. I have a description in "Physics of the Terrestrial Globe" (§ 175) referring to this principle, which makes better known the experiments known up to now on this subject. One strange circumstance is that most substances on being dissolved in water create cold. This arises from the fact that as soon as the particles are separated from each other by the water they acquire freedom to attract toward each other, according to their ability, from their immediate surroundings a certain amount of material of heat, which is contained there until a stronger attraction draws it away, and at the same time the particles so intimately combine with it that the heat can no longer affect the mercury in a thermometer. Consequently at that moment the thermometer must fall, and this differs as one salt requires more or less material of heat for its saturation than another salt. Sal ammoniac, for example, can produce 25 degrees of cold, whereas saltpeter only produces 10 degrees, etc. Just as alkalis in saturation have much the same properties as acids, so it is here with the material of heat, which in union with moleculae salinae has no effect on the mercury in the thermometer. The faster the salt is dissolved, the more quickly the thermometer falls, often 10-15 degrees in a few minutes, but in a warm room sometimes 30, 40, or more minutes are needed afterwards to regain the temperature of the room. On the other hand, heat is aroused and the thermometer rises in a solution when a multitude of crystals separate at one time; the reason is that they can mutually attract one another, let loose from themselves the material of heat which they had formerly held, and are joined into crystals.

Note 6. [Nomenclature] Neutral and middle salts, which were known from ancient times, have often acquired absurd names, mostly because of false prejudices either concerning their special use in medicine or alchemy, or concerning their composition, etc., which comes out in the various examples which follow. If we take some common denominations, for example, all others could be named and designated on the same basis. Suppose we name first the basic or alkaline part [of the compound], and then the second part which designates which acid it is united with. Thus alkali vegetabile vitriolatum could be a useful name for that compound which is otherwise called irrationally Tartarus vitriolatus; alkali minerale nitratum in place of the incorrect nitrum cubicum; alkali vegetabile salitum in place of Terra foliatum or arcanum tartari; alkali vegetabile tartarisatum instead of tartarus tartarisatus; alkali volatile acetatum; instead of Spiritus Mindereri, etc. Such names are intelligible immediately, and convey the correct comprehension [about their composition], whereas those [older] names which derived from some other reason are either false or could be more fitting.

Note 7. The pieces of information about a salt, which should be investigated before we can regard its nature as something well known, are manifold; for example: how much water is needed for its solution at various temperatures; does crystallization occur best through evaporation or through cooling; [how] to determine the shape of crystals and their taste; the proportions of water of crystallization and other ingredients; the degree of cold which arises from its dissolution, or mixing with snow; whether it effloresces or deliquesces; whether it decrepitates in heat; its behavior on mixing with other simple and compound salts, in both wet and dry methods; its uses in medicine, chemistry, industry, and the arts; whether, and how, it is found already prepared in nature; how it is usually manufactured, and whether this process can be carried out somehow more quickly and better.

Note 8. In place of the author's [Scheffer's] table of neutral salts, which is somewhat incomplete and defective, and in order to save space, the following table is presented:

Sales perfecti duplices [Perfect double salts]

Vitriolic acid

Combined with: Alkali vegetabile /	Produces: Tartarus vitriolatus
Alkali minerale	Sal Glauberi
Alkali volatile	Sal ammoniacus secretus

Acid of Saltpeter

Comb. with: Alk. veg.	Prod.: Nitrum prismaticum
Alk. min.	Nitrum quadrangulare
Alk. vol.	Nitrum flamans

Acid of Salt

Comb. with: Alk. veg.	Prod.: Sal diges-tivus Sylvii
Alk. min.	Sal communis
Alk. vol.	Sal ammoniacus

Vinegar

Comb. with: Alk. veg.	Prod.: Terra foli. Tartari
Alk. min.	Alk. minerale acetatum
Alk. vol.	Spiritus Minereri

<u>Acid of Tartar</u>

Comb. with: Alk. veg	Prod.: Tartarus Tartarisatus
Alk. min.	<u>Alk. minerale tartarisatum</u>
Alk. vol.	<u>Alk. vol. tartarisatum</u>

<u>Acid of Urine</u>

Comb. with: Alk. veg.	Prod.: <u>Alk. veg. phosphoratum</u>
Alk. min.	Sal perlatus
Alk. vol.	<u>Alk. vol. phosphoratum</u>

<u>Acid of Ants</u>

Comb. with: Alk. veg.	Prod.: <u>Alk. veg. formicatum</u>
Alk. min.	<u>Alk. min. formicatum</u>
Alk. vol.	<u>Alk. vol. formicatum</u>

<u>Sales Imperfecti duplices [Imperfect double salts]</u>

<u>Acid of Arsenic</u>

In excess with: Alkalis	Produces: Sales arsenicales

<u>Acid of Tartar</u>

In excess with: Alk. veg.	Prod.: Crystalli tartari

<u>Alkali Minerale</u>

In excess with: Sal sedativus	Prod.: Borax

FIRST SECTION, CHAPTER IV, Neutral Salts

Sales Perfecti triplices [Perfect triple salts]

	Acid of Tartar	
Comb. with both Alk.	min. and vol.	Prod.: Sal Seignetti
	Acid of Urine	
Comb. with both Alk.	min. and vol.	Prod.: Sal microcosmicus

All new names in the above table are given in Italics [underlined], when this is the only name [available] for a substance which has not been given a common name previously.

Note 9. It is known that when several salts are mixed in the same water [solution] we obtain again on evaporation a complete multitude of salts different from each other; so it is only natural to form the opinion that these, as well as other substances, attract each other with a fixed power, from which it follows that a proportionate amount of them combine intimately and cannot be separated through crystallization. Tests and experiments confirm this opinion, which will be discussed at length later on, when I have brought to a conclusion a series of investigations which I have now begun. [This was in reference to B.'s publication a year later of his "Dissertation on Elective Attractions"; see my translation of this important work, published in 1968, by the Johnson Reprint Corp., New York, N.Y. J.A.S.] Meanwhile it can be stated that triplefold and manifold salts are more numerous than we have heretofore believed, indeed for the most part the mother liquors which are obtained after the crystallization of certain salts are of just such a nature. From this we can learn that

water which is saturated with one salt can still dissolve others. If for example water is saturated with saltpeter, it can still after that dissolve a large amount of common salt; and when the solution is not able to dissolve any more common salt, it can again dissolve a [further] portion of saltpeter. Such examples seem to show an ability of salts to dissolve each other. We have believed that many salts could be dissolved in water without changing its volume, because the pores of the mass made room for it; but Mr. R. WATSON has shown that if a flask with a very small neck is filled to a fixed mark [on the neck], even 1/40,000th part of salt, compared to the weight of the water, cannot be dissolved in it without the surface rising noticably; if a similar amount of salt is put into two flasks, of which one has double the volume of the other, but which have similar necks, then the surface rises similarly in both, which shows clearly that the pores are not the explanation here.

§ 52. Tartarus vitriolatus has small, irregular octahedral crystals which dissolve slowly. They are found in the residue after the distillation of saltpeter with oil of vitriol (§ 20). Arcanum duplicatum, which is obtained after distillation of saltpeter with vitriol (§ 17), also gives Tartarus vitr., but contaminated with earth of iron [järnjord], which it contains in abundance, and through the separation of which, during evaporation until a film forms, the solution remains continually cloudy although originally it was clear. Presently alkali fixum prevails, but not sooner than when overheating causes precipitation. If this salt is to be pure, alkali vegetabile must be joined to it, but when the alkali reaches excess, which is still easily saturated by acid of vitriol.

Note 1. That which gives up earth of iron here is a vitriolic residue, and this is decomposed by boiling, which is known. Alkali saturated with acid of vitriol does not precipitate at all, and free alkali cannot be found here, since more nitrum is not attacked during distillation than the alkali is able to receive of the acid mentioned.

100 parts of Tartarus vitriolatus in crystalline form contain about 51-1/2 [parts] of alk. veg., 40-1/2 [parts] of acid, and 8 [parts] of water of crystallization.

Note 2. Tartarus vitriolatus, since it is composed of the strongest akali and the strongest mineral acid, forms a very firm union, for the reasons mentioned above. STAHL therefore reported it to NEUMANN as a very difficult problem to decopmpose this same [compound] in any manner. We did not understand correctly in the beginning, without specifying it as dissolving, how to melt the combustible substance from which sulfur is produced, since this could not be done by wet methods. BOULDUC first found the correct answer in 1774, how with the help of [their] calces, lead, quicksilver, copper, iron, etc., dissolve in other acids. When any of these solutions is mixed with dissolved Tartarus vitriolatus, precipitation occurs immediately and decomposition occurs through double attraction, in the manner which is explained in the Foreword. Mr. BAUMÉ in 1760 showed another way: if Tartarus vitriolatus or Sal Glauberi is dissolved with the aid of heat in approximately the same weight of strong acid of saltpeter, with the former [Tar. vitr.] prismatic saltpeter crystallizes on cooling, and with the latter [Sal Glaub.] the so-called nitrum quadrangulare. Consequently acid of saltpeter can by itself drive out vitriolic acid in the wet method, although the former must yield to the latter in the dry way. Mr. SCHEELE has even found that acidum salis, as well as [acidum] tartari, likewise decompose both Tartarus vitriolatus and Sal Glauberi, but not altogether with nothing left. Acid of saltpeter cannot decompose more than about 1/3 of Tartarus vitriolatus, the other 2/3 being unaltered although very newly compacted; it has attracted vitriolic acid to itself in place of the decomposed part, which protects itself in crystallization, and although it stays in solution longer than the rest, it can be driven out both by spiritus vini and by the proper amount of heat. On occasion, however, this alkali seems to have the strength to bind itself to a certain excess of acid, where if acid of saltpeter is present in this way we can imagine three attractions which react, namely: [1] the attraction of alkali for the vitriolic acid, which is necessary for saturation; [2] the attraction [of alkali] for a certain excess of the same acid; and [3] finally the attraction of acid of saltpeter for alkali; the latter two of which overpower the first to a fixed degree and produce the remarkable circumstances which are described above.

Tartaric acid dropped into a saturated solution of tartarus vitriolatus precipitates a true cream of tartar; the same thing happens with saltpeter and digestive salt [KCl], when acidum tartari is dropped into their saturated solu-

tions. The reasons are the same as those given immediately above.

§ 53. Sal mirabile Glauberi, also called Sal Umeåense, consists of the alkali of common salt saturated with vitriolic acid. It is found in the residue when common salt is distilled with vitriolic acid (§ 24), where it is observed that 5/8 concentrated vitriolic acid with common salt gives pure acid of salt; and to obtain pure Sal Glauberi in the residue more should be used, but still not enough to make it sour. The same salt is obtained if alum or vitriol is precipitated with alkali minerale. The longer the solution boils the less crystals there are, and yet this should be continued for some time after the first [crystals] begin to appear because otherwise the crystals are so wet and soft that they cannot be separated from the water. The crystals are four-sided prisms; they should not be kept in a closed place, although DIPPEL for example still maintains that this is entirely possible; they turn to flour in air; they do not precipitate silver solution if they are free from acid of salt. In the first crystallization common salt seldom appears, but it does appear in the following ones if decomposition was incomplete.

The Sal Glauberi which is obtained in the apothecary's shop is usually slightly acidic. Alkali minerale is a purgative by itself, but it harms the intestines, so it must be saturated somewhat with vitriolic acid; we are so accustomed to common salt that it does not work [as a purgative]. The dose of Sal Glauberi is 1 uns [1 Swed. uns = 29.7 g]. This [dosage] has been determined at Umeå.

Note. That certain salts cannot crystallize in vacuo seems to be a failing with acid of air, which probably makes up only a part of the composition of the salt, as is shown in the following example:

100 parts of Sal Glauberi in fresh crystals contain 16 parts of alkali, 26 [parts] of acid, and 58 [parts] of water of crystallization.

§ 54. Sal ammoniacus secretus Glauberi consists of vitriolic acid combined with alkali volatile. It has spike-like crystals. When clay is distilled with vitriolic acid, usually some sal ammoniac is sublimed into the neck of the vessel, if alkali volatile is previously contained in the clay.

Note. This salt is not known with certainty to have any special property, either in Medicine or the arts. That calx of tin may dissolve in a solution of this salt, for the preparation of carmine [dye?], has not been very successful with me up to this time.

§ 55 (a). Saltpeter (nitrum prismaticum) consists of its own acid combined with alkali vegetabile, and contains naturally almost the same amount of each.

Note 1. 100 parts of saltpeter in crystals contain about 49 [parts] of alkali veg., 33 [parts] of acid, and 18 [parts] of water.

(b) It [saltpeter] is found in fertile soil, mostly under old barns and outhouses, and is found especially wherever growth is promoted by means of the strong urine contained therein. Such soil is tested for its quality in the following manner: a known amount [of soil] is put into a vessel, water is then added, which is then filtered off, and finally the soil is poured onto the filter and cold water added as long as the filtrate tastes salty. Heat takes so much material with it that saltpeter cannot be separated. If the solution is boiled too long, pure saltpeter is not obtained, but a thick syrup filled with coarse matter, common salt, sal ammoniacus fixus, etc. So evaporation should continue only until crystallization, that is until a drop of the solution placed on cold iron crystallizes at once. When the first extraction produces no more crystals, the solution is boiled again to the same point, and this is continued until no more crystals are obtained. All the niter collected together has the appearance of the extracted soil.

(c) Evaporation until a film forms is entirely unsatisfactory with saltpeter, because a film can appear at any time, and that [film] from the extracted material may not appear before the whole mixture becomes like thick gruel when it cools. The more gently the extract is boiled, the larger and more free of water the crystals become; for this purpose, the solution is removed from the fire as soon as the first shiny particles are observed in a portion of the liquor when it is allowed to drop on iron; it [liquor] is placed first in a warm room to cool, and then in cold.

Note 2. One cubic foot of the best soil yields about 10 to 12 lod of pure saltpeter, 4 to 8 [lod] of mediocre, and 2 to 3 [lod] of poor. In Sweden 28,000-30,000 lispund [1 lispund = 20 pounds] is manufactured annually, which if it were purchased abroad would cost over 1-1/2 tunna of gold. [1-1/2 tunna, 10,000 oz.(?) of gold might be worth as much as $4 million U.S. at the present price of $400 per oz. J.A.S.] This important business has in recent years had quite a few improvements, but obviously it still needs more, not in ideas, but in indisputable experiments grounded in science, concerning the amount of its salt, before it can attain its maximum development.

Note 3. To better determine what different evaporation [methods] would do to saltpeter, I have weighed from the same batch two portions, A and B. A was dissolved and boiled until a drop of it, dropped on cold porcelain, after 10 minutes first showed some crystallization around the edges; whereas B was boiled until a drop congealed instantly. The boiling was carried out in similar vessels, which were placed near one another; B began to form crystals on the bottom usually within 5 to 10 minutes at the most, whereas this took longer with A, and the crystals were formed partly on the surface and were smaller than in B. After decanting the liquid two times, and boiling to the same point as before, the process was continued in the same manner as long as any liquid remained. When finally all the saltpeter from A and B was weighed, it was found that the product from B weighed 2 percent more, which probably came from the water entrapped during rapid crystallization. The volumes of A and B when fully evaporated were approximately in the proportion of 100 to 68.

(d) Nitrum is obtained free from common salt as follows. Most of it is separated by dissolving the saltpeter in water, boiling it down until a film forms and holding it until crystallization, when common salt crystallizes first, but when the solution becomes lukewarm it is put into another vessel wherein the saltpeter crystallizes itself; this is dissolved again and finally evaporated until a drop placed on cold iron produces visible crystals, and most of the common salt remains behind in solution, although the saltpeter obtained can still contain 3 to 5 lod [of NaCl] per 100 pounds. The solution which is left is boiled again and set aside to crystallize, when more nitrum is obtained, since the common salt requires greater evaporation, but much is lost in this manner.

<u>Note 4.</u> Different solutions of saltpeter and common salt, after different amounts of heating, give a convenient way to separate them from each other for the most part. Water, of which about 15 g dissolves one part of the former [saltpeter], near the boiling point dissolves about 4 parts, whereas for the latter [common salt], the difference is only 6/77. [The solubil- ity of KNO_3 increases from 32g/ 100g water at 20 C to 247g/100g at 100 C, whereas the solubili- ty of NaCl only increases from 36g/100g at 0 C to 40g/100g at 100 C. J.A.S.] Consequently if a solution containing similar amounts of each salt is saturated at medium temperature, and 3/4 of

the water is boiled away, then nearly 3/4 of the of the common salt will have separated through crystallization, without any of the saltpeter crystallizing out. This is a method of approximation which continually reduces the amount of common salt in solution, but always leaves some behind.

(e) If we wish to have completely pure saltpeter, it can be made from pure aqua fortis [skedvatten, dilute nitric acid] and pure alkali vegetabile; but also from the common variety from which all common salt is precipitated by silver solution, after which the solution is filtered, the water distilled off, and the residue strongly heated, by which process the silver loses its acid and drops to the bottom; then the salt is dissolved, filtered, and crystallized. This latter kind of nitrum contains so-called nitrum quadrangulare, in place of common salt.

Note 5. Obtaining pure saltpeter is a difficult task, since we must crystallize it 3 or 4 times before it becomes truly white. So the following method is of the greatest importance. Raw saltpeter from the first crystallization is dissolved in boiling water, and to this is added 3 percent of powdered alum with stirring; whereupon the solution immediately becomes murky because of the earth of alum, which is separated from its acid, and a filter retains it [along with] that which would otherwise make the salt brown. After filtering, the solution is boiled down and brought to crystallization in the usual way. The crystals should be removed immediately and spread out on heavy paper to dry, after which it is found to be fully as white as that which has been recrystallized 4 times, and causes silver solution to become less cloudy; its solution does not react with crystallized sugar of lead, by which it is shown that no appreciable trace of acid of vitriol is present.

(f) Pure nitrum does not change a solution of corrosive sublimate or sugar of lead, but since pure nitrum is so rarely obtained, some have believed that it was an essential property of nitrum to muddy these solutions. 23 lod of cold snow water dissolves 6 [lod] of purified saltpeter; BOERHAAVE requires 5-1/2 [parts] of water for 1 part saltpeter, but seems to have used a dried form [of saltpeter] which is water-free.

Note 6. Corrosive sublimate must have been mis-written for quicksilver dissolved in acidum nitri [I.E., $HgCl_2$ must have been written for $Hg(NO_3)_2$ by Scheffer] since it would be very unusual to find a solution of saltpeter which by any method would be muddied by the former salt [$HgCl_2$].

(g) Sometimes Tartarus vitriolatus is found mixed with nitrum, which can be removed by the same method as that for common salt.

(h) Nitrum melts easily in fire but does not burn unless phlogiston is present, when with vehement hissing the acid is destroyed, leaving behind pure alkali vegetabile. The hissing comes partly from the phlogiston and partly from the quantity of air which it [nitrum] contains and which is now set free. [Scheffer here was close to a method for preparing oxygen. J.A.S.]

Note 7. If a hot coal is cast into concentrated acid of saltpeter it gives indications of detonation, but far weaker than with saltpeter, which fumes and sets the phlogiston afire. Dr. HALES has shown that 1/8 of the weight of saltpeter consists of elastic fluid; when this is examined, a part is found to be acid of air. Detonation was mentioned previously in § 36 (c); the flaming is increased by the elastic matter which is set free. [Here too B. was flirt- ing with the discovery of oxygen. J.A.S.]

Through melting alone, only the water of crystallization goes off, which scarcely makes up 1/8 of the weight, but if the mass is observed to glow, some part of the acid is gradually phlogisticated, which therefore can be driven off with vinegar, and under union with its alkali it makes up a mother liquor which will not separate.

(i) Nitrum ignites all metals in fire, except gold and silver.

Note 8. White gold [platinum] still resists saltpeter, if it is free from iron.

(k) Sal prunellae is made from pure nitrum, which is melted in a crucible, and upon it is gradually cast about 1/16 to 1/10 [its weight] of flores sulfuris, which detonates, and after melting, is given off. This [sal prunellae] is used against fever, sore throat, and inflammations.

Note 9. So this, as with the following preparation, is with the same proportions entirely different in quality, depending on completeness of mixing, the rigor and length of duration of heat, the size of the crucible, etc. It just happens that one part [of nitrum] uses different proportions of sulfur.

(1) Nitrum is mixed and some sulfur is dropped on it in a crucible heated to glowing, the mixture is kept melted for one hour after detonation, it is poured out, dissolved in water, filtered, and evaporated to dryness, and so is produced Sal polychrestus or Polychrestum parisiensium, good for obstructions in ague.

Both contain nitrum mixed with Tartarus vitriolatus, but in different proportions.

<u>Note 10.</u> Saltpeter is used in incredible amounts in medications, in the ditillation of aqua fortis, in mining explosives, certain cures, glass-making, experimental fluxes, but mainly for gun powder. See Section 4. [Section Four deals with Combustible Substances.] China abounds in this salt, and the Dutch alone transport over 500,000 pounds a year from there [China], not to mention what is imported in ships from other countries.

§ 56. <u>Nitrum cubicum</u> consists of the acid of saltpeter saturated with alka- li minerale. It is obtained in the residue when common salt is distilled with acid of saltpeter; it crystallizes with difficulty after evaporation to a film, but must be caused to evaporate in an oven or in sunlight.

Its crystals are small cubes, which detonate with coal dust and give pure alkali minerale.

<u>Note.</u> Detonation occurs with a yellow flame, whereas [nitrum] prismaticum gives a blue [flame]. [Here B. is giving the "flame tests" for sodium and potassium. J.A.S.] The shapes [of the

crystals] were noted previously in § 39, Note 3. How it can be decomposed with acid of salt is noted in § 20, Note 6, and § 27, Note 1. 100 parts of nitrum quandrangulare in fresh crystals contain about 32 [parts] of alk. min., 43 [parts] of acid, and 25 [parts] of water.

§ 57. Nitrum flammans consists of acid of saltpeter combined with alkali volatile. It catches fire when heated and detonates without addition of phlogiston.

Note 1. This salt melts in a candle flame, as well as boiling and fuming, but when set on fire the border of the flame is colored vermillion; it melts easily with a blow-pipe, but can scarcely be formed into a bead with charcoal without detonation occurring; with other substances, such as glass or metal, it can be entirely fumed away without melting. If nitrum flammans is dropped on red-hot glass it detonates immediately with a yellow flame (common saltpeter only melts); if the glass is hot, but not glowing, the salt melts and becomes clear; but if it is left in the cold, and is then heated, it fumes as usual although strong heat must be applied to it. From this it should be noted how the reactions of this salt are to be corrected; that it is in no wise as dangerous as some have indicated to carry out its distillation, which I have done many times without the least difficulty.

One hundred parts of alkali volatile purum require for even saturation 132 [parts] of acid of saltpeter.

§ 58. Sal digestivus Sylvii is also called, although improperly, Sal communis regeneratus. It consists of acid of common salt saturated with alkali vegetabile. It is found in the residue when Salmiak [sal ammoniac] is distilled with alkali fixum (§ 48). The extract is filtered and evaporated until a film forms with cubic and spike-like crystals. It is obtained also from the residue when saltpeter is detonated with Salmiak.

Note. This salt crystallizes in four-sided prisms, when alakli slightly prevails, but the crystals are cubic like common salt, from which it is distinguished by its sharper appearance.

One hundred parts of Sal digestivus contain 61 [parts] of alkali purum, 31 [parts] of acid, and 8 [parts] of water.

§ 59 (a). Sal communis or common salt [koksalt] consists of acid of salt saturated with alkali minerale. It is produced by 3 methods: (1) from sea-water, which in northern Europe is let into a kind of basin and evaporated by solar heat, after which the mother liquor is drained off; this is called sea-salt (Sal marinus); (2) that which is boiled from salt springs near Luneborg and Halle [in Germany], which is called spring-salt (Sal fontanus); (3) that which is mined in solid form in Poland, Hungary, and Russia, under the name of rock salt [bergsalt] (Sal gemmae).

Note 1. Rock salt is sometimes obtained so clear that salt-cellars are made from it, which look like they were made of glass.

(b) All three kinds crackle when heated.

(c) It melts in moderate heat [804 C].

(d) It precipitates a solution of sugar of lead and silver.

Note 2. Mr. SCHEELE has discovered a special method of decomposing common salt. Litharge is pounded very finely, filled into a funnel, and salt brine allowed to run through it. In the vessel below [the funnel] is obtained a solution of alkali minerale, sometimes contamin- ated with a little common salt and lead solution; the former [NaCl] is removed with more litharge, and the latter in open air.

(e) It has cubic form.

Note 3. Cubes crystallize at the bottom, made up of four-sided cones, but on the surface only pyramids are formed.

(f) It dissolves on mixing with golden acidum nitri.

(g) It is unchanged by calcination and melting; in combination with food in our body it comes out again in the urine, whereas alkali vegetabile becomes volatile; in combination with vitriolic acid and acid of salt, alkali fixum also passes on with no volatility, which can be concluded from finding of Sal Glauberi and nitrum in urine.

Note 4. In addition to common salt, we find in urine Sal microcosmicus, Sal perlatus, Sal digestivus, and Sal Glauberi, as Mr. ROUELLE, the younger, determined, in the same amount be- fore as after putrefaction.

(h) It separates to coagulate with oily materials, which [process] is observed if soap is dissolved in the water and common salt is put into it, when the material turns itself into white lumps, and then if alkali is dissolved in the water, no lather can be produced. Acids and the greater part of middle salts have the same property, and indeed many believe that all salts serve to coagulate oily substances in water.

Note 5. This ability to coagulate, which properly belongs to neutral-salts, is very difficult to explain.

(i) Decrepitated common salt requires 3-1/2 times its weight of water to be dissolved in the heat of digestion.

Note 6. The amount of common salt which is dissolved at 15 degrees above zero, I have found to be, compared to that which the same amount of boiling water takes up, approximately as 71 is to 77. [See §55 (d), Note 4.]

Common salt is now an article of commerce, used to season food, to prevent spoilage, etc. Oddly enough, in small quantities it helps digestion, but in large amounts it prevents change. That kind which is produced from boiling saltpeter can be purified, but is usually discarded; it can be used in certain leather dressings; it is even used to advantage on accidental fires, according to the proposal of Mr. BAUMÉ, since a fiber covered with common salt does not burn.

(k) It usually contains more acid than alkali.

Note 7. One hundred parts of common salt in crystalline form contain about 42 parts of alkali, 52 of acid, and 6 of water of crystallization.

§ 60. Sal ammoniac vulgaris consists of acid of salt saturated with volatile alkali. It is manufactured in Egypt from cattle dung, which is first used for fuel, after which the soot is collected and the Salmiak is separated by sublimation in a glass retort. Acid of salt comes from soaking of manure, and alkali volatile from its putrefaction. A little alk. min. contaminates it [alk. vol.] and is sometimes extracted along with it. Since salmiak from Venice is sold more and more, we imagine that it is prepared from urine, salt and soot, but that is a mistake.

The name arises from the fact that this salt formerly was found near the Temple of Jupiter of Ammon [Egypt], coming from camels' urine; or else from ammon, which signifies sand, because the salmiak of old was found in sand, although this was not entirely comparable to ours, as can be gathered from the description of DIOSCORIDIS.

When salmiak is cast upon molten nitrum, detonation takes place and Sal digestivus is obtained in the residue.

Note. This salt has for many years been brought from Egypt to fill the entire needs of Europe, but now Salmiak factories are constructed, one in Braunschweig by the GRAVENHORSTS, and the other in Paris by Mr. BAUMÉ. I have not seen samples of either manufactured product up to now. But to anyone who lives near the sea and has any knowledge of Chemistry it would not be particularly difficult to prepare Salmiak.

This salt is decomposed in distillation with alkali fixum, burned and unburned lime, and with almost all metals, and in fact it even reacts with several of those substances in cold mixtures or digestion [in the cold]. Salmiak cast upon glowing charcoal causes the flame to become blue-green and alkali volatile is believed to be destroyed; and Mr. SCHEELE has found that if a cold iron is held over the colored flame only drops of acidum salis are condensed on it.

§ 61. Terra foliata Tartari consists of distilled vinegar saturated with alkali veget., and gives a disorderly mass of leafy crystals.

Note. To obtain this white, which is usually deired, the vinegar used ought to be that which first comes over in the ditillation, since the latter half is contaminated with a substance attacked by the acid, and the salt becomes gray with it, even if it is diluted in a waterbath. How much vinegar is required by Alkali tartari for saturation is difficult to determine because we cannot calcinate away all the water, without also driving off some of the acid; it shows such a great propensity for deliquescence that it cannot be weighed with confidence, because respiration alone makes it damp during the process. Usually 23 parts of terra foliata is obtained from 16 parts of alkaline salt. If a deliquescent portion is mixed with acid of air it separates in crystals which then do not melt in the moisture from the air. This salt is found to be already mentioned by PLINIUS, and has acquired many nonsensical names.

§ 62. Sal ammoniacus aceti con- sists of vinegar saturated with alkali volatile, and gives spike-like crystals.

Note. This [salt] is otherwise known by the name of Spiritus Mindereri, is usually used in solution, and is quite difficult to crystallize, but this can be done if the solution is evaporated to form a thin syrup.

§ 63. Tartarus Tartarisatus is purified tartar saturated with alkali vegetabile.

Note. This salt, if saturated to neutrality, cannot be made to crystallize, but if made slightly alkaline it crystallizes without difficulty. If an excess of acid is permitted, a perfect cream of tartar precipitates, which because of its insolubility cannot be held in solution, and thus all tartarus tartarisatus can be changed into crystalli [tartari] or cremor tartari. If a mineral acid is instilled we also get cremor [tartari], since the stronger [mineral] acid drives off the acid of tartar, which in a certain excess amount combines with tartarus tartarisatus. Consequently we have here an unusual salt, which when saturated to neutrality does not crystallize, but does crystallize with an excess of either alkali or acid, both of which by themselves are difficultly crystallized. It dissolves some calx tartarisatus, and also iron and calcinated antimony.

§ 64. Tartarus citratus is acid of citron saturated with alkali vegeta- bile.

Note. It should be called alkali veg. citratum , since only one constituent part of Tartarus enters into the composition. Sal absinthii is often used [as its name] when another alkali is used instead [of alk. veg.].

§ 65. Liquor cornu cervi succinatus, or Prinsen's yellow drops, consists of spiritus cornu cervi saturated with sal succini. Into the former is put some of the latter a little at a time, which is stirred and dissolves with hissing. This is continued until no further reaction is observed when fresh salt of yellow amber is added, after which it is filtered, since during saturation some oil separates from the spirit of hartshorn; finally it is preserved in a well corked flask.

Since salt of yellow amber is expensive, these drops discovered by Archiater Hjjärne are seldom found saturated, but usually prevailing in alkali, so that more salt of yellow amber should be added to make it good.

Unrectified salt of yellow amber ought to be used so that some of its oil may be present in the liquor. The drops are cooling and sweat-producing.

Note. This salt can be crystallized but is still always used in liquid form. Some prepare it from tartar and hartshorn, which are distilled together in certain proportions. This is an imitation which tolerably resembles Prinsen's drops.

§ 66 (a). Tartarus or vinsten [tartar] consists of acid, earth, and a fatty material, since it does not effervesce with other acids, causes a white precipitate with sugar of lead, gets opal colored when alkali is added to its solution, reddens litmus juice, and burns in fire.

(b) It is formed in wine casks and forms a crust all around them. Consequently it is an essential salt of fermented grape juice.

Note 1. This salt is found in the juice itself before it is pressed out, as Mr. ROUILLE [Rouelle?], the younger, has found.

(c) Tartar requires much water for its solution, but gives a purer salt after filtering and evaporation. That which forms on the surface is called Cremor Tartari, but that which crystallizes on the bottom, crystalli Tartari. A stronger solution is formed if the water is heated, but the crystals become yellow with oiliness. Red tartar can be similarly purified, but gives less than the white.

Note 2. Most tartar is purifed in Venice, as far as I know, and next most in Montpellier; in the former place eggwhite is used, and in the latter place some kind of clay.

(d) If a glass retort is almost filled with pulverized red tartar and a receiver is sealed onto it with flour paste and heavy paper, and this is cautiously distilled on a sandbath, an acid phlegm comes over called spiritus Tartari, and at the same time a burned oil, which can be separated in a separatory funnel. The residue is discussed above (§ 35 (a)).

Note 3. The acid and the oil ought really to be separated completely before this, because otherwise the former absorbs the latter, changing it thereby and causing a bitter taste.

One hundred parts of purified crystalli tartari gives on burning 32 parts of alkali vegetabile, whereas almost 6 parts are acid of air; 3/100 earth is obtained from the same quantity. White tartar gives 21/100 alkali and 8/100 earth; the red [tartar] contains a little less alkali and more earth. That the essential consituent parts of purified tartar are alkali vegetabile and acid of tartar, in the foregoing proportions, the latter [acid of tartar] in excess to a certain degree, or more abundantly than is required for saturation, can be shown clearly by both analysis and synthesis.

§ 67 (a). Sales essentiales are acidic and are obtained from plants, especially sour [plants], without heat in the following manner: the fresh plant is crushed, the juice pressed out, mixed with double its amount of warm water, filtered, and evaporated to the consistency of honey or thin syrup. This is put in a stone vessel, covered over with oil to prevent mold, placed in in a cellar where after 8 months is found around the brim a body of salt which, freed from juice and impurities, is called Sal nativus or essentialis plantae ex succo crudo. Tartarus, Sal acetosellae, etc., belong here.

Note 1. Cremor tartari and Sal acetosellae, when they are dissolved in water, and quicksilver solution [in acid of saltpeter] is added, can be wholly and completely decomposed by double attraction, the quicksilver being joined with the acids and precipitating, whereas the alkali takes the acid of saltpeter to itself.

(b) Sugar is also an essential salt, concentrated from the juice pressed from sugar cane, but which does not change the color of syrup of violets, but contains its own acid; if an alkali is also found it has not been identified. Sugar can easily be burned to charcoal but scarcely to an ash. If silver solution is precipitated with this it can be ambiguous, if the alkali is to be deduced, because sugar cane grows in salt water. Sugars detonate with charcoal and nitrum lunae [silver nitrate], after which alkali ought to be found in the crucible, and is probably of the same nature as mineral [alkali] from the salt water in which the cane grows.

Note 2. If acidum nitri is used several times to extract sugar, some entirely acidic crystals usually crystallize out at last, which can be saturated with alkalis, earths, and certain metals to form different neutral- and middle-salts. This [acid] can be called acid of sugar, which is found also in gum arabic and various other substances. Acid of saltpeter is so far the only known means for bringing this [acid] out of sugar, which is dephlogisticated during extraction as much as required.

To Drive Sal Succini Out Of Yellow Amber.

§ 68 (a). Yellow amber [bernsten] is broken up into small pieces and a re- tort is half filled with them. Over this is spread one tum [inch] of dry sand; then a receiver is cemented on with flour paste and this is distilled carefully on a sand-bath, partly so that no yellow amber may boil over, and partly to retain the oil which dissolves more of the salt by means of stronger heat. We continue to heat strongly, not quite to the end because a thick black oil then covers the salt and to some extent reacts with it.

(b) During the distillation phlegm passes over first, then salt, and finally oil. The oil has a sour smelling odor, is used in Medicine against pleurisy and similar illnesses, comforts the stomache, acts as a purge when other oils obstruct, is thick and brown, but becomes clear, light yellow, or almost water-clear when it is redistilled with water.

Note 1. When yellow amber is distilled, Mr. SCHEELE has found, that a liquid goes over at the same time which behaves very much like vine- gar. This gives us reason to believe that yellow amber is derived from the vegetable kingdom.

(c) The salt condenses in the neck of the retort, is long needled, and yellowed by oilyness. By means of sublimation in a mixture with sand, it can be made much lighter, but it never becomes as white as other salts; it is used only in medicine.

Note 2. If the salt is mixed with lime- free clay and sublimed it becomes purer than in the other usual method.

(d) This salt is acidic in its nature, but mixed with oily material and earth like essential salts. We could say that alkali vol. is also an oily material, but it does not burn without the help of acidum nitri, and leaves nothing like an earth behind; pure alkali vol. produces no oily material which is not salt, nor any salt which is not oily, whereas these two materials are inseparable here. On the other hand, an oil is obtained from Sales essentiales and sal succini which [oil] is not a salt, and which after burning off leaves an earth. Yellow amber itself seems otherwise to resemble a mineral product more than a vegetable one.

Note 3. Purified Sal succini, dissolved in water and spiritus vini, effervesces with mild alkalis, allows itself to be saturated with all of them, and these neutral-salts do not make aqua regia from acid of saltpeter, do not precipitate silver solution, but sugar of lead certainly, yet are precipitated as completely different lead salts. Consequently we cannot regard salt of yellow amber as containing acidum salis, although Mr. BOURDELIN suggests this; his reasons are that yellow amber is found near the sea, and after detonation with pure nitrum gives a residue which tastes salty, crystallizes in almost cubic form, crackles on [hot] coals], gives gray fumes with concentrated acid of vitriol, and precipitates quicksilver and silver from aqua fortis with a white color. This salt is entirely volatile, but requires a much stronger heat than alkali volatile; it is very expensive and consequently is often likely to be adulterated with added sugar, cremor tartari, salt of hartshorn, acid of vitriol, etc. In Königsberg it is prepared largely from bits and chips which are collected from the processing of yellow amber. This salt has a place here among the closest relatives of the essential salts.

§ 69. Borax is produced from nature in Persia, the East Indies, and Transylvania, looks red-brown, and is called Tinkal before it is refined in Holland and Venice. It then shows the following properties:

(a) It requires 18 times as much water to dissolve it as its own weight.

Note 1. It dissolves also in spiritus vini, and when this [solution] is set afire, it gives a green flame.

(b) It makes syrup of violets green.

<u>Note 2.</u> Borax combined with a like amount of sedative salt [boric acid] does not make syrup of violets greener.

(c) It precipitates alum solution, silver solution, and other metals dissolved in acids.

<u>Note 3.</u> These precipitations arise from the abundance of alkali minerale.

(d) It does not effervesce with either acids or alkalis when the solutions are cold, but does when they are warm and abundant.

(e) It melts easily with swelling in heat.

<u>Note 4.</u> The clear glass which is obtained from melting can be dissolved in water again and crystallized.

(f) Borax combined with acid, so that syrup of violets is not altered in color, causes litmus juice to become red, and if more borax is added, syrup of violets becomes green. There would seem to be evidence in this of both alkali and acid.

(g) If borax is dissolved in syrup of violets and the solution boiled somewhat, crystals slowly form which resemble Sal Glauberi.

(h) If borax is saturated with acidum vitrioli and distilled, a white salt condenses in the neck of the retort which is called Sal narcoticus, [Sal] anodynus, or [Sal] sedativus, which provides the way for this compound to be identified and resolved.

<u>Note 5.</u> Borax dissolved in boiling water but not an excessive amount, and saturated with acidum vitrioli, deposits Sal Sedativus, and when

the remainder is crystallized Sal Glauberi is obtained; with acidum nitri we obtain nitrum quadrangulare, and with acidum salis common salt. These behave entirely as they should if they were entirely free from Sal Sedativus, which readily reacts with them. Consequently, borax is Sal Sedativus with an excess of alkali minerale.

(i) Borax saturated with acid cannot be boiled down to a scum unless the time is almost past in which it can be prevented from crystallizing.

Note 6. One hundred parts of purified borax contains about 17 [parts] of alkali min- erale, 34 of sedative-salt [boric acid], and 49 of water. It effloresces in dry air.

Since Mr. GRILL ABRAHAMSSON brought home from the Orient some raw earth, it seems that this salt can be obtained in natural and not artificial form. Mr. BAUME´ experimented with vitrified earth macerated for a long time with alkali minerale and animal fat in free air, and obtained borax. Mr. CADET continues to insist that copper enters into this salt.

(k) Borax is used as a flux for melting metals, particularly in the purification of gold, wherein it is called Chrysocolla.

Note 7. Borax has a strong ability to vitrify earths, and metallic calces, without reduction.

§ 70. Sal Seignetti [potassium sodium tartrate; Rochelle salt] is pre- pared from Crystalli Tartari, which is dissolved, saturated with Sal sodae, boiled for some time, and allowed to crystallize.

Note. This salt carries the name of an apothecary in Rochelle, who sold it at a good profit for many years and kept the method of preparation secret. LEMERY mentions it already in 1675. Since then several attempts to make it have been tried, but to no avail until 1731 when two French Academicians, BOULDUC and GEOFFROI, at one and the same time, disclosed its composition. It is by itself a three-fold salt, composed of acid of tartar saturated in part with vegetable [alkali], and partly with mineral alkali, or more accurately it is Tartarus Tartarisatus [K_2Tart. 1/2 H_2O] combined with alkali minerale tartari- satum in such proportions that it can have an ex- cess of an acid like crystalli tartari. Mr. SCHEELE has devised the following directions for a method for preparing Sal Seignetti. Two pounds of crystalli tartari are saturated with potash, and afterwards 5 ounces 2 drachmas of common salt is dissolved with it [1 ounce (uns) - 8 drachmas - 29.7 grams]; when this [solution] cools and tartarus vitriolatus [K_2SO_4?; B. must be in error here] settles out, it [the solution] is strained through wool and evaporated to a scum. The first and second crystallizations produce beautiful Seignet-crystals, the third has a little of Sal digestivus [KCl] in it, and the fourth consists entirely and wholly of this [KCl]. The reasons are that tartarus tartarisatus and common salt exchange their components; the acid of the former goes to the base of the lat- ter, and vice versa.

Seignett's salt dissolved in water is decomposed not only by mineral acids, but also by the juice of citrons, vinegar, decoction of Tamarinds, etc. What happens is that alkali is removed by the excess of acid, which should be in excess in order to produce a really pure tartar, with the ability to precipitate.

FIFTH CHAPTER

About Middle Salts

§ 71. Neutral or Middlesalts of other classes are those whose base is an alkaline earth.

Note. The following chart shows the principle middlesalts known up to now:

Earthy Middlesalts With Acids.

Acid of Vitriol [H_2SO_4]

Satd. with calcium, gives Selenite [$CaSO_4$].
Satd. with magnesia, gives Bittersalt [$MgSO_4$].
Satd. with clay, gives Alum [$Al_2(SO_4)_3$].

Acid of Saltpeter [HNO_3]

Satd. with calcium, gives calx nitrata [$Ca(NO_3)_2$].
Satd. with white magnesia, gives magnesia nitrata.
Satd. with clay, gives argilla nitrata.
[Note that Bergman here is careful to specify "white" magnesia, to differentiate it from "black" magnesia or manganese, which he and Gahn had just discovered.]

Acid of Salt [HCl]

Satd. with calcium, gives Sal ammoniacus fixus.
Satd. with magnesia, gives salt ash.
Satd. with clay, gives argilla salita.

Acid of Fluorspar [HF].

Satd, with calcium, gives Fluorspar.

Acid of Vinegar [Acetic Acid]

Satd. with calcium, gives Sal cretae, corallior, etc.

Satd. with magbesia, gives magnesia acetata.

Satd. with clay, gives argilla acetata.

Acid of Tartar [Tartaric Acid]

Satd. with calcium, gives calx tartarisata.

Satd. with magnesia, gives magnesia tartarisata.

Satd. with clay, gives argilla tartarisata.

Acid of Urine [Phosphoric Acid]

Satd. with calcium, gives calx phosphorata.

Satd. with magnesia, gives magnesia phosphorata.

Satd. with clay, gives argilla phosphorata.

Acid of Air [Carbonic Acid]

Satd. with calcium material, gives calxspar, or usually calx earth.

Satd. with magnesia, gives the usual magnesia alba.

Satd. with clay, gives argilla aerata.

Earthy Middlesalts With Alkali

These are still little known. On calcium, alkalis do not seem to exert any dissolving power, since this base is really by itself a distinct alkaline salt. Magnesia seems to be attacked more, or so I have found at least, that when this earth [magnesia, Mg] is precipitated from acid with caustic volatile alkali, it deposits in liquid after a few days small round clots, which presumably contain magnesia combined with volatile alkali. When we use mild alkali to cause precipitation, a large amount always dissolves in acid of air; Mr. MARGGRAF believes that this solution is due to the alkali itself, but it is not difficult to find the correct composition. How pure fixed alkali behaves with magnesia has not yet been investigated. The behavior of alkalis with clay is also unknown, but it seems that they should have strong attraction. Does not the capacity of clays to decompose in the distillation of saltpeter and common salt bear witness to this?

§ 72. Gypsum, Selenite, or Alabastrite consists of acid of vitriol saturated with calcium earth. Nature produces crystals in gypsum deposits, but science has not yet produced it in such [crystalline] form.

It is insoluble in water. Calcium can be freed of acid either by means of red-hot charcoal, whereby the acid is sulphurated, or by melting it together with fixed alkali in a crucible, when, after leaching of the salt, a sizzling calx remains behind.

Gypsum precipitates from fixed sal ammoniac [$CaCl_2$?] when acid of vitriol is dropped into it (§ 78). It also hap- pens with finely pulverized limestone when it is spread out thinly on a dish, sprinkled with acid of vitriol, mixed well and allowed to dry, where we find that if we use just the right amount of acid, which is not easy to do, a product is obtained which after burning takes up water just like natural gypsum.

The usual method of burning gypsum is to heat it with wood in an oven, and then the gypsum which has broken down into fine pieces is swept out; openings are made again and the whole lot is left undisturbed until it has cooled down. In small quantities, gypsum can be burned quickly in a pan over a small fire.

Note. One part of Montmartre gypsum is fully dissolved by 550 parts of boiling water, and is consequently a true salt. 100 parts of selenite contain about 34 [parts] of lime, 44 of acid, and 22 of water.

In addition to the foregoing method for decomposing gypsum by the dry method, it can also be done by the wet method using fixed alkali. Gypsum, heated below red heat, gets hot with water, but less than lime, is coagulated in a few moments and expands into the same dry volume, which always makes it suitable for shaping, molding, etc.

§ 73. English Salt, or Epsom Salt, consists of calcium earth dissolved in acid of salt, and then mixed with some acid of vitriol. It is best prepared in small amounts from Sal ammoniacus fixus (§ 78) which is precipitated with acid of vitriol, and then when nothing more separates, it is filtered, [the filtrate] evaporated until a film forms and then allowed to crystallize, when the long pointed crystals separate, consisting of the lime which the acid is not able to precipitate.

At Epsom it is prepared from a mother liquor at the boiling point, mixed with colcothar of vitriol [?], strained, boiled and crystallized, but must be made in small amounts, since it is moderately expensive. It deliquesces in air.

Note. the author [Scheffer] is not describing here the well-known English- or Bitter-Salt, of which the base is certainly not calcium, but rather White Magnesia, earths which hardly resemble each other except that both fizz in acid. Those crystals which are described in the text as English Salt are nothing else than gypsum which dissolves with difficulty and is first brought to separation by evaporation. An excess of acid makes it more soluble than it usually is. When the sea-water mother liquor is boiled with colcothar of vitriol, acid of vitriol is separated from earth of iron, drives off the acid of salt from the magnesia and unites itself with it [Mg], producing a bitter and slightly soluble salt, entirely different from gypsum. Commercially Glauber's Salt is sometimes found in mixed crystallization, instead of English [Salt], but in addition to some difference in taste, the latter is decomposed by limewater, but not the former.

[Colcothar of vitriol, from the description in the foregoing Note by Bergman, could be iron sulfate. J.A.S.]

§ 74. Alum is made up of acid of vitriol and earth of clay, but is not fully neutralized, which both taste and fizzing with alkali reveal, when both solutions are strong enough.

Some believe that earth of alum is a kind of lime, but it does not form gypsum with acid of vitriol, does not fizz with acids, does not turn syrup of violets green, and does not dissolve after strong heating, in either acid or water, which are all characteristic of lime. Earth of alum does not melt by itself in fire, and even [melts] sluggishly with borax, and consequently is next to tin calx in difficulty of melting. Little of the mineral is dissolved by vegetable acid, even after heating.

Note. Ordinary alum contains an excess of acid, which by the test given in the text is detected by its ability to turn tincture of litmus red; but this excess cannot always be neutralized, when alum solution is heated with earth of alum so that the salt becomes saturated with earth, until it loses most of its taste, solubility, and the rest of its properties, as Mr. BAUMÉ has reported. 100 parts of crystallized alum contains approximately 18 [parts] of pure clay, 38 of acid and 44 of water.

[If Bergman is referring to potassium alum, $KAl(SO_4)_2$-$12H_2O$, the calculated percentages should be: Al, 7.1%; Al + K, 17.5%; SO_4, 25.4%; H_2O, 57.1%.]

To Produce Alum from Alum-shale

§ 75. (a) Broken pieces of shale are placed in a crucible, over which another [crucible] is placed,. and the whole put in a forge [heated level area equipped with a bellows.]

(b) When the crucible begins to glow the shale burns with a flame and the smell of sulfur. We let this continue ubtil it all glows gently throughout and no more flame is to be seen, after which the crucible is allowed to cool. When it all is completely burned it must be mixed once or twice in order that the part underneath comes to the top and the part on top is turned under. The fire should be forced cautiously [with the bellows] until no fumes of acid of vitriol come off.

(c) The burned shale is then put into boiling water, which after several days begins to taste of alum, and this is continued as long as any more solution occurs, the heating is stopped and [the solution] preserved; after this some fresh boiling water is added, proceeding as before until no further taste of alum can be detected.

(d) After the shale is heated to glowing 3-4 times, soaked in lye as before, it is ground until fine, and burned and washed as long as it yields more alum; it gradually becomes completely red at the end, when after two further burnings it is entirely black inside.

(e) All extracts are combined, filtered, boiled to film formation, and then held until crystals form, but ochre must be strained off.

(f) Alum-shale is nothing else but petrified clay intermixed with sulfur and iron.

Note 1. Alum-shale is a petrified mixture of clay and flint-sand, intermixed with rock material which makes it black, and visibly sprinkled with pyrite. These substances are found in all alum-shale that I have examined, but in addition both lime and magnesia are often encountered as well. It never contains alum before burning if weathering has not decomposed any of the pyrite; it [alum] is produced during burning, phlogiston is dispelled and acid of vitriol is retained partly by the iron, and partly by the clay, from which vitriol and alum are produced, and without which both gypsum and bittersalt are produced when these bases are present.

(g) In an alum factory shale is burned in large stacks underlaid with wood, whereupon it is ignited and allowed to burn itself out. Afterwards it is leached and then reburned. As long as more raw hale is available it is added and burned two or more times as long as more alum is produced, presumably because water can penetrate better the more soluble the shale becomes. The process is stopped when on adding more wood for further burning over new shale, more than twice as much wood is required as in the first burning.

The burning is carried out in lead pans, since other metals are attacked by the alum acid.

Most of the alum in Sweden, England, and Germany is manufactured in this manner, and all are of about the same quality, but for dyeing of wool the Swedish [alum] is best. All of these contain iron and are unsuitable for dyeing silk and for other delicate uses.

Note 2. Director and Knight RINMAN at the refinery in Nerke has built ovens such that pans can be used in heating the shale instead of wood. The stacks in which the shale is usually burned are called beacons [fyrar]; only brushwood is generally used for igniting.

(h) Roman alum which is manufactured at Civita Vecchia [Italy?] and at Solfatera is iron-free, and so is the best of all. The ore is like lime or gypsum, but is of an entirely different nature. It is burned first with wood, spread out thinly on the ground, sprinkled with water for a few weeks, washed and simmered in the usual way, but without wood at Pozzouli, where the pans are buried in the earth and the boiling is produced by underground fire.

Note 3. The district where the ore is quarried for the so-called Roman alum is called Tolfa. The ore is broken open by stoney hard rocks. Lava, which is found in the district, gives reason to believe that it was exposed to underground fire for some time. The stone is white, does not fizz with acid, and is often intermixed with veins of quartz. Alum does not seem to be formed from it before burning, since I have boiled many pounds of it in water, after first grinding it to the finest powdered form, without being able to produce the least trace of this salt [alum].

At Solfatera lava rocks become entirely white with acid of sulfur; pieces of it are collected together, heavy fumes rise up which provide the necessary acid thereby, so that it can then be leached and with boiling give alum. In Braunschweig a red alum is produced which has a quality opposite to Roman [alum]; I have tested it and found that when it is precipitated and the earth is melted with borax, a blue gas is formed, and clearly reveals that the [blue] color comes from cobalt. I have iron-free alum from China, like that manufactured in Garphytten, in another way which I proposed in a 1767 dissertation; it is free from earth and appears to be 5 percent better than the Roman [alum].

To Produce Alum from Clay

§ 76. (a) Finely dried and unburned Cölnic [from Cöln or Cologne] pipe clay is put into a retort, enough acid of vitriol added to thoroughly drench the earth, a receiving vessel attached and [the contents of the retort] distilled over an increasingly hotter fire until the retort begins to glow.

(b) In the receiver is found a little acid of vitriol, some sal ammoniacus secretus in the neck [of the receiver], and the clay left in the retort is found to be completely dried out.

(c) Water is added to the residue [in the receiver] and allowed to digest for several days, [the mixture] is filtered, boiled to film formation, and allowed to crystallize. The excess acid is neutralized with as much alkali as the solution requires without precipitating any earth of alum. Between each addition of a few drops it is stirred, and as soon as the least cloudiness appears, the addition of alkali is stopped. One should scoop a little of it out and test it in this manner ahead of time, since this helps to judge afterwards how much more alkali to add [to the bulk solution.]

Note. Alum can be precipitated, not only by alkali and lime, but also by zinc and iron; dephlogisticated iron is precipitated however by clay. When alkali is added to an alum solution it first attaches the excess acid to itself, and part of it becomes difficultly soluble, and precipitates by itself, taking with it the portion of the acid which it possessed. If a mild alkali is used we find also a small part of earth of alum dissolved in acid of air, which again is separated, if the strained solution is warmed; at the moment of precipitation the earth has the greatest possible surface, so it dissolves and then the weak acid of air separates the substance, which otherwise it would not be able to do.

If we add a lot of acid of vitriol to an alum solution it attaches the water to itself, and the salt precipitates for lack of solvent; but if smaller portions of acid are added crystallization is prevented; and to this degree the acid often prevails over the lye in an alum factory; to correct this I made the proposal in 1767 to add clay, which by relieving the prevention [of crystallization] also increases alum [production].

§ 77. Acid of Saltpeter is added to lime until saturated, so that neither syrup of violets or litmus juice is then changed by the solution.

The solution is evaporated at digestive heat [until] finally a brown pellucid mass is formed, but no crystals, because lime can be brought [to crystallization] in nature only by acid of vitriol.

The brown mass dissolved in water changes neither syrup of violets nor litmus; and it releases the lime to give alkali. If it is distilled, pure acid of saltpeter passes over, but in order for all of it to be driven off, not only the retort but also the mass must be heated to glowing, or else some nitrum calcareum is left behind, which when dissolved in water is also found to be neutral.

The solution is evaporated by allowing it to stand until dry, when it becomes like syrup in heat, and like wax in cold, but it deliquesces. In a crucible over fire it fizzes strongly, but does not melt until the bellows is used, after which it can be poured out leaving a clear mass which deliquesces in air, and detonates with charcoal like nitrum but more slowly and more feebly.

Note. If a solution of calx nitrata is strongly evaporated and quickly plunged into severe cold, crystals separate, but transitorily, which crystals once again deliquesce readily. The water from calx nitrata, which is evaporated to dryness, was called by BALDUINUS of the previous century Spiritus mundi, and it was believed at that time to be a deadly medicament. If this salt is exposed to some heat, until it is calcinated to a soft yellow, it attains the capacity to attract fire or sunlight so much that it can then produce a faint glow in a dark room; it is usually called Phosphorus Balduini, and the chalk from it seems to be more suitable than other lime.

§ 78. Sal ammoniacus fixus [$CaCl_2$] is gotten from distillation of Sal ammoniac with lime (§ 49). The residue should be freed from all alkali volatile by heating, whereupon it also melts together. It is then leached with hot water, filtered and boiled to dryness. It can also be obtained in the form of loaf sugar by evaporating to film forma- tion, gathering the crystals together, which are however quite deliquescent, which arises from the acid of salt, since all of this combined moisture forms with- out [the presence of] Corrosive Sublimate [$HgCl_2$].

Acid of vitriol precipitates the calx of gypsum from this salt with heat and fizzing; but not all yet, since the rest comes down if alkali fixum or [alkali] volatile is added, preferably the former, since it takes a long time with the lat- ter.

The calx can be precipitated by al- kali volatile, which seems strange since sal ammoniac is decomposed by calx. But it is noted that burned lime displaces alkali volatile by its very nature; the slaked [lime] also has some effect, since it still has causticity. The clear [solution] which is left after precipitation gives sal ammoniac. Alkali volatile is not able to precipitate all the calx, which is not- ed when the non-volatile [alkali] is elim- inated.

Note. Fixed Sal ammoniac [$CaCl_2$] or- dinarily reacts with the alkali which comes from burned [unslaked] lime, and apparently can be used to make a solution of cream of calx [cremor calcis]. Acid of salt just saturated does not give the kind [of $CaCl_2$] for a reagent either for acid or alkali; evaporated to dryness and allowed to deliquesce it has the name of oil of calx [oil of lime?]. Iron, encrusted with molten Sal ammoniacus fixus, gives off light, if it is broken or rubbed in a dark room; it is Phosphorus Hombergianus. The precipitation with alkali volatile is mentioned in the Foreword.

§ 79. Lime is also dissolved in vinegar, and thus is produced Sal ammoniacus fixus aceti, or vegetabile.

Note. This salt separates in needles which are inclined toward beautiful growth. It is usually named from its source, Sal cretae, oculorum cancri [crabs' eyes], corallorium, etc., but should be called calx acetata.

On Middle Salts of [Other] Metals

§ 80. Every metal is attacked by one or more acids, producing with them salts, which are often deliquescent, and a whole lot of them can be crystallized.

Note. The following Table shows the most noteworthy compounds which have been made between metals and the principle salts [acids] known up to now. A large number of these cannot be produced directly without requiring either that the metal have been previously precipitated with alkali from another medium, or at least have been dissolved, after which they can be precipitated by acids which otherwise cannot react with them. These are marked with the letter 'p'.

[This Table, given on the following pages, could be the earliest use by Bergman of his new chemical nomenclature.]

Metallic Middlesalts with Acids.

Acid of Vitriol [H_2SO_4]

Saturated with:	Gives [B.´s new names]:
gold (p)	gold-vitriol
quicksilver	quicksilver-vitriol
lead (p)	lead-vitriol
copper	copper-vitriol
iron	iron-vitriol
tin	tin-vitriol
bismuth	bismuth-vitriol
nickel	nickel-vitriol
arsenic	arsenic-vitriol
cobalt	cobalt-vitriol
antimony	antimony-vitriol
zinc	zinc-vitriol

Acid of Saltpeter [HNO_3]

Saturated with:	Gives [B.´s new names in Latin]:
gold	<u>aurum nitratum</u>
silver	<u>argentum nitratum</u>
quicksilver	<u>mercurius nitratus</u>
lead	<u>plumbum nitratum</u>
copper	<u>cuprum nitratum</u>
iron	<u>ferrum nitratum</u>
bismuth	<u>vismutum nitratum</u>
nickel	<u>niccolum nitratum</u>
arsenic	<u>arsenicum nitratum</u>
cobalt	<u>cobaltum nitratum</u>
zinc	<u>zincum nitratum</u>

Acid of Salt [HCl]

Saturated with:	Gives:
gold (p)	aurum salitum
silver (p)	argentum salitum
* quicksilver (p)	mercurius subl. corrisivus
"	mercur. praecipitatus albus
"	mercur. dulcis
lead	plumbum salitum
copper	cuprum salitum
iron	ferrum salitum
tin	stannum salitum
bismuth	vismuthum salitum
nickel	niccolum salitum
arsenic	arsenicum salitum
cobalt	cobaltum salitum
* antimony (p)	Butyrum antimonii
zinc	zincum salitum

[* In the cases of quicksilver and antimony, Bergman was not yet ready to give up the old names. In the case of quicksilver he had three different salts to confuse him.]

Acid of Vinegar [Acetic Acid]

Saturated with:	Gives:
silver (p)	argentum acetatum
quicksilver (p)	mercurius acetat.
lead	saccharum saturni
copper	Viride aeris; Spanish green
iron	ferrum actetatum
tin	stannum acetatum
bismuth	vismuthum acetatum
cobalt	cobaltum acetatum
zinc	zincum acetatum

Acid of Winestone [Tartaric Acid]

Saturated with:	Gives:
quicksilver	mercurius tartarisatus
lead	plmbum tartaris.
iron	ferrum tartarisat.

Acid of Urine [H_3PO_4]

Satyrated with:	Gives:
copper	cuprum phosphoratum
iron	ferrum phosphorat.
tin	stannum phosphor.
arsenic	arsenicum phos.
zinc	zincum phosphor.

Acid of Ants [Formic Acid]

Saturated with:	Gives:
silver	argentum formicatum
lead	plumbum formicatum
copper	cuprum formicatum
iron	ferrum formicatum
zinc	zincum formicatum

Acid of Air [Carbonic Acid]

Saturated with:	Gives:
lead (p)	plumbum aeratum
iron	ferrum aeratum
zinc	zincum aeratum

Metallic Middlesalts with Alkalis.

Alkali Fixum [NaOH and/or KOH]

Saturated with:	Gives:
gold (p), by tincture gallarum	<u>alkali fixum auratum</u>
quicksilver (p)	<u>alk. fix. mercurisatum</u>
copper	<u>alk. fix. cupratum</u>
iron	<u>alk. fix. ferratum</u>
cobalt	<u>alk. fix. cobaltatum</u>
antimony	<u>alk. fix. antimoniatatum</u>
zinc	<u>alk. fix. zincatum</u>

Alkali Volatile [Ammonium Hydroxide]

Saturated with:	Gives:
silver (p)	<u>alk. vol. argentatum</u>
quicksilver (p)	<u>alk. vol. mercurisatum</u>
copper	<u>alk. vol. cupratum</u>
nickel	<u>alk. vol. niccolatum</u>
cobalt	<u>alk. vol. cobaltatum</u>
zinc	<u>alk. vol. zincatum</u>

These salts for which alkali acts as a solvent, are very little known up to now, and remain a new field for investigation. Alkali imparts saline properties to eight metals much as acids do, and some of these compounds separate in beautiful ctystals. So far the true nature of blood lye has not been investigated, so I have not ventured to introduce the metals which dissolve in it; these unite seemingly more quickly to produce three-fold or even more complicated salts.

Three-fold Middlesalts.

Aqua Regia

Saturated with:	Gives:
gold	aurum regalisatum
platinum	platina regalisata
antimony	antimonium regalisatum

Alkali Volatile [NH_4OH]

With: copper vitriol Gives: Sal antiepilepticum puerorum

Tartarus Vitriolatus [Potassium sulfate?]

With: iron vitriol Gives: Tart. vitriolatus ferratus

Sal Ammoniac [NH_4Cl]

With:	Gives:
merc. subl. corr.	Sal alembrot
copper	Sal ammoniacus cupratus
iron	Sal ammoniacus ferratus
tin	Sal ammoniacus stannatus
antimony	Sal ammoniacus antimoniatus
zinc	Sal ammoniacus zincatus

Purified Winestone [Na or K Tartrate?]

With: iron	Gives: globuli martiales
antimony	tartarus emeticus

Bittersalt [$MgSO_4$]

With: iron vitriol	Gives: Sal amarus ferratus

Iron Vitriol [$FeSO_4$]

With: copper vitriol	Gives: Salzburger vitriol
nickel vitriol	Native nickel vitriol
zinc vitriol	Gosslar's green vitriol

Zinc Vitriol [$ZnSO_4$]

With: copper vitriol	Gives: Gosslar's blue vitriol

Fourfold Middlesalts.

Sal Ammoniac [NH_4Cl]

With: copper diss. in acid of saltpeter	Gives: Sal ammon. cupro-nitratus
iron [ditto]	Sal ammon. ferro-nitratus

Purified Winestone [Na or K Tartrate?]

With iron vitriol	Gives: Tartarus martialis

Iron Vitriol

With: copper and zinc vitriol	Gives: Fahlu Vitriol [§164]

The three- and four-fold salts are without doubt much more numerous, but up to now no one has given attention to them. Mr. MONNET has made us aware of a lot of compounds between iron-vitriol and other salts, which cannot be further separated by crystallization; I have also named these previously, putting them among the neutral salts. Among the three-fold salts above I have not included other solutions with aqua regia, other than those which are only dissolved by that complex fluid; because others, which can be attacked individually both by acidum nitri and [acidum] salis, by chance produce not one, but two separately. Probably these are combined in a specific case, but it has not yer been determined how this occurs.

§ 81. Water-free acid of vitriol [fuming sulfuric acid] dissolves silver and shows no cloudiness with pure water. Volatile alkali precipitates it well at first, but it dissolves again if more is added. Acid of vitriol precipitates it [silver] and silver dissolves in acidum nitri; the precipitate dissolves in [excess] water.

Silver-vitriol is precipitated by many metals, since acid of vitriol dissolves zinc most readily, then iron, then tin, copper, and silver, and last quicksilver. Iron requires the least acid for saturation, next tin, copper, etc., in the order of being attacked [activity].

§ 82. Quicksilver can be reacted with water-free acid of vitriol in a different way. If one part of the fluid [dissolved

?] metal is distilled in a retort with two parts of the acid until dryness is reached, a little acid of vit- riol is found in the receiver, and in the retort a white corroded mass of quicksil- ver, which is not dissolved by oil of vit- riol, but when stirred with cold water de- cays to a gold [colored] powder, called Turpetum minerale by apothecaries. If the color does not deepen, it indicates a need for acid, and it helps if the mass is ground up in a glass mortar with a little oil of vitriol, when afterwards water may be needed, but too much will dissolve it altogether; the solution releases a red powder, when fixed alkali is added, as usual. If some acid of vitriol is added to the Turpetum in a retort, and [the retort] placed over a fire, a thick brown solution is formed, which becomes clear with the addition of a little water, but still re- tains its color. It can be boiled until crystallization, but if it begins to form a film it is taken away from the heat, and so are obtained very small crystals diffi- cultly soluble in water. The low solubili- ty comes from the fact that the acid en- ters into the crystals in small amount.

The white mass, heated to red heat in a crucible, is washed in cold water and well sweetened [i.e., washed free from salt]; when it becomes bright golden allow it to dissolve in oil of vitriol, and keep it dissolved, even if some water must be added.

We might expect that addition of phlogiston to Turpetum minerale at the proper temperature would produce cinnabar [HgS], but when sublimation continues with colophonium and Turpetum used, a volatile acid of sulfur goes over first, then a fluid quicksilver, and the residue consists of earth from colophonium. [Colophonium is a pine resin; colophonite is a garnet-like mineral consisting of a calcium iron silicate. J.A.S.]

Turpetum minerale is both a purgative and an emetic, can also increase salivation, wherefore it is also used against lues venerea [venereal disease?]. It can be ground up well with both water and oil, which protects its color, but it is expensive to employ.

Note. Acid of vitriol does not dissolve quicksilver except in the form of its calx; therefore it is also known that when they are put together in a retort over fire, that a very highly volatile acid of sulfur is formed. If the distillation is continued until nothing more comes over, if there was 1 part quicksilver, we find approximately 1-1/2 parts of vitriol; if the heat is increased until the bottom of the retort glows, white vitriol sublimes, which produces Turpetum in water as usual. The wash water holds a lot of vitriol dissolved, and if the acid is increased enough, the whole is dissolved and held in solution.

§ 83. Lead is corroded by acid of vitriol and is precipitated by it, when it has been first dissolved in acid of saltpeter.

Note. The precipitate is vitriolate of lead, which is not soluble in distilled acetic acid, unless infinitely slowly, and is thus differentiated from lead combined with acid of salt; this difference serves very well as a test in solution. [Bergman is giving here the test for distinguishing $PbSO_4$ from $PbCl_2$, by using the solubility of the latter salt in acetic acid.] Lead vitriolate can still be entirely dissolved and held in solution, but it requires much water, without which it again separates under evaporation. Mr. GROSSE has already reported that Plumbum salitum precipitates the [lead] vitriolate with acid of vitriol, which indicates a closer affinity.

§ 84. Copper vitriolate is copper ash dissolved in acid of vitriol. Copper [metal itself] is attacked slowly by acid of vitriol, which ought to be well diluted with water, but if the metal is in the form of its calx it proceeds much more easily. The solution is measured, boiled until a film starts to form and crystallization gives a bright blue vitriol, like that from Cyprus.

In most cases blue vitriol is prepared from water from certain copper mines, by a process which drives the phlogiston out of the sulfur of the ore, whereby the acid remains which dissolves the copper essence. It can also be produced very pure from filings from kopparbleck [copper-tin?] which is calcinated with sulfur and slaked in water, which [solution] is finally boiled to a film in a copper kettle, and is brought to crystallization by the addition of alkali, lime or putrified urine, which absorbs the excess acid. In this way a blue vitriol is produced in Turkey and at Tumba where the Southern Crescent [Söderelja] is situated. Such [blue vitriol] cannot be obtained from copper ore, if it is not iron-free, like is found at Sunnerkog in the Kållberg mine.

Whether blue vitriol is iron-free can be determined by the fact that its [iron-free] solution does not turn black with astringent water; and with enough spiritus salis ammoniaci [spirits of ammonia] it gives a blue color without precipitating any green powder. That from Cyprus meets this test.

Alkali fixum and [alkali] volatile precipitate copper vitriolate with blue color, and the precipitate is dissolved again by the latter [alkali volatile] if a lot of it is added. The precipitation takes place with effervescence when both solutions have the proper strength.

The largest consumption of this [blue] vitriolate is for making paper from linen, and Cattun's printing, which produces white and blue in the dyer's tub; it must be iron-free, otherwise it produces rust flakes in the white product. Blue vitriol will not do for all paper. It is sometimes called bluestone and is sometimes used for eating away dead flesh [from wounds]. The Germans call it "kupferwasser" or "kupferrauch", from which many of us get the name kopparrok [copper smoke] without knowing what this means.

All vitriol has a sour and metallic taste, which is not at all like the neutral salts.

<u>Note.</u> Since copper is attracted by acid of vitriol more weakly than is iron, we see shiney iron, which is placed in a solution of blue vitriol, immediately covered over with a skin of copper, because just as some particles of iron dissolve, copper is let loose in its place. This was taken in ancient times as apparent proof that metals could be transformed from one kind to another. Zinc also precipitates copper.

§ 85. <u>Iron vitriolate</u> is acid of vitriol saturated with terra martis, which separates in green crystals of rhomboidal shape. It can be easily prepared from iron filings, onto which oil of vitriol is poured in a retort, and then diluted with at least four times as much water. It should dissolve immediately with strong effervescence, but to bring it to completion the retort is set over a slow fire and watched carefully, so that no solution boils over, which happens easily. We then note that a grey powder precipitates, if the acid is too strong, and it must be further diluted, but if nothing dissolves it is too weak and should be strengthened. When all the iron dissolves, except for a black powder, which always precipitates and is the phlogiston from the iron, it is allowed to settle until a sediment appears; then it is heated until a film forms in the iron pan, but is not allowed to crystallize into a glass. With boiling we meet the difficulty that an ochre forms on the surface which prevents proper observation so that evaporation is easily under- or overdone. The acid should also be well saturated, or else the vitriol becomes sour and less fit; calx of iron does not react with it in the least.

It is precipitated by alkali fixum and [alkali] volatile with green color, which upon drying becomes brown-red. A green vitriol is precipitated with lime, which fades to light gold, and with this linen can withstand washing and acquires a very beautiful appearance.

Iron vitriolate turns black with astringents, and is therefore used to blacken fur, wool and silk, but for calico-printing the iron is dissolved in vinegar. The ancients called it vitriol Calcanthum and the Latins called it attramentum sutorium. Roman vitriol cannot be obtained now, or is at any rate obtained under the name of either Cyprian or Hungarian, and so forth. The fermented [?] or English [kind] is better than the Dylta vitriol, which is very sour and watery, [a property] which reduces the amount of the desired material. Free acid, especially vitriolic, removes the black which is made with iron; when sour vitriol is used therefore, more of it must be used, which can be bad for the goods and weaken the color. Since fresh iron precipitates dissolved copper, so must the former [iron] first be dissolved entirely from an ore which contains both [metals] together, and [then] no more acid put into the vitriol than that amount which can react with the copper; whence it follows further that iron vitriol contaminated with copper always contains excess acid. So therefore the leather colored [iron vitriol] from Salzburg, if iron is to be added with advantage, gives more of this [iron] than might be expected from the weight. Copper contaminated [iron vitriolate] is identified quickly if the vitriol is moistened and rubbed on shining iron.

Green vitriol calcinated to redness is called colcothar vitrioli, and also Crocus martis astringens, and is a very good styptic or method of staunching blood. If sweet spirits of salt is poured on it we get a gold tincture, which is said to be good for strengthening weak viscera, but doctors of our time refuse it and maintain that it is very dangerous to the lungs; it is called Tincture martis aurea.

Note. One hundred parts of green vitriol in fresh crystals are obtained with about 52 [parts] of water, 16 of acid, and 32 iron, which becomes somewhat dephlogisticated. That a lot of phlogiston is separated during solution can readily be learned from [the amount of] inflammable [gas] given off, which ascends as long as any [iron] remains and is attracted by the magnet, until both [solids] are fully dissolved in the acid; later precipitation occurs with alkali, and the precipitate being washed and put in a dry closed vessel, and acid of saltpeter being added thereto, it begins immediately to give off red fumes. The vitriol itself, sorted out with a long and nimble magnetic needle, keeps its appearance still almost unchanged. One lod [one-half ounce] of iron, upon dissolving in vitriolic acid, produces 207 cubic inches of inflammable air. Zinc precipitates iron from solution made with acid.

Mr. DE LA FOLLIE has melted colcothar vitrioli with half as much burned lime and obtained thereby a hard black substance, which shows marked polarity with a magnetic needle, but does not attract filings [of iron].

Preparation of Green Vitriol from Pyrite

§ 86. (a) The pyrite is broken into small pieces, put into a crucible which is tightly closed, placed on a forge and kept in a flame until no more fumes are visible. Then the cooled pyrite is put into an iron pan, moistened with a little water and allowed to stand for a while. Pyrite contains sulfur and iron. If it is roasted on an open fire it gives off both phlogiston and acid; in a closed crucible the iron retains its phlogiston well enough to be able to be calcinated, but the acid is retained and then this sulfurous state is likewise decomposed when water is added, whereupon the acid condenses and phlogiston escapes separately. The acid attracts water very strongly, and metals, even phlogiston, which is learned from iron filings mixed with sulfur and a little water, from which vitriol can be leached after a few days. More water may not be added, even that which comes regularly from burned pyrite, since otherwise it does not decompose so well; when the water is first driven out, it is not taken up, and it continues like this until it all falls apart, which requires about a half year or more.

(b) The decomposed pyrite is boiled with water and this is continued as long as the filtrate forms a precipitate with alkali fixum; it is filtered and boiled until a scum forms [on the surface], whereupon a lot of ochre continues to separate, which cannot be dissolved. Straining is of no help here, since no separation shows itself immediately.

In a sulfur-works [smelter] the residue [caput mortuum] from the distillation is put out to dry in the air, and acquires in the process so much internal heat that it bursts into flames; finally it is leached in a vat and boiled in a lead pan.

Note. The solution of iron vitriol is not decomposed in a filled and closed vessel, but in the open ochre separates continuously, from which we conclude that the air attracts phlogiston and so more of this inflammable [element] separates, whence more acid is required to keep the iron in solution. This goes on still more in warm solution, and most strongly during boiling. A solution of very dephlogisticated iron is a red-brown mother liquor, which will not crystallize by itself, but can be brought to do so by the addition of a little alkali. The crystals appear a bright white, and it is known that the vitriol becomes weaker the longer it is boiled; and the more often it is crystallized the more this process decreases the inflammability; even the taste is changed and becomes somewhat like alum. These circumstances easily confuse those who are fond of confusion, since some alum can be obtained from all green vitriol; but if one goes to the trouble to examine it closely it becomes quickly apparent that this base does not exhibit the least odor of earth of alum, but rather that of pure earth of iron. Sometimes true alum is found mixed with vitriol, but not always, and only when the pyrite contains some clay, which we know happens sometimes; it is never obtained from vitriol which is prepared from iron and pure vitriolic acid, although it forms alum-like crystals from the mother liquor.

Vitriolum Romanum is prepared in the district of Monte Fiascone, and contains some copper; it is seldom found in Sweden.

§ 87. Tin is dissolved in water- free vitriolic acid. The solution becom- es brown and is obtained in the same small sulfur granules, which look brown during melting, but golden afterwards. Strong heat is required before the acid begins to boil and attack the metal, so it [the me- tal] melts but the solution succeeds any- way. The tin can be completely and wholly precipitated with water.

Note. During solution inflammable air sep- arates.

§ 88. Bismuth does not dissolve in vitriolic acid.

Note. If vitriolic acid is distilled away from the bismuth in a retort and the residue is leached, we find that it turns brown and nothing precipitates immediately with alkali; but if the solution is allowed to stand 6-8 hours, a fine white powder settles out; Mr. ROUELLE (the elder) assures us also that the solution in aqua fortis is precipitated by 8 times as much water, and if the clear [filtrate]is then concentrated, dilute vitriolic acid, and all salts that contain it, can precipitate a bismuth vitriolate.

§ 89. Nickel dissolves in vitri- olic acid.

Note. It dissolves only in the form of the calx and separates in bright green crystals, which are like alum, with two points symmetrical to each other.

§ 90. Arsenic is dissolved in the form of the calx in acid of vitriol.

Note. The regulus is fully dissolved, and from this solution I have obtained small round divided crystals. Mr. SAGE must have obtained long undivided pyramids.

§ 91. Cobalt-regulus is dissolved in water-free acid of vitriol, both in the form of the calx and the metal, with a rosy red color, and it appears that it can be freed from iron in this way when both are in the form of the calx. The solution separates cold, but the crystals melt, so easily that they only need warming. Alkali volatile precipitates the metal gray, but compacted it dissolves again with red color, but this darkens if more alkali is added.

Lime, clay, and roasted cobalt [ore], melted together to a glass, are dissolved by dilute acid with brighter color than cobalt alone, and perhaps the lime is responsible for this.

Note. Cobalt vitriolate is red-brown and I have some that has stood seven years in an unheated room without deliquescing. In France cobalt must be contaminated with nickel, because Mr. CADET and Mr. BAUMÉ have obtained green crystals; such [green crystals] would not be expected from a red solution, and it [cobalt] dissolves with such [red] color in all acids, unless careless handling prevents it.

§ 92. Antimony-regulus is dis- solved in oil of vitriol over a low fire and then separates in white crystals, as soon as the solution is cooled, which crystals melt in mild heat as easily as butter. The regulus mixed with lead does not dissolve, unless ground to a white powder.

The solution is precipitated by water to yield a white powdery precipitate.

Antimonium crudum dissolves similarly, whereupon sulfur falls to the bottom [of the container].

§ 93 Zinc-vitriolate is white, is called Gallizenstein by the Germans, and is produced only at Goslar. The material is taken from roasted Rammelsberg silver ore while it is still hot, and is thrown out in large amounts to be leached. The leachate is then boiled and allowed to crystallize. Usually it is, in loaf-sugar form, about as expensive as blue vitriol, and about 10 to 12 times more expensive than green [vitriol]. To keep the price steady in the pure form it is produced then only every seven years. White vitriol can easily be freed with zinc from all contamination by iron and copper, and green [vitriol] freed from copper with iron, but zinc can be separated from neither of these when they are present.

It is used in medicine, and a little in rims [for barrels or wheels?], and in metal working.

Note. One lod [1/2 oz., 13.3 g] of zinc when dissolved gives off 180-1/2 cubic inches of inflammable air. It was not known until 1735 what substances occurred essentially in Gallizenstein, but then it came out in Nova Acta Upsala, from Mining Counselor BRANDT, and at the same time in the French Memoires by HELLOT, that it was made up of zinc and vitriolic acid.

§ 94. Acid of saltpeter [nitric acid] dissolves silver, and if the solu- tion is strong it produces crystals, which are called for no good reason vitriolum lunae. It is very bitter and caustic, so it is used to remove hard growths [warts?] and the like.

The crystals are dried, melted over a weak flame and after effervescing, yield a black material which appears metallic at a fracture, and is called Lapis infernalis. This is more corrosive than anything up to now, and is therefore used to remove warts, dead flesh in wounds and in venereal diseases; and to process and blacken hides. It should be kept well enclosed so as not to deliquesce. Lunar [silver] crystals detonate like niter with a glowing coal, but more quickly; and also lapis infernalis, which therefore gets the name of nitrum lunae ardens.

If to a silver solution is added a strong saltpeter solution, and then it is evaporated to dryness under continuous stirring, we get Luna purgans Angeli Salae, which is used as a purging solution in dropsy.

Note. Silver solution is precipitated by [both] acidum vitrioli and [acidum] salis, and likewise by all neutral and middlesalts, [or solutions] which contain both of them; also by alkali fixum, lime water, and alkali volatile crystallizatum, but according to Mining Counselor BRANDT'S research, only one boiling with burned lime does not produce it completely. Quicksilver, lead, copper, iron, bismuth, and zinc precipitate silver in metallic form, showing strong attraction for solution.

A copper-free solution [of silver] in pure acid, which without being filtered or stirred at all, and from which phlogiston can be obtained, when evaporated and fused with glass becomes entirely white, so Mr. SCHEELE has found. He believes that the consistent power of corrosion which this salt exhibits to a greater extent than that of the constituent acid arises from the fact that the silver, during solution, is partly dephlogisticated, and then with its last strength seeks to restore this deficiency. It is known that during solution it gives off a red vapor; that earth of silver becomes highly corrosive, and that it also attracts phlogiston in the wet method, becomes black and can be again dissolved with red vapor. It seems to be of like nature to quicksilver solution, mercurius sublimatus corrosivus, and other corrosive metallic salts, fixed in unequal degree. Acid of salt always attracts the phlogiston of these metals, yet more or less depending on the circumstances.

§ 95(a). Quicksilver is dissolved readily in acid of saltpeter over mild heat, the solution becoming in the beginning somewhat milky, but then clearing up.

This solution is very corrosive and is usually called by the name of Aqua

—

divina Fernelii. When well saturated, the solution produces white crystals, spreading like radiating beams, when kept in a cold room. It can also produce greater quantities after each additional evaporation, in order to get as many crystals as can be obtained from the solution.

These crystals are incorrectly called Vitriolicum mercurii; they readily deliquesce in air, so they ought to be kept in a tightly closed container; they melt easily in heat; they have a disagreeable taste; and are among the most corrosive and poisonous substances.

The solution is precipitated with alkali volatile to give a grey powder which is nothing else but quicksilver itself which becomes runny [liquid] upon grinding in a mortar made of glass.

Note 1. Quicksilver solution has different natures depending on how it is treated, either with or without heating; in the first case [with heating] a red vapor arises, the precipitate with alkali volatile is white, and the solution produces a gold powder or Turpetum [upon dilution] with water; but in the latter case [without heating] it has a greater tendency to crystallization, the precipitate with alkali volatile is dark, produced by finely dispersed quicksilver, and the salt itself has little or no corrosiveness. The crystals from either of these solutions, well drained on absorbent paper and dried, cannot then be completely dissolved in water unless a small amount of nitric acid is added, and finally they [the crystals] turn yellow and decrepitate.

Copper and iron precipitate quicksilver in the metallic form, which generally happens when the metal is the means of precipitation, giving up its own phlogiston. This precipitate obtained in metallic form does not increase in weight, whereas alkalis and earths augment it in weight and produce some dephlogisticated precipitate.

Warm urine precipitates the so-called <u>mercurius praecipitatus incarnatus</u>. It is compacted by alkali volatile to obtain <u>mercurius praecipitatus niger</u>.

(b) <u>Mercurius praecipitatus ruber</u> is obtained from quicksilver solution which is evaporated to dryness and cal- cinated. The substance at the bottom at first turns white, then golden, and final- ly red. We may obtain the highest degree of coloring finally by using stronger heat than glass will tolerate, so consequently the vessel must be changed or small a- mounts used. When the solution begins to dry, it should be stirred vigorously with a glass rod, to get the heat distributed evenly, and this should be continued dur- ing the entire operation. Metallic vessels and instruments must be avoided here since they are attacked, especially copper and iron.

If the material which remains in the container is lumpy, it should be ground up in a glass mortar. During calcination the acid is observed to escape in the form of red fumes. The redder they are the better the reaction is believed to be. It is used for the same purposes as Mercurius cosmeticus, but it is stronger and cannot be taken internally. Since it is not precipitated in any case, it would be better named coagulatus or inspissatus [evaporated].

In the apothecary's shop, after complete burning of this material, it is extracted seven times with distilled vinegar and then likewise with rectified spirits of wine, when the color becomes brighter and it is called aurum horizontale or Arcanum corallinum Paracelsi, and it is considered better and more gentle [in its action]. Usually it is not given so many distillations.

Note 2. If acidum salis is poured on this red calx of quicksilver it first becomes white and is then entirely dissolved; vitriolic acid also makes it white, or transforms it to the vitriolate, since if water is afterwards added, Turpetum minerale is obtained; also the red color is changed to white by alkali volatile; nitric acid dissolves it completely.

Since quicksilver is strongly attracted by nitric acid, the Alchemists believed that through repeated extractions they could convert it into gold, and SYLVIUS assures us that it was successful. When quicksilver is well dephlogisticated in this way, it does not let itself be reduced without additives; Mr. BAUMÉ has found that it is sublimed with sufficient heat and produces ruby-red cubic crystals.

(c) If eight parts of quicksilver are dissolved in aqua fortis and one part of copper is added, and then it is mixed and evaporated to dryness, a green powder is obtained, which is used under the name of Mercurius praecipitatus viridis against Gonorrhoea of particular virulence. Also, as the flame is applied the color is altered. When much acid remains it is entirely green, but with less whiteness than when finally burnt, if most of it [acid] is driven off.

§ 96. Lead is dissolved by acid of saltpeter both in metallic form and in the form of the calx, and at the same time white crystals separate, which have an astringent taste and are soluble in water. These are consequently called saccharum saturni [sugar of saturn], which name properly belongs to its solution in vinegar.

Note. Lead dissolves readily in acid of saltpeter and at the same time precipitates a gray powder which is calcinated lead, and not quicksilver which the Alchemists imagined it to be; Messers MACQUER and BAUMÉ have investigated it very closely, and 25 pounds of lead have produced not a trace of quicksilver. Plumbum nitratum separates in octahedral crystals.

§ 97. Copper in metallic form forms a blue solution in aqua fortis, but does not crystallize.

Note. When an excess of finely divided copper is added to acid of saltpeter all at once, it gives off noxious and very red fumes, but the solution acquires a beautiful bright green color. The salt can hardly be crystallized, and is slightly soluble in spiritus vini with a dull green color.

§ 98. Iron in metallic form dissolves without heat in acid of saltpeter. The solution becomes red-brown and forms a red precipitate with alkali, and the large quantity of phlogiston is imagined to come from the acid used here. The solution will not crystallize by itself, unless evaporated to a dry syrup by boiling; if phlogiston is added and the material is dried, it takes fire on a heated stove, and begins to fume in your hands [i.e., at body heat].

Note. When a small amount of iron is gently dissolved without heat the solution be- comes greenish, but as it approaches saturation, and if a lot is added at a time, or if heat is added, the color becomes red-brown. Acidum nitri can also dissolve calx of iron, but it is at- tacked more sparingly than the metallic iron, and is also precipitated by the latter [metallic i- ron]. If a solution of alkali vegetabile is added a little at a time to iron solution which is acidic in the beginning, some precipitation oc- curs, but this is dissolved immediately by acid of air with mild heat; this four-fold saline mix- ture when allowed to stand produces beautiful vegetation known by the name of arborescentia martis.

§ 99. Tin is attacked but not dissolved by acidum nitri.

Note. In this manner a white and well de- phlogisticated calx of tin can be obtained most quickly; serviceable as an enamel, especially if the acid is strong, heat is used, or much [acid] is added all at once; if these conditions are avoided some solution can be obtained.

§ 100. Bismuth is dissolved by acid of saltpeter, which [solution] is then precipitated with water [to produce] a white powder called magisterium marcas- itae or Blanc d'Espagne, which when well sweetened becomes entirely white and is used as a paint [pigment]. They have tried to give it a complexion color [for use in cosmetics?] by burning it, but it will often turn to gold; yet it seems that some means has been found to give it the desired complexion reddening because I observed this once to give a test for bismuth. For complete precipitation it is necessary to use 40 to 60 times as much water as solution.

Note. The solution can be caused to separate in small, clear, prismatic crystals, which when dropped into water become white instantly, and opaque as well, and precipitate out in the form of a powder.

§ 101. Nickel is dissolved in acid of saltpeter with a green color.

Note. A saturated solution allows itself to become dark green, some blue crystals precipitating, which deliquesce in moist air.

§ 102. Arsenic in the form of its calx is dissolved in acid of saltpeter.

Note. The regulus itself dissolves also.

§ 103. Calx of cobalt is dissolved by acid of saltpeter with a rosy red color. If alkali volatile is added to this solution it finally precipitates like iron, which can be present, as a gray powder, and at the same time the solution becomes purple- or columbine-colored, particularly if the alkali is in excess. Pure water also precipitates something white, like bismuth, lead, etc.

Note. A saturated solution gives a red writing [ink], when moderately warmed, but in strong heat becomes brown and finally black. If acid of salt is added, free or combined, Sympathetic [invisible] ink is obtained. See § 119.

§ 104. The regulus of antimony is dissolved by acidum nitri.

Note. **Mr. MACQUER says that if the regulus is pulverized and aqua fortis is added to it until it is half covered, it takes it up and fine transparent crystals separate in the interstices. Or it is calcinated with more acid and in this manner we can produce antimonium diaphoreticum or Benzoardicum minerale.**

§ 105. Zinc is dissolved by acid of saltpeter.

Note. **This solution crystallizes easily. Upon solution, the acid becomes very fuming and red, but neither with zinc nor with any other metal is inflammable air produced, which seems peculiar.**

§ 106. If to silver solution acid of salt is added, or common salt, or sal ammoniac, a white powder precipitates, which when completely neutralized, dried, fused and poured out, is called Luna cornea. It is not soluble in water, is quite tough under a hammer, but otherwise brittle when broken; sometimes brownish, other times opal in color. It is sometimes found in nature when it is called Horn-erts [like horn].

Mercurius sublimatus corrosivus when distilled with silver filings produces luna cornea in the residue, according to Dr. POTTS´ research.

The precipitate with alkali fixum again produces luna cornea.

Note. **acid of salt does not dephlogisticate metals, but does attack them in the form of the calx. The precipitate, which has formed at the same time, resembles a slime, whereas if acid of vitriol is used the silver precipitates in more granulated form. The slime is soluble in a**

large amount of boiling water, and after fusing in a flame is likewise called Luna cornea. That acid of salt has a stronger attraction for silver than acid of vitriol seems clear from this, and that the former [acid of salt] precipitates a true argentum salitum from a solution of silver vitriolate. This salt [silver chloride] darkens in daylight and gets still darker in sunshine; it is decomposed in via humida by copper, bismuth, brass, iron, and zinc, particularly the latter two; it is volatile in flame; it contains 3-1/2 parts of silver to one part of acid; it is sometimes used for silver plating, but with damage; it can be reduced with alkali minerale; with lead; by grinding with quicksilver, alkali vegetabile, and sal ammoniac.

§ 107. If acid of salt or common salt is put into a quicksilver solution a white powder precipitates, which is called Mercurius praecipitatus albus, consisting of quicksilver combined with acid of salt. When nothing more precipitates with the aforementioned acid, still much more is precipitated by alkali volatile, which precipitate has the same name as given above, and is of similar nature. The precipitate should be well neutralized. It is also an emetic and purgative, taking 4 to 8 grains. It has been used against lues venerea; also against worms; but if use is continued for several days it arouses salivation.

When common salt forms no further precipitate, what remains is a weak quicksilver solution, which is called Mercurius cosmeticus, and has been used to beautify the skin and take away blemishes, but must then be well diluted. Some have proposed Merc. praec. albus itself for the same purpose, but it produces black spots. Others cure eruptions and venereal sores [with it], but some purging should also prevent these illnesses that affect the body.

> <u>Note.</u> Mercurius praec. albus is quicksilver combined with acid of salt, which [acid] has a stronger attraction for this metal than acid of saltpeter, but cannot attack it [quicksilver] directly, unless it has been dephlogisticated to some degree. This salt precipitates like a slime, but is still soluble in sufficient water. That which is precipitated by alkali volatile is by no means of the same properties, but rather [has those properties] of a calx of quicksilver, occasionally mixed with fine metallic grains, which during self-precipitation become reduced. I have found quicksilver vitriolate to be precipitated as a white slime by acid of salt, whose closer attraction consequently is manifested. Merc. praec.albus darkens in lime water.

§ 108. <u>Mercurius sublimatus corrosivus</u> is a compound of quicksilver and acid of salt, with an eruginous taste and is among the strongest poisons, so that it must only be used externally with much water, for removing dead flesh, cancers, scabs, and venereal sores. It is precipitated red with alkali fixum, and white by [alkali] volatile. The solution, gently evaporated, separates into long radiating crystals, which end more abruptly and shorter, if quicksilver is attacked by acid of vitriol (§ 109 (b)), than they are if acid of saltpeter is used (§ 109 (a)).

Note. Merc. subl. corr. is a salt which RHASIS first named, and which was being manufactured in China when Europeans came there. It runs [liquefies?] [in] litmus juice. Boiling water dissolves over half its weight, which then separates in four-sided prisms with points, building up on top of the smallest sides. Spiritus vini also dissolves it. Three parts of water mixed with sal ammoniac dissolve 5 parts of the sub- limate: spiritus vini mixed with sal ammoniac takes twice as much: this substance is combined intimately to a triple salt, which used to be called Sal alembrot. Mr. MACQUER has published noteworthy research on this substance. Merc. subl. corr. is precipitated by acid of vitriol, but the sediment is a corrosive sublimate which is deprived of the water of solution, and not vitriol, since it is dissolved in water and gives no Turpetum [residue?]. Merc. subl. corr. turns yellow in lime water.

To Prepare Mercurius Subl. Corros.

§ 109. (a) Quicksilver dissolved in acid of saltpeter is poured into a retort and a similar amount of common salt is added, an alembic is sealed on with milk cement, and the whole apparatus heated carefully on a sand bath. As soon as the salt begins to have a vigorous reaction, which ought to be watched so that it does not boil over, the fire should be moderated with much caution until the water is distilled off.

Note 1. It is usually customary to evaporate the quicksilver solution to dryness, then to mix the mercurius nitratus obtained with common salt and calcinated vitriol. In Holland where this salt is most often prepared, quicksilver is used, along with vitriol and common salt, equal amounts of each, which are ground together and put into several retorts. Mr. MONNET has given a method to prepare this salt by a wet method, from calx of quicksilver precipitated by alakli which is then dissolved in acid of salt.

(b) When the fire is then increased the acid of salt combines with quicksilver and is sublimed to corrosive sublimate, alkali minerale combines with acid of saltpeter, and the excess acids together make up an aqua regia which is distilled over.

(c) When all is dry in the retort the heat is increased still more, and when the mercurius sublimatus corrosivus mounts in earnest, sticking like other sublimates, remove the surrounding sand immediately. By now the sublimates are driven off over the sand so that a clear glass appears on it, which is a sure sign that nothing more is reacting, whereupon the vessel should be allowed to cool down.

Note 2. If the heat is strong the corrosive sublimate forms white very dangerous fumes, which fill the entire room when the operation is carried out in a retort.

(d) The ascending salt [vapor] is entirely white, radiating and transparent; if it continues long it becomes gray, brittle, and opaque. An entire pile of merc. subl. corr. can collect in the form of a flour.

(e) The residue consists of common salt with nitrum quandrangular, if acid of saltpeter was insufficient or too weak to decompose all of the common salt, since there is also some red nitrum mercurii present with red color, and likewise some sublimed running mercury.

<u>Note 3.</u> TACHENIUS reports that at the great works in Venice 280 pounds of quicksilver, 400 of common salt, 600 of vitriol, 50 of col- cothar, and 20 of corrosive sublimate were taken, which were mixed by a certain machine, divided into 15 large glass retorts, and sublimed with 5 dung fires, producing altogether 360 [pounds] of sublimate. It follows from this that the quicksilver needed is to the acid as 3-1/2 is to 1.

(f) Apothecaries trouble to add, in addition to the substances mentioned in (a), a quantity of white calcinated vitriol, to aid the decomposition of the common salt, because they usually use much weaker acid of saltpeter.

(g) Some people prepare M. s. corros. and dulcis from altered Merc. praec. with common salt, which still is not very successful because so little acid of saltpeter is present that hardly 1/4 the Merc. subl. is obtained, but if altered Merc. praec. is taken along with the common salt and calcinated vitriol, and these materials ground together thoroughly in a glass mortar, some Merc. dulcis is obtained but no corrosivus, since quicksilver is present in the red precipitate already reduced through the evaporation of the acid.

<u>Note 4.</u> The quicksilver is in no way reduced by the departure of the acid of salt- peter.

(h) If quicksilver is attacked by oil of vitriol [vitriolsolja] we obtain fine white crystals (§ 82), which are dried over a mild flame and ground together with double their amount of common salt to produce on sublimation a proper Merc. subl. corros.

Note 5. This is KUNCKEL´S discovery, by which we obtain, in addition to acid of saltpeter, a perfect Merc. subl. corr., more readily than by any other method.

(i) Merc.subl.corr. becomes purer through repeated sub- limation.

(k) It [Merc. subl. corr.] is sometimes found adulterated with white arsenic, which can be revealed by paper, colored blue by litmus, which becomes red if a solution of good Merc. subl. corr. contacts it, but black in the case of the adulterated form.

Note 6. That such Merc. subl. corr. adulterated with arsenic becomes black with alk. fix. or vol., I have not observed, although this test is given by many, especially more recent chemists, as certain.

§ 110. Merc. subl. dulcis con- tains less acid of salt and consequently is milder [than M. s. corr.].

(a) It is prepared from M. s. corrosivus by grinding it in a glass mortar until the mixture becomes evenly gray colored, which is very difficult, and must at this time be well observed for rising vapors. The mixture is then sublimed in the usual manner, when Merc. subl. dulcis vaporizes similarly to corrosivus, but opaque, without taste, insoluble in water, and what is more peculiar, [insoluble] in acid of salt.

Note 1. One part mercurius dulcis requires 750 parts of boiling water to dissolve it; so when a little water is used the solution gets somewhat grayish. Quicksilver seems to produce 3 different compounds with acid of salt; Merc. subl. corr. contains a maximum of acid, Merc. dulcis a minimum, and Merc. praec. albus less than the former but more than the latter. Merc. dulcis is less volatile than corros., turns black in lime water, becomes yellowish on grinding, is distinguished readily from corrosive sublimate by its slight solubility, tastelessness, etc., and contains about 6 parts quicksilver to 1 part acid.

(b) All mercury compounds which are used the most, and which are good for constipation and for purging the blood, ought not to be taken too often, since they cause salivation. It is odd that Merc. dulcis, which is insoluble is water, can overcome pain so in the body. It is known that acidum nitri most easily dissolves quicksilver; that [quicksilver also reacts] with vegetable acid is known from much common evidence and separate experiments; does the animal acid in our body also dissolve Merc. dulcis?

Note 2. Merc. dulcis is seldom free from all mixing with Merc. subl. corr.; and this salt should be used internally as proposed by Mr. BAUMÉ, i.e., ground with hot water and a little sal ammoniac until it dissolves; it is better purified [in this way] than by many sublimations, and less Merc. dulcis is lost, since sal ammoniac does not make it soluble.

(c) Merc. dulcis sublimed 9 times becomes purer and purer, and some call it Mercurius Calomelas [Calomel].

To Convert Merc. Dulcis Back To Merc. Corrosivus.

§ 111. The following research was carried out to accomplish this purpose:

(a) Merc. dulcis, decrepitated [common] salt, and some white calcinated vitriol, but a little more of the latter two [materials] to be certain of reaction, ground together and sublimed in the usual manner, produce Merc. dulcis [should be Merc. corrosivus?]. This always occurs when the procedure is followed according to these directions.

<u>Note.</u> Mr. BAUMÉ describes [the procedure] as follows: from 3 oz. Merc. dulcis he obtains 3 oz. and 1 drachma of corrosivus, of which 4 grains dissolves in the usual amount of water; he has mixed Merc. dulcis with common salt and the vitriol on a grinding stone. Can common salt be volatile?

(b) To determine if quicksilver volatilizes before acid of salt has time to dissolve, Merc. dulcis is sublimed with free acid of salt, but nothing is obtained other than Merc. dulcis, externally contaminated with the acid, which is strange because the acid ought to dissolve it.

(c) To investigate whether or not acid of saltpeter was needed, some of this [acid] is poured on Merc. dulcis in a glass [vessel], and the mixture evaporated on a slow fire, while the sublimate reacts. After which the decantate is diluted, more [acid] added, until almost as much is used as required to convert quicksilver into Merc. dulcis. Then it is heated to dryness and slightly burned, after which the mass is ground to pieces, mixed with 2 times its amount of fine salt and sublimed. But still nothing is obtained but Merc. dulcis. Consequently it appears that acidum nitri is not needed, since it is plainly evident that no reaction occurs (§ 109 (b)). In this example it is quite necessary that the quicksilver be attacked [by acid], but this can happen as well with acidum vit- rioli as with [acidum] nitrosum.

The residue left after the above described operation is white; some of it is dissolved in water and precipitated with alkali fixum, whereupon a white quicksilver-powder settles out, which also afterwards retains its color; and this is very strange, since both vegetable and minerale alkali form a red precipitate with quicksilver. Similarly a white precipitate was produced with aged potash.

(d) In the final experiment 1 part Merc. dulcis was mixed with 1 part common salt and 2 parts uncalcinated vitriol, which upon sublimation gave a perfect Merc. subl. corrosivus.

From all this we may deduce that the desired mutual decomposition and recombination needs the help of some phlegm [moisture?].

§ 112. <u>Acid of salt</u> dissolves only a little <u>lead</u> in metallic form, but a lot if it is calcinated, such as minium, whereupon the solution becomes yellow at first and small crystals separate out, which in a crucible over a fire are melted to a horn-like mass, called plumbum corneum, which requires much water and boiling in order to dissolve it completely. From [solution in] other acids, acidum salis immediately precipitates lead as a white slime.

Note. This salt is volatile in fire. Acid of salt has a special ability to volatilize metals; even gold itself, the densest material that we know in nature, can be volatilized when sal ammoniac is added to the solution mixture.

§ 113. Copper is dissolved by acid of salt, both in the metallic and the calcinated form. In the former case [metal] it takes a long time and finally produces a yellow color, which indicates how difficult solution is, since with a greater amount the color is green. Copper calx is attacked more easily after boiling into deliquescent crystals.

Copper is precipitated by both alkali volatile and [alk.] fixum, but the point of saturation must be closely observed, because if too much is added the precipitate redissolves, especially with alkali volatile, with which the solution also becomes [more] bluish, and at the same time the precipitate grayish.

Note. When the acid of salt is concentrated and aided bu heat, copper gives with it a brown solution, which upon dilution becomes green; it produces green crystals which deliquesce slightly, are readily soluble in spiritus vini, and give a bright green color to a flame with white and red striations. Mr. BAUME´ in 1757 made a gold sympathetic ink in a skillful manner with copper, but [the ink] will not disappear in cold, which Mr. MACQUER then perfected, when he noticed the property of this salt to become yellow-brown on drying out, and when water is added again becomes olive colored, then green, blue-green, and blue, all after taking up water. Writing made with this [ink] and dried in the air becomes invisible, but bright yellow on warming; it completely disappears in cold when calx salita is added and a little acid of salt, which makes the ink more deliquescent. Similarly with copper vitriolate, cuprum nitratum, etc., only acid of salt, free or combined, reacts, which result indicates that the copper is attracted to it [acid of salt] from other acids. If the writing is held over alkali volatile causticum it becomes bright blue, but then in air becomes green, unless it disappears.

§ 114. Iron dissolves, in both metallic and calcinated form, in acid of salt. In the former case [metal] the solution becomes somewhat greenish and it forms a green precipitate with alkali fixum, but in the latter case [calx] the color is orange and the precipitate red, becoming yellow in air. Consequently the difference comes from the phlogiston.

A solution of iron in metallic form, after boiling, produces spike-like crystals, and it is still difficult to determinbe whether it is best to boil for a short time or long. If they are evaporated strongly, the crystals become smaller than usual, but after deliquescing in air they shoot out larger rays. The solution in water gives a black color when tincture of gallapple or tea is present, which indicates the presence of iron, which is partly precipitated by tea, but not as much by gallapple; most of the iron here is from vegetable substances, which have so much iron in themselves that they become black. The clear solution after precipitation contains still more iron than the precipitate. If gum is added until saturated the black dissolves. The mother liquor which is not crystallized becomes brown with astringents, a sign of an excess of acid and a lack of iron.

Iron can actually be precipitated with strong lime water, although those great chemists, Prof. POTT and Mining Counselor BRANDT deny this; but the truth is that a large amount of lime water is needed before any precipitation can be observed, which is not to be wondered at, because one grain of burned lime can hardly be dissolved in one pound of water, and that acid of salt must be added to it, if the iron is to be separated from it. The precipitate becomes green and in this manner almost all the iron is removed, although some still remains, but so little that alkali fixum cannot detect it, since the precipitate becomes white with lime. The liquid, when boiled after precipitation and filtering, produces a little green-gray substance, which indicates the iron.

If the iron could be kept in solution after the calx is formed, a good medicament could be obtained from it for all kinds of illnesses which require the use of an astringent.

Note. The solution of metallic iron has a bituminous odor and sweetish taste, but more astringent if made from the calcinated [iron]; if metallic iron is added to the latter [calcinated iron] both the taste and color are altered. The vapor given off is inflammable. If pure quicksilver is added to the iron solution it quickly acquires a viscous film which deprives it of much of its liquidity; this surface film is lightly gray when metallic iron is in the solution, but darker with calx of iron, and arises from the iron powder which forms and reddens in fire; and in the liquid itself I have found some trace of quicksilver. This is a case of double decomposition; the dephlogisticated iron takes some phlogiston from the quicksilver, so that this latter metal can react with acid of salt.

Metallic iron when dissolved separates in spar-like light green crystals, combined to form four-sided hollow cones. When acid is much in excess they become pointed like needles, which I have found to occur with various neutral, earthy, and metallic middle salts.

§ 115. Tin is dissolved by acid of salt. It is pre- cipitated by alkali and also vitriolsolja [oil of vitriol], which has a stronger attraction for tin than acid of salt, but when this latter acid is mixed with water the vitriol brine cannot remain in solution but precipitates the tin.

Note. This salt allows itself to crystallize. If the liquid is smoking it stops as soon as the tin begins to be dissolved. If 4 parts of tin are amalgamated with 5 [parts] quicksilver, and mixed and distilled with 9 [parts] of Merc. subl. corr., there first passes over Spiritus fumans Libavi, which gives thick, stinging and long-lasting fumes; also called fumigatorium perpetuuum Joviale. At the end we get a thicker fuming material called butyrum stanni. Application of still more heat produces quicksilver. Since an amalgam is used, combined with Merc. subl. corr., we obtain in the neck [of the distilling flask] white flocks called Barba Jovis [Jove's Beard]. What causes the smoking spiritus fumans is not established; fuming acid of salt, when it dissolves much tin in the usual manner, does not smoke.

§ 116. Bismuth is not dissolved by acid of salt.

Note. In metallic form it reacts very slowly, but some can be precipitated with alkali from acid of salt digested with this metal; calx of bismuth dissolves more readily. From the solution produced with aqua fortis, first precipitated with 8 times as much water, according to Mr. ROUELLE the elder's research, when afterwards the clear [solution] is well concentrated, a bismuth salitum can be precipitated with acid of salt and leached salt, although with 20 times as much water nothing precipitates.

§ 117. Nickel's reaction with acid of salt has not been studied.

Note. Nickel is dissolved with a green color and gives a yellow-green deliquescent salt.

§ 118. White arsenic is dissolved in acid of salt.

Note. This solution proceeds easily by itself with the help of heat, but ought to be carried out in a retort with a receiver attached, because the vapor carries arsenic acid and is very dangerous. If arsenic is distilled with Merc. subl. corr., Butyrus arsenici is produced. The regulus is scarcely attacked by either dilute or concentrated acid of salt, although boiling has been used; Mr. BAUMEˊ reports however that a yellow powder deposits, insoluble in water.

§ 119. Calx of cobalt is dis- solved by acid of salt.

Note. The solution is red in the cold, but turns greenish at 70-80 degrees of heat; it separates in beautiful red crystals, which turn into blue, with the same shape as table-stones [tafelstenar]. The crystals when dissolved give writing which becomes bright blue in heat and does not disappear in cold, but if more acid of salt is added [the ink] becomes green in heat and invisible in cold. HELLOT first observed how to obtain a similar sympathetic ink; it is obtained from cobalt dissolved in whatever acid it is that is obtained whenever acid of salt or common salt is added, since acid of salt attacks calx of cobalt and produces this substance. The inventor clarified the strange material thus obtained to produce moist writing in the cold which became invisible on dilution, but when the heat dissipated, part of the cobalt salt becomes dense enough to be visible. Mr. BAUMEˊ believes that the moisture plays a part in this, since an invisible writing becomes visible immediately upon being dipped in hot water; but that the heat is not necessary can be concluded from the fact that the writing is visible in cold and even dry air, if held in a flask over concentrated acid of vitriol or calx viva [unslaked lime], as Mr. SCHEELE first observed.

§ 120. Antimony is not directly dissolved by acid of salt, but yields itself [to solution] readily on distillation with corrosive sublimate.

Note. Calcinated antimony is attacked instantly by concentrated acid of salt.

To Prepare Butter of Antimony.

§ 121. (a) One part Antimonium crudum and 2 parts Merc. subl. corrosivus are ground together in a glass mortar (some take Ant. crudum alone, but then 1/3 is more than required) until rising vapor can be detected by nose or mouth. The mixture is heated over a slow fire in a glass retort with a receiver cemented on, and this is continued until a red sublimate is observed.

Note 1. The author [Scheffer] follows STAHL´S proportion of 1 to 2; LEMERY recommends 3 to 4, and when the regulus is used 3 to 8. STAHL proposes that with antimonium crudum, common salt and calcinated vitriol, [we can also] distill butter [of antimony], and this succeeds also with antimony alone.

(b) The butter [of antimony] vaporizes and condenses completely dry in the neck of the retort. It is now very brownish, but after redistillation it becomes pure and is called rectified [butter of antimony]. It melts with little heat, like butter, and produces a dense vapor which makes different combinations [compounds?] with acid of salt.

Note 2. The ancients used the name butter for all dry metallic compounds with acid of salt, produced under distillation with Merc. subl. corr.

(c) The basis of the process is as follows: acidum salis will not combine with antimony without the aid of a third substance, for which purpose Merc. subl. corr. serves best. Under distillation the acid of salt is combined with the reguline part of antimony, which is attracted more strongly than the quicksilver; it can be concluded then that the regulus could be used instead of Ant. crudum, and through the combination described the quicksilver ought to dissolve, but something prevents this, something contained in the butter that is, which acts like sulfur, for which the quicksilver has a stronger attraction than for the acid, and therefore produces cinnabar.

Note 3. Acid of salt does not attack the regulus [of antimony], but since the quicksilver in the Merc. subl. is in the form of the calx, double decomposition occurs here; phlogiston in the regulus [of antimony] reduces the quicksilver [calx], and so the antimony thereby becomes calcinated to a degree, so that the acid of salt can combine with it.

(d) This Butter [of antimony] is very corrosive, so that surgeons use it for eating away warts, hard growths, and the like. One drachma is thinned with 1/2 quart of brandy and this is applied with a woolen or linen bandage to the swelling, in cases of pain or gout.

(e) When the butter deliquesces, a white powder sep- arates out, which is called Mercurius vitae. By solution in water we obtain something similar which when washed carried the same name, although it ought to be called Mercurius mortis. Two to three grains cause strong vomiting. The best consists of the regulus combined with as little acid as can be dissolved in water, exactly like Mercurius dulcis, luna cornea, plumbum corneum, etc.

The water which first runs off, as the mercurius vitae settles out, is called Spiritus vitrioli philosophicus, and was formerly thought to dissolve gold, but in vain, because it only contains acid of salt.

If Mercurius vitae per se is melted down until it produces strong glowing fumes, it disappears completely, but with phlogiston it produces the regulus.

Note 4. Mercurius vitae, so named by PARACELSUS, is now usually called Pulvis Algeroti after a doctor in Verona, VICTORIUS ALGEROT, who brought it into vogue and gave it the name Pulvis angelicus. If it is calcinated until all acridity goes away, we get Arcanum Reverii, and if this is heated strongly to melting we get vitrum antimonii succineum, etc.

(f) The butter is corroded by acid of saltpeter with strong effervescence to a white powder. If the liquid is distilled off from this, an aqua regia is obtained, which gets the name of spiritus nitri Bezoardicus; when gold is dissolved in it, it becomes bright red and a white calx is precipitated. The Alchemists called the red solution anima auri, and the precipitate auri corpus album, or luna fixa, imagining that this precious metal had been decomposed and this one part precipitated, but deceived themselves, since the white calx is the regulus of antimony, which passes over during distillation; and if the red solution is evaporated to dryness a red gold is obtained, the usual color of which can be restored by calcination.

Note 5. The acid of saltpeter dephlogisticates antimony further here, which [metal] consequently precipitates fully calcinated.

(g) The residue after the foregoing distillation (f), well heated, is called Bezoardicum minerale, of which BASILIUS VALENTINUS is the originator. This substance, considered of little use, can be used for fevers. Bezoard is a Persian word meaning anti-venom.

Before calcination and before the acid is driven off, it [the residue] is yellowish, and is nothing other than the usual crocus antimonii, but without volatility in fire, which arises from acidum salis which accompanies acidum nitri here, having a stronger attraction for it than for Mercurius vitae. It is most extraordinary that saltpeter and this acid take away the emetic power. Therefore Bezoardicum minerale and antimonium diaphoreticum both have the same nature, and antimonium diaphoreticum, Bezoardicum minerale, and antihectici Poterii, although cast out the window, do the patient no harm.

§ 122. Zinc is soluble in acid of salt.

Note. During solution much inflammable air is given off, as with acid of vitriol. The solution is clear and can be caused to crystalllize.

The black flakes which appear do not burn in a flame, and are soluble in both acid of vitriol and acid of saltpeter, according to HELLOT'S research, but Mining Counselor BRANDT says they are sulfur. In distillation with Merc. subl. corr. we get butter of zinc.

§ 123. Distilled vinegar does not dissolve silver.

Note. As long as silver has all of its inflammability [phlogiston] left it is not attacked [by vinegar], but precipitated with alkali it is completely soluble, and separates in glistening crystals. The precipitate, when silver solution is dropped in until it dissolves, [is called] Terra foliata tartari.

§ 124. Quicksilver is not dissolved by vinegar.

Note. Quicksilver, precipitated with alk., or calcinated by acid of saltpeter, is greedily dissolved [by vinegar] and gives beautiful silvery crystals.

§ 125. (a) Lead is dissolved by vinegar in both metallic form and in the form of the calx, but better in the latter form. The solution is usually called Lac virginale metallicum, but also acetum Lithargyrii. After moderate boiling it separates into crystals, but the proper degree of evaporation is difficult to find because no film appears. Some recommend that 9/10 ought to be boiled away, but that seems to be excessive.

The crystals are usually called Sal vitriolum or magisterium Saturni, but more correctly saccharum saturni crystallisatum.

Sugar of lead is usually prepared in this manner: the solution is first evaporated to dryness, then dissolved again in vinegar, and then afterwards boiled until crystals form. It is then dissolved two [more] times in vinegar for no particular reason; if not, when white lead is used, which must be dissolved in a certain way, it is easier to crystallize than when litharge, red lead, or lead [metal] is used.

Out of this solution lead can be precipitated with alkali fixum; although something precipitates in the beginning, it dissolves again when more alkali is added.

Lac virginale metallicum is believed to be good for all inflammations, but is a slow-acting poison. Used as a paint, for example, it causes consumption. Internal use of sugar of lead spoils the stomache, destroys the appetite, makes the skin pale, and finally causes deadly consumption. It is very powerful for extinguishing all sexual desire, and in this respect is a good medicine for monasteries.

Note 1. Sugar of lead is insoluble in spiritus vini and oils. It is precipitated by several substances not enumerated in the [Scheffer's] text, such as all salts which contain acid of vitriol or acid of common salt, by acid of winestone [tartaric acid], and acid of air [carbonic acid]. If Sal Glauberi crystals and this salt decompose while being ground together like porridge, copious water of crystallization is produced in this manner. Since lead neutralizes vegetable acid, and this without changing the clarity or color, which result cannot be achieved with any other known material, so lead is used sometimes in the form of litharge or other saturnina [lead compounds] to repress excessive acidity in wine. This criminal use is discovered if liquor vini probatorius [wine testing solution] or other dissolved liver of sulfur is dropped into it [the wine], since the precipitate gets dark instead of yellowish.

(b) If we wish to prepare white lead, a retort containing vinegar is placed on a sand bath, thinly beaten or rolled lead [sheet] placed in the mouth of the retort, a cap placed on top and cemented on with milk cement. Then the distillation is driven [by fire] underneath as much as possible so that vapor converts the lead [sheet] to a white calx, which is scraped off, and is called white lead or cerussa. The distillate contains sweet dissolved lead and causes severe choking.

In commerce, wooden pails or Waldenburgian pots are used to hold glass containers of vinegar, and over these is placed a thin rolled sheet of lead. On the pot also is laid lead sheathing as a lid, after which the pails are set in manure, packed into a walled space, which is [then] closed up. After some time the lead is found to be corroded.

White lead is used to absorb acidic liquid in sores. That which is usually used in paint causes consumption. In North Holland where cerussa is made, people can scarcely endure 3 years in the works without dying. Dusting oneself with this substance is very dangerous.

Note 2. White lead is useful as a white oil pigment, than which nothing better is to be found, and which was in demand because of its tendency to be changed by phlogistic vapors. In Sweden there has never been more than one such factory, which was established in Norrkoping some time ago. This substance has little if any taste or acidity. Vinegar, when mixed with it, is in many respects different from acetum lithargyrii. This latter substance [acetum lithargyrii] gives off a white powder in air; it deposits a soluble mucilaginous sedative salt and corrosive sublimate; this does not separate without much difficulty, which circumstances are opposite to those found with vinegar and white lead. Mr. SCHEELE has noted this difference, which seems to occur with lead calx free from acid of air, with the help of the oily material in vinegar.

§ 126. Copper in the metallic form is only slightly souble in vinegar, but very soluble in the form of the calx.

(a) Verdigris is produced in France from pressed grapes, which are laid in layers with copper plates [in between], which are squeezed together and left some time undisturbed, and in the meantime the grapes get hot and give off acidic vapors, which corrode the surface of the copper to a greenish-gray crust, which is scraped off and is the common variety of verdigris, which is not soluble in water. When the plates have been scraped off, they repeat the process as before.

Note 1. Verdigris, to my knowledge, is prepared nowhere else but at Montpelier [France}. It is not entirely insoluble, since both reaction and precipitation can be used, but much water is required and the dried material is perhaps insoluble to some extent. Mr. MONTET reports that it contains 5 parts copper to 27 parts acid.

(b) Common verdigris, dissolved in vinegar and crys- tallized, has the name distilled [verdigris] and is soluble in water. It can also be made from copper calx dissolved in distilled vinegar, filtered, boiled and crystallized. Iron produces no precipitated copper from [solution in] vegetable acid.

Crystallized verdigris, distilled in a retort, lets loose a sharp acid which is called acetum acerrimum.

If acid of salt is poured on the crystals and distilled, the distillate dissolves gold, which shows a relation between acid of saltpeter and the vegetable [acid].

Distilled verdigris is strongly emetic. Six to ten grains in a quart of milk can be taken internally, if something poisonous has been swallowed, which [then] is quickly thrown up again through vomiting.

Note 2. Both common and the so-called distilled verdigris pre- cipitate copper very slowly on iron, but require however one whole day; if a little vinegar is added to the solution the precipitation proceeds more strongly and much more plentifully.

That the distillate from verdigris and acid of salt dissolves gold, I have not succeeded in repeating, unless this precious metal be precipitated with alkali from aqua regia, when distilled vinegar with a little common salt is best.

§ 127. If to iron filings 20 times as much spiritus aceti is added, and [the mixture] is allowed to stand 24 hours in digestion with seething up, we obtain Tinctura martis made with spiritus aceti, which is good against caterpillars, and against phlegmatic illnesses.

Rectified wine also dissolves iron, but gives a black- ish tincture.

Note. If the vapor from the digestion is collected and studied, it is found to be inflammable. The solution has a stiptic taste when it is allowed to stand in open air, all the iron being in the form of the calx; iron calx is not soluble in vinegar. JUNCKER says that the solution can be crystallized

§ 128. Tin is dissolved by concentrated vinegar, but only slightly.

Note. Tin calx is more easily attacked than the metal itself. The solution can be crystallized, precipitated by alkalis, acid of vitriol, common salt, etc.

§ 129. Bismuth has not been tested with vinegar.

Note. It dissolves on boiling in strong vinegar.

§ 130. Nickel has not been tested with vinegar.

Note. It is soluble in the form of the calx with boiling.

§ 131. White arsenic dissolves in vinegar.

§ 132. Cobalt has not been tested with vinegar.

Note. It dissolves in the form of the calx, the solution becoming red and giving a sympathetic ink.

§ 133. One drachma of vitrum antimonii gives emetic power to 16 times as much white wine for 10-12 hours.

Formerly also goblets were cast from the regulus [of antimony], in which wine, after 6-10 hours, formed a vomative [emetic]. Its size should be such that it holds 2 ounces of white wine.

Note. Wine contains the same acid as vinegar. The old calices vomitorii and pillulae perpetuae need more careful study, since in the reguline form no weight [of antimony] is lost when vegetable acids are boiled with it.

§ 134. Zinc is dissolved by vinegar; also flowers [of zinc] and calamine [gallmej].

§ 135. Lead is dissolved by alkali fixum when it [lead] is precipitated from vinegar. See § 125 (a).

Note. These metallic salts, which require alkali to dissolve, are little known. Authors who described them in various places I have reported here in their proper place. Usually metallic salts are precipitated by alkali, but here acid is needed to separate the metal [from solution].

Lead is attacked by alkali fixum in both the dry method and the wet method.

§ 136. Copper is dissolved by alkali fixum, yet it is precipitated out of acids [by alk. fixum also].

Note. When caustic lye is boiled many times to dryness on copper filings, we finally get a beautiful blue solution; the same thing is obtained more easily by digestion of the [copper] precipitated from some acid. Copper is attacked also in the dry method.

§ 137. Cobalt is dissolved by alkali fixum.

§ 138. That antimony in the form of the calx is dissolved [by alkali] can be concluded from Tartarus emeticus, which is prepared from 1 part vitrum antimonii and 9 parts alk. fixum, boiled with water to dryness, after which new water is added, filtered and boiled. This must be used cautiously, since otherwise it could cost one his life.

Note 1. Here [Scheffer] describes Tartarus emeticus prepared as ZWELFERS directs, which was published in Pharm. Holm. for the year 1682. This method is now hardly used at all.

Note 2. If we omit the solution with alkali fixum referred to, there are still a few other examples. Precipitated gold is attacked some by caustic lye during boiling, I have found, and if the precipitation has been carried out with tincture of gallapples, the lye becomes yellow without heating. Precipitates, with [tincture of] gallapples, of quick- silver, iron, cobalt, zinc, etc., are similarly soluble in caustic lye, but we do not yet know the true nature of these, not one of them, nor if they could be caused to crystallize; perhaps if they are not double salts they may be some kind of combined salts. Various metals can be dissolved by alkalis at the monment of precipitation, when they are then in their finest state of division, but not otherwise.

§ 139. Silver is dissolved by alkali volatile (§ 81).

Note. Silver precipitated out of acid of saltpeter is dissolved by alkali volatile and separates in crystals which are very soluble in spiritus vini. The solution is precipitated immediately by acid of salt and common salt, by acid of urine, but not by Sal microcosmicus, nor by vitriolic acid, as Mr. MARGGRAF has determined. If a little quicksilver is added we obtain arbores Dianae, which form more easily if 2 parts silver are amalgamated with 1 part quick- silver, then dissolved in aqua fortis and this solution diluted with about 8 times as much distilled water, since when a small globe of amalgamated silver is put into half a wineglass of the above described solution, in 10 or 15 minutes a silvery bush forms with branches and twigs of very beautiful appearance.

§ 140. If bits of copper are put into spiritus salis ammoniaci they dissolve with a dark blue color, but air must be present or otherwise the solution becomes cloudy and less colored. When a film is observed to cover the surface it can be collected and separated in small bright blue crystals from which the excess alkali escapes. The crystals decompose in air, becoming green and losing their alkali. If they are ground fresh with lime and a little water they give off a pungent alkaline odor. The solution is used to color maps, because it does not blot out the writing.

Note. The crystals are four-sided, like spar, but the edges of two planes come together from the pointed corners to form a gable.

When copper filings are put into caustic alkali volatile, the flask can either be left open or corked; in the former case the solution becomes blue, but in the latter case the reaction differs according to the circumstances. If absolutely no vacant space is left in the flask and it is tightly closed immediately, then when copper is put in, no solution at all occurs, even after long standing, but if it is opened even a little [to the air], or if some empty space is found between the cork and the liquid, a colorless solution is obtained which becomes blue when exposed to air, and becomes colorless again as often as this is repeated if it is merely kept closed over fresh copper filings. These conditions Mr. SCHEELE has carefully determined to be as described. It follows from this entire series of reactions that copper is not dissolved unless it loses a little phlogiston which is taken up by free air, which attracts the inflammable element strongly; but when it occurs to some small part, and air is excluded, it dissolves without color, and [the solution] becomes blue just as soon as air is admitted to remove the residual phlogiston.

This copper calx is precipitated when the blue solution reacts with the required amount of dephlogisticated copper and can be made colorless thereby in a closed container as often as you please.

The colorless solution is precipitated by zinc, by acidum vitrioli, but not by iron; the taste is almost sweetish and it does not smell of alkali volatile; the blue [solution] on the other hand always has a pungent odor and can be precipitated by distilled water.

§ 141. Cobalt in the form of the calx is dissolved by spiritus salis ammoniaci, which becomes red thereby. See § 91.

§ 142. Zinc is dissolved by alkali volatile.

Note. In addition to this [i.e., the metals listed in §§ 139-142], nickel is dissolved by alkali volatile with a blue color.

§ 143. Aqua regia dissolves gold [with] bright yellow [color], better when acidum salis rather than acidum nitri prevails [in the aqua regia]. Upon evaporation the solution produces crystals. It is precipitated by quicksilver, iron, copper, tin, tin and iron solution, and alkalis. The solution becomes green after precipitation with alkali fixum; it is noted that the gold is mixed with copper.

Spiritus nitri Glauberi also dissolves gold, but only in a closed vessel and with the help of fire, and the solution precipitates by itself as soon as air gets to it by shaking.

Metals can precipitate each other from solution, and always those which have the stronger attraction for the liquid than the other [cause the precipitation], but the precipitating metal must dissolve at the same time. [Bergman followed up on this statement of Scheffer's, determined the amount of phlogiston moving from one metal to the other in the precipitation, and thereby made his discovery of the equivalent weights of the metals. J.A.S.] Up to now gold, which has the least attraction for the acids [of all metals], is the exception; it can be precipitated by various metallic solutions. When tin and gold solutions are mixed, both metals precipitate; quicksilver dissolved in aqua fortis (but not when it is dis- solved in aqua regia) precipitates gold; iron solution precipitates gold and holds itself in solution, but always precipitates a little ochre.

If the aqua regia is made from acidum nitri and Sal ammoniac, the precipitate is yellowish, which is obtained with alkali fixum after neutralization and drying, to the extent that it discharges in a weak flame with a sharp report, and is called <u>aurum fulminans</u>. If the aqua regia consists of acid of saltpeter with common salt, or both acids [nitri and salis] mixed, the precipitation must be produced with alkali volatile, at least at first, otherwise we get no fulminating gold [knallguld]. If it is rubbed in a glass mortar, or in some other manner so that heat is produced, it catches fire with incredible violence, and is consequently very dangerous to keep closed up. If we wish to detonate it without loss, it should be put into a retort with a tubular hat on a sand bath, and when the sand is heated to the point where the hand cannot stand it, knallguld is put in, a little at a time, and the stopper put on at once to prevent the fumes from reaching the air. As soon as the powder is put in a small explosion occurs and white fumes arise. In this way the entire quantity may be detonated. The retort ought to be somewhat large, and small portions put in at a time, otherwise [the retort] would burst. That which is obtained in this operation is called <u>flores auri</u>; it is soluble in alkali volatile, and melts in a crucible to gold, which can best be collected if a little borax is added.

Note. Although gold, which has all its phlogiston intact, is dissolved with difficulty by some simple acids (which can be deduced from the above described research by Mining Counselor Brandt concerning acidum nitri which is the only exception up to now), it is still attacked readily after precipitation by all mineral acids. It has also been shown previously (§ 23) that acidum salis can be dephlogisticated, and such an acid of salt robbed of its combustible element [phlogiston] attacks gold; we then have good reason to believe that acid of saltpeter in aqua regia does little else but take up phlogiston from acid of salt; and at least aqua regia has the smell of dephlogisticated acidum salis. Double attraction seems to take place therefore in the dissolution of gold; namely dephlogisticated acid of salt which has lost some of its combustible part, and an ordinary acid which attracts the weakened de- phlogisticated calx of gold. May not strong and boiling acid of saltpeter dissolve gold on the same basis?

The solution is precipitated by alkalis, lime water, iron, copper, tin, quicksilver, and white arsenic, which same [precipitations] from solution are described in the [Scheffer's] text, by silver and lead dissolved in aqua fortis, which takes acid of salt to itself and consequently decomposes the aqua regia. Tincture of galls also precipitates gold, as well as other metals, which shall be described further in the last section. Wine, vinegar and tartar solutions precipitate gold with metallic color, but very slowly; concerning the action of oils and ether, we will have more to say later (§§ 221 and 214).

Knallguld is a strange material. We have had opinions about the reasons for its violent explosion at a certain degree of heat, the most general opinion having been that some nitrum flammans has so attached itself to the precipitate that it cannot be separated from it by water, and that this [nitrum flammans] is prevented from detonation by the density of the gold, until the heat becomes so strong that all at once the violent assemblage bursts into flame. I have on other occasions examined the behavior of aurum fulminans, and I know that it can be obtained without the least amount of acidum nitri, from which [fact] I will now conjecture that gold dissolved by dephlogisticated acid of salt and alkali volatile produce knallguld, even though no acid of saltpeter has come in contact with the gold from the beginning. This proposal, which assumes metamorphoses, can still of necessity require a slight change of some sort in acid of saltpeter, but these suppositions, I admit, do not carry with them any particular authority. Meanwhile it is agreed that something in alkali volatile shows itself in the process in question, since a precipitatum non-fulminans obtains similar power through digestion in alkali volatile, and what is worthy of note, also in digestions with solutions of all neutral salts which contain alkali volatile.

If gold leaf is ground together with alum, saltpeter and enough common salt, and then boiled sufficiently in water, the gold dissolves; the acid of vitriol in alum decomposes the other two salts, thus forming aqua regia, and so [this preparation] has been named <u>menstruum sine strep-</u>

ita, because no alkali volatile is contained in it, although when it is used for precipitation, knallguld is obtained.

Gold precipitated with tin solution is called Purpura mineralis from its color, and is used in glass making to make a ruby-colored flux (§ 179).

Gold solution mixed with saltpeter, held [soaked] in linen rags, which mixture has then been burned dry, is used for cold gilding in this manner: that a wet cork dipped in the burned linen residue is then rubbed on pure silver [to produce gilded silver]. For gilding wheels and small pieces of steel, dipping in gold solution has been used, then rinsing in water, and finally polishing them, but excess acid always attacks the finest edges and points; so Mr. BAUMÉ has proposed bringing the solution to crystallization, letting the crystals melt and dissolving them in pure water, whereby all the excess acid is removed. So-called Greek gilding is carried out using Sal alembrot, or Sal ammoniac, and as much Merc. subl. corr. as is present in acid of saltpeter, and in which gold is then dissolved, the solution concentrated and filtered, and the silverware dipped in it, whereupon it becomes entirely black, but on heating becomes gilded.

§ 144. Platina or white gold also is dissolved by aqua regia. The solution becomes red, is precipitated by quicksilver in the same manner as gold, by alkalis into a red powder, or even by pure water, which precipitate dissolves again if more aqua regia is added. Iron vitriolate does not precipitate Platina, and therefore can be used to separate [platinum] from gold.

Note. Platinum solution is precipitated by Sal ammoniac, [and by] gold and tin dissolved in aqua regia; likewise by silver, quicksilver and zinc dissolved in aqua fortis, but not by lead, copper, iron or bismuth solutions; and what is unusual, not by alkali minerale, which Mr. MARGGRAF discovered. Consequently, if gold and platnum are mixed in solution, the former [gold] can be separated by green vitriol or alkali minerale, and the latter [platinum] by Sal ammoniac.

Platinum precipitated from aqua regia is soluble in mineral acids.

§ 145. Silver is precipitated by aqua regia. [Meaning is obscure here. See § 147, Note. J.A.S.]

§ 146. Quicksilver is dissolved by aqua regia.

§ 147. Lead is dissolved by aqua regia.

Note. This and the next preceding [§ 146] are taken to mean that Mercury and Lead salts dissolve in aqua regia. [By analogy then, § 145 could mean that silver salts are precipitated by aqua regia.]

§ 148. Copper is dissolved green by aqua regia.

§ 149. Iron is soluble in aqua regia.

Note. Dr. ZIMMERMAN dissolved 1 part of iron filings in 12 parts aqua regia (made from acid of saltpeter with 1/8 Sal ammoniac) without heating. The solution is filtered twice, leaving a yellow white earth on the filter, which after heating to glowing with coal dust is attracted strongly by a magnet; I have let it [the solu- tion?] stand for 2 years, after which a bright white earth falls to the bottom.

§ 150. Tin is soluble in aqua regia, which should consist of equal amounts of acid of saltpeter and acid of salt; the solution takes a long time, and becomes pale yellow in color, though clear. If the metal is rubbed with a little acid of salt it corrodes to a white powder. If any lead is present in the tin it precipitates. One part tin requires 5 parts of pure aqua regia, and 6 of impure [?]. At first it is poured on the corroded tin, and it does not [all] dissolve if not enough aqua regia is used.

Note. Aqua regia, made of 2 parts acid of saltpeter and 1 part acid of salt or Sal ammoniac, dissolves its own weight of tin clear, and if it [tin] is added in small pieces no heating is required. The reaction here does not seem to be completed, since if the solution is heated an internal reaction takes place, after which it often stiffens like gelatin. In regard to acids of unequal strength, the weighing can be less than certain; Mr. EXLEBEN proposes therefore to take 4 parts acid of saltpeter to 1 part acid of salt, but to test [the solution] with a small bit of tin to see if it dissolves completely; if a black powdery residue remains, increase the acid of saltpeter, but if the residue is white, increase the other [acid of salt]. Tin solution is used to revive the color of cochenile and Purpura mineralis .

§151. Bismuth is dissolved by aqua regia.

Note. Nickel is dissolved also and with green color.

§ 152. White arsenic is soluble in aqua regia.

§ 153. Cobalt is soluble in aqua regia with red color.

§ 154. Regulus of antimony is dissolved clear by aqua regia, which consists of 1 part acidum nitri with 5 of acidum salis; but if this solution, and also the butter [of antimony] is precipitated by water, it is necessary to use the strongest acids we have. Antimonium crudum is dissolved similarly to the regulus [of antimony], but sulfur falls to the bottom, and is the purest, like that from a puff of flame, that can be obtained: it is called Sulphur antimonii per aquam regis.

If excess acid of saltpeter is in the aqua regia, the solution becomes turbid and calx antimonii precipitates. If this solution is precipitated with water and [the precipitate] washed, we obtain Flores antimonii purgantes. If the dried corroded calx is mixed with a similar amount of sal ammoniac and sublimed in a retort with a cemented-on cap, flowers vaporize which when diluted with sufficient water are called Flores antimonii emetici, which when taken internally in amounts of 1/2 to 1 grain produces strong vomiting.

Flores antimonii emetici, mixed with 3 times as much saltpeter and heated to burning in an iron pan, loses all its emetic power and finally becomes sudorific; it is called antimonium diaphoreticum Helmontii. According to the directions given by the discoverer we ought to get flores antimonii rubri, which settles on top when antimonium crudum and sal ammoniac are sublimed together in a retort with a long neck.

It is written that antimonium diaphoreticum edulceradt, mixed with 1/2 resina scammonii and 1/6 cream of tartar, is called Purgans diaceltacesson Helmontii, or Paracelsi, which is good for all agues and intermittent fevers, if 8 to 36 grains are taken internally 3 hours before any paroxysm.

§ 155. Zinc is soluble in aqua regia.

§ 156. Solutio lunae, mixed with a lot of alkali volatile, is a good blood purifier, used along with mineral waters, either as it is or made into a solid pill. It tastes indescribably bad, but it is not corrosive or dangerous.

Note. This can be regarded as a solution of silver in nitrum flammans, or the resulting triple salts such as solutions of metals in sal ammoniac.

§ 157. Sal ammoniac dissolves copper.

§ 158. Sal ammoniac dissolves iron. If haematite and enough finely powdered sal ammoniac are well mixed and sublimed in a retort, yellow flowers are obtained. The residue is richer in iron and is called oleum martis per deliquium.

Note. This sublimate is usually called Flores calendulae mineralis; if iron filings are used for bloodstone we get Stroma philosophorum or Aroph Paracelsi, and with colcothar vitrioli Ens Martis.

§ 159. Sal ammoniac dissolves tin.

§ 160. Sal ammoniac dissolves regulus antimonii.

§ 161. Sal ammoniac dissolves zinc.

§ 162. Arcanum duplicatum, which is not purified by precipitation with lye salt, always contains iron, which is detected from the cloudiness, the precipitation, and the reaction. Through evaporation it is brought to separation into five-sided, tangled crystals.

Note. In addition, the aforementioned triple salts also produce ordinary Tartarus emeticus here, which consists of the calx of antimony dissolved in purified tartar. But rather it ought to be prepared from Pulvis Algeroti, about which I have written extensively in another place.

§ 163. Iron vitriol and tartar, mixed well and boiled in water to the consistency of honey, then digested for 24 hours in spiritus vini, produce Tinctura martis Ludovici, which is a good medicine for scurvy, dropsy, and childbirth pains.

Note. Four parts of purified tartar dissolves up to 1 part vi- triol, so that it does not separate in crystallization, but together produces a quadruple salt. It is noteworthy that green vitriol, which is not attacked in spiritus vini, is dissolved better the more it is dephlogisticated by boiling or burning, an obvious proof of the falsity of the conclusion that homogeneous [materials] attract each other most strongly.

§ 164. Fahlu vitriol, which forms in mines, is light blue, and contains zinc, copper and iron in the proportion as 16, 4 and 1 have to each other. It can well be used as a black coloring, but when iron is added, which precipitates copper, [the ratio] becomes 16/21, and it is useless, so that for blackening cloth 8 pounds are required, whereas Salzberger [dye] requires [only] 5. It serves better as a brown fashion coloring, than a black. Four pounds of it cost the same as 1 pound of Salzberger, so that it can be used with profit.

Note. I will mention only two quadruple salts. If sal ammoniac is dissolved to saturation in a hot iron-containing aqua fortis, we obtain on cooling red cubic crystals, which KUNCKEL calls ruby sal ammoniac [Salmiac rubiner]; it is not successful without iron. If sal ammoniac is dissolved in copper solution with aqua fortis, we obtain blue crystals, which become yellow on heating, and a solution of these gives a sympathetic ink, which was described previously in § 113.

To finish our discussion of salts, we introduce the preparation of Prussian blue; this is carried out with the help of salts and the product is probably saline itself.

To Prepare Prussian Blue

§ 165. (a) Soda is finely ground to pass through a sieve, then put in an iron pan with water and boiled down about half, when the pan is filled again and the solution boiled a while longer, after which the solution is filtered as soon as it thickens. Water is added to the residue again and boiled, and this is continued as before as long as any salty taste can be detected. All the solutions are combined and evaporated until just before the solution thickens and dries up; a very small fire must be used so that no phlogiston is burned off from the Sal soda, which is most important in this process. For the same reason the salt is not calcinated but is dried very cautiously, after which it is ground and put through a sieve.

(b) With the soda salt is weighed out some finely ground oxblood, either dried in the summer sun, or in a tile oven in winter, and these are mixed together thoroughly. We have found no better phlogiston than that in oxblood.

Note 1. To better saturate the alkali we can take with [one part of] it 2 to 3 parts of blood. The blood is dried in an iron pot with stirring until it loses all its volatility, when it is divided into portions and placed on an iron sheet in a moderately warm oven; it can also be done altogether in the pot, but the fire should be reduced then very much, so as not to burn. Erlanger Blue is made with fine soot.

(c) One medium-sized crucible is placed on the forge and heated; the mixture from (b) is put into it. The blood should ignite immediately, and as soon as this is finished a bellows is used so that the necessary heat may be obtained more quickly, since the longer it burns the more the combustibles are destroyed. The material does not melt quickly because of the abundance of phlogiston, but when it is like mush it is taken from the crucible, and the material finely ground.

Note 2. To give the mixture just the right amount of combustion is one of the most difficult operations. Most of it does not melt, and to take the mixture out from the flame as soon as it glows makes it show only a little white flame on the surface, like a [hot] coal. This indication is fairly noticable, but not always dependable; most dependable is Mr. MARGGRAF'S suggestion, to dissolve a little of the material in water and add a few drops of silver solution, since if it gives a brown precipitate, it has burned long enough, but otherwise not enough. We then try to work the hot mass out of the crucible into water without having to grind it. After the combustion, the weight of alkali is found to be hardly increased by 1/8.

(d) The thoroughly ground material is boiled with water in an iron pan until it all has time to separate, then it is allowed to settle and is filtered.

Note 3. This solution is called blood-lye.

(e) Meanwhile weigh out 1 part of iron vitriol and 3 [parts] of alum, which are dissolved together in water over a flame.

(f) This solution (e) is filtered immediately into the lye (d), when both solutions are still warm, the mixture is stirred, and a yellowish precipitate forms continuously, and should be watched carefully until it changes to green completely, since if we mix insufficient amount of (e), witout allowing it to settle, neutralize, and dry, the whole mass becomes blue.

Note 4. Usually we take 8 parts of alkali to dissolve 3 parts of vitriol and 4 of alum, and this is mixed with lye (d).

(g) Prussian Blue consists of Terra martis and phlogiston, since most of it burns away in bits and leaves only a bit of red earth; the trick therefore is to combine them correctly. Phlogiston must be dissolved, because it does not mix otherwise with iron during precipitation; for this reason alkali is required and makes the rest unnecessary, since it does not go into the color. The iron must also be dissolved, so we use vitriol, because during burning this acid takes the alkali to itself, whereupon the phlogiston and the iron have nothing to hold them together so they precipitate together. But since phlogiston takes very little iron, there is not enough acid for the alkali, and so the phlogiston is not completely separated, alum is added to it, which with its earth increases the color, but still gives it some whiteness. The stirring should cease as soon as the precipitate turns green, since if it is continued until it becomes dark blue, the color becomes very pale after drying, from the excess earth of alum. We can also, if we so desire, leave out the alum entirely and use either free [acid of] vitriol or acid of saltpeter instead, but acid of salt turns the color to green. When free acid is used the vitriol solution is first poured into the lye and then acid is added as long as foaming is observed, after which the precipitate becomes very blue even after drying. We could rather have used acidum vitrioli and in the amount which alum contains. If by using excess earth of alum we get a pale color, we can correct it by adding dilute acid which dissolves some of the earth.

Note 4. [B.'s mistake. Should be Note 5.] The alkali used for Prussian Blue is believed to be loaded with phlogiston during burning. There is no doubt that pure essence is found in such alkali, but if the alteration of the alkali is just being completed, it might not yet be done. It seems to me, as it does to Mr. SAGE, that some animal acid is chiefly the agent here. If caustic lye is poured on Prussian Blue, it takes the coloring material to itself and leaves the earth of alum and earth of iron behind; if after this it is poured on a new portion, it can finally be saturated and allowed to come to crystallization. The crystals are transparent, yellowish, (or green if they contain the dissolved color); they do not taste like lye; they do not effervesce, nor are they decomposed by mineral acids, but precipitate a little blue powder, which is partly dissolved in them, partly caused by the iron contamination which is always found in the mineral acids; the blue is not changed by vegetable juices; it does not precipitate the middle salts of earths, but all metals dissolved in acids, and this precipitate cannot then be redissolved by more acid. This behavior is clear enough to be used as a test for neutral salts. I have also noticed an apparent effervescence when not very caustic alkali is poured on Prussian Blue. If the saturated bloodlye is digested with iron we do not get a blue color, even if the process is prolonged; the coloring material cannot be separated from the alkali, unless we use double attraction, in which case the iron must be dissolved in acid, so that it attracts the alkali while the metal [attracts] the color, which instantly makes it blue, but is not attracted by a magnet. Other metals are precipitated similarly, and bloodlye might be the best method to detect and separate them, if they clearly showed a different color. Usually alkali forms a green- ish precipitate with [iron?] vitriol, but after- wards the precipitate turns yellow-brown. When the bloodlye is not saturated, it precipitates from solution (e) partly Prussian Blue, partly earth of alum, and partly earth of iron, which [latter] soon turns into ochre; the last named [ochre?] which damages the color most, dissolves when acid is poured on the precipitate, indeed can even be prevented if part of the bloodlye which is not saturated with coloring matter is previously saturated with acid. Acid of salt is quite as good here as any other [acid]. Distillation of Prussian Blue yields alkali volatile and an oil; the residue is black and has lost its power.

(h) Commonly white flux (§ 37 (a)) is used, or otherwise vegetable alkali, but this is much less costly than refined potash, and though alkali nitri causes the separation better than the former [vegetable alkali], it gives in addition a little Prussian Blue itself, because this alkali contains some phlogiston. From one pound of alkali vegetabile we get finally a half ounce [1 lod] of Prussian Blue, and something more of soda salt.

Note 5 [6]. Prussian Blue boiled with water and unslaked lime is decolorized and the coloring material gives with the lime a deliquescent salt, the solution of which is used like bloodlye to precipitate metals. I make use of more of this than of ordinary bloodlye, because the latter is obtained saturated and less contaminated with dissolved color, so it should be avoided for its acid of vitriol content, since although a bit of gypsum is produced thereby, we cannot be certain whether or not it might be mixed with the precipitate and change its color.

(i) Before the dye can be used, after the first drying, it must be ground in a glass mortar with water to make a thick gruel, and then dried [again]. Without the water it becomes too hard. If it is too dark, you can add some white lead, but this makes it bad, foul, and lumpy.

Note 6 [7]. This dye was discovered in the beginning of our century by DIESBACH, when he used Florentine lacquer instead of bloodlye with alkali; thereafter DIPPEL had abstracted animal oil; his solution contained cochenile, alum, and a little iron vitriol, and from this he obtained, with the last named alkali, not red but rather his marvelous blue lacquer. DIPPEL then shortened his procedure but kept it secret. WOODWARD reported the procedure first in 1724, after various attempts to improve it and establish its composition, but up to now no one has been more successful than Mr. MACQUER who gave an enlightened account based on his research.

SECOND SECTION: CONCERNING EARTHS.

FIRST CHAPTER: On The Nature Of Earths.

§ 166. An earth is a tasteless and odorless substance which is insoluble in water. [Earths] occur in all of Nature's kingdoms, but in the vegetable and animal [kingdoms] they are mixed with spirits, water, oil, and salt, which must be removed in order to get them pure; this is done with strong heat to produce a glass.

Note. It has already been noted that the mineral earths, which occur on the surface of the [planet] earth, are saline, but more or less difficultly soluble in water (§ 1, Note 2.). On the other hand, if some single earth is found with the vegetable or animal [kingdoms] which is not found in the mineral kingdom, it can with reason be called into question. After all the research which I have carried out up to now, vegetable earths are a blending of several mineral [earths], and likewise with animal earths, which are better illustrated in the following [paragraph?]. On the other hand it should be noted that the author [Scheffer] in this Section discusses the Earths only briefly and generally; a complete discussion of these [earths] belongs in Mineralogy.

§ 167. The mineral earths in general appear to be of two kinds, namely, lime and clay. Marble and alabaster belong to the first kind [lime], but paints made from them contain metals which can be separated from them with skill. The nature of flintstone can be deduced from the chalk crust which deposits on its surface; whereas pure clays are of the same nature as quartz, but which ever of them is produced is not so easily settled; it is also possible that clay is finely divided quartz which is converted into coagulated clay.

Note. That which is quoted in this paragraph [§ 167] is found in AHLSTRÖMER'S handwriting, but is missing in all the other later [lecture notes], which were also reviewed by Mr. SCHEFFER. I should note this because the author [Scheffer] probably found reason later to have other thoughts [on this subject], or at least to call this section into question. At this time only 5 earths are known in nature, according to investigations made up to now, which cannot be derived from each other or from anything simplar. [This could be considered part of B.'s definition of an element. J.A.S.]. These are: lime, earth of heavy spar, [barium], magnesia, clay [aluminum], and flint [silica]; they are identified most conveniently by reaction with acid of vitriol, since the first gives gypsum with it, the second heavy spar, the third bittersalt, the fourth alum, and the fifth does not react at all.

Up to now we have grouped together lime and magnesia, although they scarcely have any other similarities than that they both fizz with acids; on the other hand they are greatly different. The first [lime] becomes soluble in water through burning, but not the second [magnesia]; lime contains 43 percent acid of air, magnesia 25; lime has greater attraction for acid of air than all other alkalis, which [acid of air] is consequently neutralized by it, but magnesia has lesser [attraction], and other differences besides, as Dr. BLACK (1755) and Mr. MARGGRAF (1759) have observed. Earth of heavy spar seems to be more closely related to lime; both dissolve in acids with fizzing, and likewise in water after burning, make creams [like milk of magnesia? J.A.S.] in free air, causticize alkalis, dissolve sulfur when boiled with it in water, precipitate corrosive sublimate yellow, make mercurius dulcis black, etc., but they differ in many other essentials from each other. Earth of heavy spar produces firm crystals with acid of saltpeter; with vinegar, deliquescent gummy [crystals]; with acid of fluorspar and tartaric acid salts which are much more soluble in water than the salts with lime; it [barium] loses acid of salt in distillation; after burning it decomposes Tartarus vitriolatus in the wet method; etc. We must therefore consider them [lime, magnesia, and barium] to be different substances, until evidence shows us otherwise. Alkali minerale is different from [alkali] vegetabile, although in many cases they are similar in behavior. The new earth which Mr. SCHEELE mentions in his article on Brunsten [manganese], is of the same strength as heavy spar, which [fact] Mr. J. G. GAHN has just discovered. Concerning clay and silica, more is given occasionally in the following.

Mr. CRONSTEDT found occasion in 1758 for good reason to name nine distinct earths, but since then research on the subject has brought us instead down to the above mentioned five [earths]. Fluxing shows unmistakably that lime is of the same basis as gypsum; garnet, shirl, and zeolite are similarly made up of flint, clay, and lime, but in different proportions; mica consists of flint and a little magnesia; asbestos, amianth, soapstone, and serpentine reveal on analysis flint and magnesia; but brunsten contains a new semi-metal, concerning which more is given later. See "Physical Description of the Earth", § 171.

§ 168. (a) Lime effervesces with acids, but can lose this property in several ways, and form gels with them instead at all times. Lime, mixed with clay and then burned but not melted, becomes like lime which is heated strongly with quartz, because the mass is so soft that it can be rubbed to pieces between the fingers.

Note 1. The source of the fizzing is acid of air, which fact can be seen from § 51, Note 1. Consequently when this [acid of air] is expelled, no other reaction occurs; this happens during strong burning, and goes faster the larger the surface of the lime, a condition favoring all evaporation. Consequently, when the lime is pulverized in the aforementioned mixtures, it gets much firmer. A lime which is burned like this, so as to lose all its acid of air, does not grow hot with water, and can be called dead-burned, since it slowly falls apart, whereas any kernels which remain [unburned], heat up strongly with water, whence the water taken up turns into vapor, and loosens the entire mass into a very fine powder. Such distinction is necessary when [the lime] is to be used for masonry, because dead-burned [lime] should not be used [for masonry]. When lime is precipitated from acids by means of fully caustic fixed alkali, the precipitate is found not to fizz with acids. When allowed to lie in open air, burned lime again takes up acid of air and becomes just

like unburned [lime] when it is again fully saturated [with acid of air].

(b) Lime really contains a substance many times more soluble in water, which BOERHAAVE called Sal calcareous. We can test this idea if one pound of burned lime is boiled with 1-1/2 stop [1 stop = 3 pints] of pure snow, or rain water, until half of it boils away, it is removed from the fire, allowed to settle, filtered through heavy paper, then poured into a glass retort to which a condensing head is cemented, and to this tube a flask [is attached] as a receiver. It is then distilled over a sand bath until clear water is received and no more of the silver solution, none of the dissolved sugar of lead, nor of blue vitriol, because 2 drops of spiritus salis ammoniaci makes this bright blue: nor does syrup of violets change color with it. The taste is flat like that of distilled water, and it has a bad odor like that which water distilled from lyesalt has, perhaps arising from the phlogiston in the wood, and which sticks to the calx during burning. On Gothland ovens are so set up that the residue turns to stone, but then as much combustion can follow as the work requires.

If pure snow water is poured on the powder in the retort it dissolves most of it. The solution is filtered and turns syrup of violets green; some of the lime water is first neutralized with acid of salt, and then the earth can be precipitated by lye salt, and the greater part of the calx settles out by itself.

Note 2. Burned lime is really a peculiar kind of alkali, which will dissolve slowly in water, about 1 part [lime] in 300 [water], but turns vegetable juices greenish-blue, precipitates metals and other kinds of earths from acids. Cf. § 31. This alkali becomes more difficultly soluble by combination with acid of air, just like lye salt, which deliquesces in moist air, but with acidum vitrioli produces a many times more difficultly soluble neutral salt. So when lime water stands in free air it attracts an overlying layer of acid of air, thus forming on the surface a skin of lime, and this continues until it is all separated. In a well sealed vessel lime water can remain unchanged for many years. If someone wants to know the correct composition of lime, they should dissolve raw lime in a flask with aqua fortis, and to the neck [of the flask] tightly cement a bent glass tube, the other end of which leads down into a flask containing lime water. As soon as the elastic vapor appears, which comes off with fizzing through the tube into the other flask, we notice a turbidity because the lime substance dissolved in the water takes the acid of air to itself and settles down, because it is no longer soluble if an excess of this delicate acid [acid of air] is present. This cannot be the result of some acid of saltpeter distilling over, since this acid dissolves lime; but most of the vapor is the product which comes from the fizzing which is so much more evident, as the lime water itself does not fizz, but will now precipitate a powder like raw lime.

In burning, this material [lime] undergoes no other change but the loss of about 15 percent water and 43 [percent] acid of air; when this is replaced, it returns to its original condition. If some material of fire attaches to it which then, when the water is poured on, separates and produces heat, I do not know but that a dead-burned lime, which is always observed in fire, forms heat with the water, as I have many times not- ed. A dead-burned [lime] shows thereby that this remaining kernel shows not the slightest indication of fizzing with aqua fortis. And the lime precipitated by alkali fixum is also like a burned [lime], but slaked, because it is obtained in a liquid medium, but is free from combined [water]. It is alleged that both alkalis and acids owe their causticity to fire, but it is not easy to comprehend how the acrid acid of vitriol can be made so mild by phlogiston, which also consists mainly of fire, that the sulfur [in the acid of vitriol] does not show the least corrosiveness; or how burned lye can be changed into a mild soap with oil, and with a corrosive acid forms a harmless neutral salt. Numerous experiments show plainly that alkalis and acids are sweetened [neutralized] by phlogisticated compounds, and not made caustic. They are of their own nature more caustic, the purer they are; and the more they are bound up with other substances the milder they become.

(c) Some people maintain that limestone contains alkali volatile, but in that case this substance ought to reveal itself on distillation (see (b)). With more reason the French chemists proclaim the presence of acid of vitriol. We have always believed that natural French lime was usually mixed with gypsum, and that [lime] from Gothland was of similar quality. Alkali fixum also precipitates lime water to a degree, and this indicates some acid present and that it is vitriolic acid which is detected when milk is boiled and lime poured into it until it becomes insoluble, since a yellowish whey is obtained, from which true sulfur can be precipitated with acids. If we wish next to investi- gate whether a lime contains acid of vitriol by means of precipitation, it is best to use alkali minerale, since Sal Glauberi reveals itself most plainly by its large crystals.

Note 3. When limestone is replete with fossils, or laid down under seawater in which swarming creatures are found in great abundance, it is not unusual to find a little alkali volatile. There is no doubt that acid of vitriol makes up a part of limestone, aand that gypsum also occurs in the sea, with which it [limestone] could be contaminated: but the idea that this [gypsum] makes up an essential part [of limestone] is not corroborated, as many investigations show, since most [limestone] is always found to be free [of acid of vitriol]. Whether alkali fixum precipitates something from limestone depends on its compolsition. Caustic alkali does not precipitate aanything, if it is not added in large amounts so that the water [solvent] becomes insufficient to dissolve both: but when some mild alkali is added,the lime material takes the acid of air from it and precipitates as it can.

In densely layered lime[stone], which is laid down under seawater, we often find traces of acid of salt, but this is not anymore an essential ingredient [of limestone] than is vitriolic [acid].

Limestone from Gothland I have not yet subjected to such study, but other [kinds of] lime coagulates milk causing a yellowish precipitate to separatefrom the whey, producing a material like sulfur. To combine acid of vitriol with phlogiston in the wet way to saturation is up to now quite an unknown art, and if it happens with Gothland [limestone] it must have different properties than those which belong to all other limes, and not depend on the addition of acid of vitriol, since sulfur is never obtained by mild heating of gypsum when it is boiled in milk.

(d) Other limestone often contains contaminating materials which do not belong with this substance [normally]. Thus in Skåne we find a kind [of limestone] which when burned with coal gives off an odor of orpiment [orpiment is As_2S_3, J.A.S.]; this without doubt contains arsenic and acid of vitriol, which get phlogiston from the coal, and then produce orpiment.

Note 4. Limestone, which contains known amounts of clay, is excellent for mortar, since during burning the clay hardens and attracts limewater by which it is bound much better than to smooth quartz. Burned and rotted [bokad] clay is for the same reason a useful material.

(e) Swinestone is a limestone combined with phlogiston, which when sparingly distilled by itself gives some traces of fatty material; on opening the vessel, a harsh odor is noticed, the residue is found unchanged, only somewhat sooty, and the receiver a little greasy, but no oil. Swinestone does not burn by itself with a flame unless mixed with coal, from which it can be concluded that it does not have more phlogiston than that which can be held in a closed vessel. When the fatty substance is burned away it can be used, and really is used like ordinary lime in the southern part of the kingdom [of Sweden] where it occurs in abundance.

Note 5. Archiater HIERNE relates that he has obtained from 4 pounds of swinestone a little oil. The amount of fatty material is not always the same, so consequently it cannot be estimated, and unless an account is also made of the abundance when it is examined, the result will vary.

(f) Clay mixed with lime is usually called marl [jord-mårg], but it does not fizz with acids sufficiently to be noticable in the least, a useful test for good soil, although experience shows that many clays effervesce with aqua fortis although they contain varied amounts of acid of vitriol, which damages vegetable substances everywhere in proportion to the quan- tity of it present, in particular if it is combined with phlogiston since this is released and something else taken up in its place.

Note 6. The author [Scheffer] notes with good reason that fizzing with acid is a deceptive test for soil suitable for cultivation. If its usefulness for agriculture is to be as perfect as possible, and this is to become a science, the farmer should undertake to study his soil to determine the proportion of the basic substances which make up its composition, and this can be done in a very short time. When something of its composition is known, gathered over many years of study, and how it [the composition] is best maintained in wet or dry [weather], hot or cold, then we get a firm basis for improving its fertility. It is a prejudice held by many that manuring and draining are most important; it should be beneficial to alter the fertility to make it like the soil of a garden, if both could be watered similarly when needed, but otherwise it might spoil it.

§ 169. (a) Clay is a kind of earth which hardens in fire, but when unbaked forms a milky fluid with water. The pure kind is always fireproof; the other kinds are so diverse, from the foreign materials mixed in with them, that they cannot be classified any more than water [can be classified because of the impurities it contains]. All [kinds of clay] contain alkali volatile, and most [kinds] also contain iron.

Note 1. Clay, insofar as we know up to now, is a distinct kind of earth which never is found entirely pure on the surface of the [planet] earth, and nearly always occurs mixed with sand, and although it can often be separated as much as 70 percent by washing, the best part is still so infiltrated [with sand] that it cannot be separated mechanically without dissolving most of the clay in acid in order to separate it (see § 175, Note 2).

Clay is among the most useful kinds of earth that we know. Its reluctance to release water by evaporation, and its refusal to take up more than a certain amount [of water], makes it valuable in arable soil when it is available in the proper amount, useful in both dry-lands and wet-lands. Its pliability with a certain amount of water, and its ability to harden in fire, give us a means for making bricks and all kinds of useful vessels, from the coarsest pottery to the finest procelain. Common and fusible clays can be put to many uses, according to circumstances, using higher or lower degrees of heat to prevent fusing. Fireproof [clay] is either coarse, which takes on color in fire, or the kind which becomes white. The former, used for [making] crucibles, is mixed either with flint or burned clay, to obtain the desired rigidity without becoming oversensitive to heat or cold. The higher the clay content the more rigid the product, but also the more difficult it is to prepare and use. The latter [white] kind is used for tobacco pipes and porcelain. The last named material [porcelain] must have, in addition to the hard- ness of flint, whiteness and durability in fire, and also to be half transparent, properties to which the clay by itself alone cannot be brought. Consequently additional ingredients are required here, which produce a half-fusion; the Chinese call this [additional ingredient] Petunsé, which seems to be either gypsum, or a kind of fluorspar, or feldspar. The sample [of porcelain clay] that I got from China is feldspar, but contains iron, and consequently is [practically] useless.

(b) Good Fuller's earth [valklera] should : (1) be free from sand; (2) be so fine that it feels slippery and oily between the fingers, but due to its fineness and not oiliness; (3) attract fatty materials; and (4) soften readily in water. On the other hand, Fuller's earth is of different varities of internal composition.

Note 2. The genuine English Fuller's earth is related to stone marrow, consisting of magnesia, clay, flint, a little iron, and acid of salt, as I have found by chemical analysis. So-called Meerschaum [Sjöskum, i.e., sea scum], from which tobacco pipes are carved, also seems to be a stone marrow, but I have not yet had time to investigate it.

To Make Fireproof Splints [Ceramic Oven Lining]

§ 170. (a) From kiln or pipe clay is selected that which feels very fine between the fingers and is obtained in thin scales. On the grating of a blast furnace is laid down first a layer of coal, then clay, and so on layerwise, until a large enough quantity has been laid in, a layer of coal on top and on this some burning coals. The furnace is sealed both above and on the sides, so as to burn slowly. When all the clay reaches the grating, the oven is allowed to cool, but before this is completed more coal is put in, as long as the burning continues.

When the clay is taken out it should be separated according to whether it is much or little bvurned. That which appears black or gray inside is insufficiently burned; that which fractures like flint is to be used for hearths; that which is entirely white inside is suitable; the first two kinds are not suitable [for oven lining], particularly the first kind.

(b) That which is chosen is pounded and ground to the finest flour, and 1 part of this material is taken with 1 part of unburned clay which goes through a fine but sturdy sieve, and in addition 1 part of crucible sand, ground to the finest flour. These materials are mixed well, first dried in a trough, then a little pure water is added until it forms a thick dough, but not too soft; after this the mixture is worked with a wooden stick until it becomes quite viscous and fine in texture, which takes a fairly long time. Then it is allowed to dry for about half a day. We must take care that no splinters, paper, or other similar material gets into the mixture, since these would be consumed in burning and leave holes in their place, and then the material cannot flow out by itself, and ruins the furnace.

(c) The mold which consists of two parts, called Monk and Nun, is lightly greased inside with olive oil, so that the material will not stick to it; linseed oil is less suitable for this purpose. Then the Nun is filled and the dough is pressed firmly with the Monk, so that it becomes somewhat concave; roasted splint is pressed still deeper, although the bottom of it ought to be thin, but that which is used in art school is pressed much less. In this way a splint piece is formed and taken out of the Nun, set aside, and a new one prepared. When the dough begins to stick to the mold it is oiled again. The formed splint is allowed to dry until it does not spread when squeezed together in the hands.

(d) The moderately dried splint pieces are placed, one after another in the Nun which is greased, and the greased Monk on top, which is then beaten down with a wooden club until it reaches down over the edge of the Nun. Then this is completed so that the splint bottom does not bend up, which would cause it to break. After this it is allowed to dry until it becomes entirely white.

(e) Burning occurs in this way, that the fully dried splints are put in the muffle furnace of a testing oven, which is first heated gently, but then heated strongly at the end so that the splint glows white.

To Make Mufflur [Ceramic Muffle Furnace Lining]

§ 171. (a) Two parts of black French clay, which turns white on burning, are measured out, and to this are added two parts of a meal made from unused crucibles, and one part of red sand, but if meal of crucibles is scarce take instead two parts of raw clay, and the same of burned clay, one part of crucible meal, and one part sand, all pounded fine.

(b) The materials are mixed and worked well with a very little amount of water which is not salty, until the mixture becomes very fine and smooth; it is then left for a day to become firm.

(c) The dough is then rolled out to a thickness of 1/4 to 1/3 inch to the size of the muffle stock, and bone ash is strewn on both the stock and the roller until nothing sticks. The rolled out piece is cut to size, and to the cut end another piece is stuck fast before the ends dry; otherwise it will break off; sticking fast occurs when the dough is the same in both pieces. Then on each side two oblong holes are cut on the lower edge, and one on the back piece, which [holes] are used for pulling.

(d) When enough has passed through, the piece is allowed to dry, during which process something is laid against the sides to hold their shape. After one day the muffle is lifted a little from the stock and props are placed underneath, so that it does not sink down; after another day it is lifted higher, and so on, until it is completely dry. If it is not supported in this way it breaks on drying and is ruined, because it cannot be patched.

(e) Once it is fully dried, the muffle is laid in a blast furnace along with some coal, heated very cautiously at first, but finally heated to white heat.

(f) Muffle leaves are made of splint pieces (§ 170), cast and fitted to the bottom of the stock, dried well and burned in the same way as the muffle.

(g) Finally the leaves are laid in on the grate of the test oven and cemented fast to the mouth.

(h) Sudden changes of heat and cold crack the muffle, and therefore heating and cooling should be done cautiously.

If during testing or cupelling something is spilled on the muffle bottom, some bone ash should be spread on it immediately which will take up the [spilled] material which would otherwise eat a hole [in the muffle].

§ 172. From ashes we get vegetable earth. To get it free from alkali fixum we should boil it, and then wash it on a filter with hot water until the filtrate no longer turns syrup of violets green; but if this experiment is made with ordinary ash from a stove, we find that many days are needed before it passes this test. When ash so prepared is introduced a little at a time into aqua fortis, it partly dissolves with strong fizzing. Finally this reaction stops when new ash is added, but the acid is still not saturated, and still colors litmus red, but it cannot be made to react further, even if much ash is added.

The solution, filtered and precipitated with alkali, produces a fine white earth which is called Magnesia alba by apothecaries; much of that which we come across is from saltpeter works, especially the ash used in crystallization; but more that we buy is from abroad.

The solution, filtered and evaporated until it forms a film [on the surface], yields no crystals when cooled, although the small amount of liquor remaining forms a gel in a hour.

Note. Vegetable earth is usually regarded as different from other familiar earths, since it both fizzes with acids and melts by itself; but this is an error in that [vegetable earth] is regarded as homogeneous or, what amounts to the same thing, in that all particles are identical, which is quite a mistake. I have investigated several [kinds of veg. earths], with ashes carefully prepared in separate crucibles; burning is the only method known up to now for freeing earths from contaminating materials; all of them were found to be mixtures of lime [calcium], earth of heavy spar [barium], magnesia, clay [aluminum], and flint [silica], the same principal kinds of minerals [earths] that were named previously (§ 167). The proportions are different enough, and sometimes one or another [earth] is missing. At the same time it seems evident why vegetable earth fizzes with acids and also is more or less fusible. Sometimes we also encounter earth of bones (§ 173) in ashes, and undoubtedly this was the case in the investigation cited above by the author [Scheffer], since all the others which are mentioned above can saturate an acid so that it does not react further, but not in this case. The same kinds of earths can be obtained from rotted vegetables, about which further details will be given elsewhere. Mr. MODEL has confirmed the separation from rhubarb root of 1-1/2 ounce of gypsum per pound, by grinding in water. The ashes always show traces of iron, and often of brunsten, but seldom of other metals, unless the ashes are from mountainous regions. Tartarus vitriolatus, saltpeter, sal digestivus, and common salt are the usual neutral salts found here.

§ 173. Animal bone, which has been exposed to air for a long time, is freed by degrees more and more from fatty materials and salts; whatever remains at the end is an earth which, although it resembles lime, still differs in various ways from it. We use this [earth] in cupellation.

Note. Egg shells and oyster shells, which have been well ignited to burn out the spongy matter, which then leaves a skeleton of the stoney material, are then dissolved in aqua fortis without residue. The solution releases pure lime with alkali fixum, and gypsum with acid of vitriol, but yields no trace of magnesia. During the solution of ignited oyster shells a hepatic odor [H_2S] is detected, which comes from within the shells and not from the disturbed coal. Some gypsum must therefore be present here [in the shells]; this [odor] is not observed with egg shells. Newt and other animal bone, horn, hooves, and insect hide leave an earth after ignition which is very different, and Mr. J.G. GAHN in 1769 found by both analysis and synthesis that it is lime combined with acid of urine.

To Make a Cupel

§ 174. (a) Ash from bones of various animals is useful for this [making cupels], except bone ash from swine, which is not fire resistant. Instead we choose skulls of horses, sheep, and cattle, since these contain a finer earth than other bones; They should be aged and bleached white, not dark or black, because these are too expandable in fire. The selected bones are ground and ignited in a blast furnace like clay [for splints] (§ 170).

(b) From the ignited bones we select those which become entirely white all the way through, grind them asunder to make a flour, put this in a linen bag and boil it in water for a time; after this the bag is taken out and the water allowed to drain, leaving a fine powder which is pressed through the linen and is called klâra; this is colllected, dried well and set aside.

(c) The ash is boiled to free it from all salts, although some always remain, which must be remembered when it is put to some other use. The ash is taken out of the linen bag and placed on a piece of sackcloth or other filter, where it is first washed with cold water, then with boiling water, until the filtrate no longer changes [the color of] syrup of violets; this ash is then dried. If it should now appear gray, this is a sign that we did not start with a completely ignited bone, and that the ash must again be ignited in a crucible, boiled and extracted as before, because after a new ignition new salt is found.

(d) When the cupel is to be formed, the bone ash is moistened with a very little pure water and mixed well, so that the moisture is spread out; no more water should be used than is necessary for the bone ash to barely hang together when it is pressed between the fingers. The nun [mold, see § 170 (c)] is filled and packed down well and the monk is placed upright on it, held steady, and driven down with a wooden club as long as it gives; it is then taken off, klâra is sifted into the cavity, which sticks when the monk is put on again and driven down with a blow; if it is beaten on more than once the klâra sticks to it. Finally the prepared cupel is pressed out on a board, a little bone ash sprinkled on it, and left for several days to dry.

(e) When a cupel is to be used it should first be heated to white heat in a test oven, where it is placed upside down. As the oven gets warm the cupel is not placed in the strongest heat, but is first warmed in the mouth [of the oven].

SECOND CHAPTER

On The Fusing of Earths

§ 175. Alkalis have the property of bringing fireproof earths to fusion, indeed so that different earths which withstand the strongest flame will often fuse readily when they are mixed together [with alkalis]. The following investigation shows readily what has just been described.

(a) Pipeclay for kilns does not melt by itself, but if it is ground fine and mixed with double its weight of black flux (§ 37b), put into a crucible and sprinkled with common salt, which draws all the excess alkali to itself during fusion, and when the salt starts to crackle [the flame] is blown higher with a bellows until the mixture gets clear, and we get a transparent greenish glass. Silica [also] fuses with alkali, as we find in the following paragraphs (§ 76-78) [176-178 ?].

<u>Note 1.</u> Earths are either <u>fusible</u> or such that they cannot be melted by themselves, which we call <u>fire-resistant</u>. In the latter class are included lime, clay and silica, when they are free from contaminating materials; earth of heavy spar [barium] and magnesia can be melted in strong flame. Mixtures of only two of these [three earths] show little or no tendency to melt, but when all three are mixed together they generally melt easily. A burning glass or [burning] mirror produces the strongest heat

known up to now with which flint and many other fire-resistant substances melt; usually nothing can stand this, i.e., being brought to this high temperature. Diamond is volatilized.

One circumstance ought to be well noted in investigations of fusion, that often the crucible makes the material in it more or less readily fusible, and when the mixture melts all through, yes even dissolves the entire crucible through in some cases, such material is called skarande [cutting].

(b) Quartz and lime do not melt together but produce a brittle mixture; but if alkali fixum is added, a clear glass is obtained which, if the flint separates, is not soluble nor does it form a gel with acids, but becomes firm. There seems to be noted some difference between quartz and clay, because the latter melts with lime without other additives; the cause seems to be found in the fine earth which is left behind when the clay is dissolved in acid of vitriol.

Note 2. The author's [Scheffer's] conjecture has later been confirmed by Mr. MACQUER. Wholly sand-free clay or earth of alum does not melt in any proportions with lime, but if ever so little of flint is added, the property of melting immediately shows itself, and this [melting] is greatest when about one part of each of the first named [clay and alum] are taken with 2 to 3 parts of flint. Now all the clay found on the surface of the earth is mixed with flint (§ 169, Note 1), so it is not to be wondered that clay melts with lime. That the residue after solution of fire-resistant clay is flint has been confirmed; also that it is not attacked by acid of vitriol and gives a clear glass with half its weight of alkali fixum, which occurs with no other earth. But though clay so freed from flint does not melt with lime, it still prevents many other occurrences, so that quartz and clay appear to be the same substance (§ 167, Note; § 178 (b), Note). [B. is not saying that he believes that flint and clay are the same substance, just that it appears so.] Two parts lime with one of quartz melts in a strong flame.

(c) Lime clay, and flint or feldspar produce glass, yet most of the flint is taken up through the other two. The glass becomes clear, but looks somewhat yellowish; if it is pounded fine and mixed with dilute acid of vitriol it reacts vigorously, and in a little while becomes milky colored, congeals like gelatin, is insoluble in water, and crumbles in air. The cause of this lies in the lime, which Mining Counselor VON SWAB discovered in a collection of bottles which, on a ship going from Sweden to the East Indies, were corroded by wine.

A kind of zeolite, which is found at the gold works at Adelfors, has the same property of forming a gel with acids.

Note 3. Many kinds of feldspar and flintstone, fluorspar, gupsum, etc., I have found to melt clay by themselves when heated with a blowpipe. With kiln clay feldspar melts completely, and can serve for preparation of porcelaine if it is iron-free, but should be so proportioned that the mixture can only be half melted. Counterfeit porcelaine has similar appearance to the genuine porcelain, but it melts in fire, and consequently looks like half-transparent glass.

Fayance is nothing else but a common pottery material entirely covered with an enamel crust, which gives it a nice appearance. Proper-flint porcelaine or stone-porcelaine consists of a fire-resistant mixture which is either white or half glistening without any other glazing than a glassy material on the surface from a base of common salt. The most beautiful genuine porcelaine is now made a Seves in France, of which I have samples from Mr. MACQUER. It has all the inner beauty of the Saxon or East Indian [porcelaine], but excels all the rest in the decoration fashioned with peculiar whiteness equal to that of snow.

One hundred parts of Adelfors zeolite contains about 80 [parts] of flint, 11 of clay, and 9 of lime; an excess of flint prevents it from melting like the other kinds of the same genre.

To Make Crystal-Glass

§ 176 (a). White refined potash, 1 part, is mixed to- gether thoroughly with 2 parts of ignited flint, placed in a crucible, over which another crucible is overturned and fastened, the whole burned with a flame to ashes, as in the test for iron. If we want the glass free from cracks it should be allowed to remain on coals to cool.

Soda polutes fine glass.

Note 1. Flint is not turned into glass, and since alkali fixum makes it melt, there must always be some of this [alkali] in the composition of the glass, although with the purpose in view [of using that which is] more or less pure. The transparent fused mass is called glass, but if opaque, gets the name of enamel or slag, from the uneven quality of its composition.

(b) This impure glass is a little greenish, which comes from the potash, which always has some phlogiston in it. To remove this greenish color it is placed with window glass in a filled glass pot with only a knife's daub of brunsten, which is only enough to make it turn red, as in amethyst. With borax we get a proper black magnesia [manganese] hyacinth tint, and with alkali fixum purple or violet.

Note 2. Mr. SCHEELE has discovered that brunsten attracts the phlogiston in the glass mixture, and when it is evenly distributed the glass becomes quite colorless, but if more phlogiston is present than the amount of brunsten added can attract, the color weakens but does not go away completely, so more brunsten must be added than can be reacted with, to give it its own color. Iron with its highest content of phlogiston gives a greenish solution, both in the wet way and the dry way, but yields a yellowish [solution] if its inflammability [phlogiston] is lacking, which latter color black magnesia [manganese] cannot remove.

(c) We use also 1/10 saltpeter with the above-named in- gredients to remove phlogiston; too much makes the glass discolored. Sometimes along with brunsten we use white arsenic, which enhances the colorlessness, but the heat must be controlled since arsenic turns black on ignition and diminishes the clarity; too much arsenic gives a milky color.

Redlead is the best material for decolorizing [glass], taken in 1/7 part of the entire mass, although a little brunsten must still be used. We can use red-lead, brunsten and saltpeter together; flint and potash are calcinated beforehand, and when they are put into the glass pot the aforementioned mixture is produced.

<u>Note 3.</u> Sulfur, gupsum, white arsenic, calx of lead, and calx of tin in the proper amounts can destroy the red color of brunsten by [their] phlogiston, which nothing by itself pos- sesses or adds to the mass.

(d) Too much lead causes the glass to get streaky in the sun. This happens with the English [glass] which is made from 7 parts lead, 8 nitrum, and 24 flint; the glass becomes very pretty, but also is very expensive. The pots are covered over during the melting, so that the coal smoke cannot make contact with the glass and ruin its clarity. Therefore no very strong heat is applied to the glass, which therefore must have its fusibility increased by [the addition of] much lead.

<u>Note 4.</u> Crystal glass made with just salt bearing substances is lighter, clearer, sharper melting, harder and more fragile than that which contains much calx of lead; Venetians and most other Europeans made the first kind of glass, and it is called <u>Crown-glass</u> in England, while the latter kind is called <u>Flint-glass</u>. DOLLONDS found that in adapting them to optical use, each of those [glasses] considered above corrects the refraction of the other. To be able to make flint-glass properly, free from veins, we should bring this discovery to its completion, for which

also the Parisian Scientific Academy offers a prize of 1200 pounds.

To Make Cheap Glass.

§ 177 (a) Glass for ordinary use is made from 3 parts quartz and 2 parts soda or potash, whichever costs the least. This composition is not suitable for retorts and crucibles since it fuses too readily. Lime and chalk also make good glass, provided the quartz is properly in preponderance.

(b) Bottle-glass is made from furnace ash and beach sand, somewhat more of the latter, otherwise acids attack the glass. The sand should not contain quartz [?], since the lime in the ash does not fuse with it. Pine- and fir-ash can also be used here.

(c) In the retort are placed 2 parts dry red beach sand with 1 part ash. The more sand the better the glass, which should be tested with large furnaces, since they produce higher heat, and consequently can fuse [the glass mixture] with less salt than smaller [furnaces].

Note. Mr. DE REAUMUR found a kind of porcelaine, which is now usually named for him, but which he himself called "porcelaine through vitrification". Mr. LEWIS has taken this noteworthy and useful investigation much further. Glass which is made with soot, coal-dust, sand, bole, clay, bone-ash or gypsum, exposed to the proper heat, becomes harder, more difficult to melt, less fragile to changes of heat and cold, more opaque, sometimes half transparent, black, brown, gray or white, resembling cement. Metallic and earth-containing glasses do not withstand such changes [of heat and cold]; and those which contain the most saline flux to make them melt, suffer them [changes of heat and cold] the least. Bottle glass is the most suitable. Heating to redness is insufficient, but when the heat applied is near the melting point the texture becomes opaque on the surface, and fibrous, so that finally both components mix within the mixture. If the heat is applied too long the texture gets grainy, and then coaser still, and finally like a mass of sand clinging together. For retorts, cooking vessels, mortars, grinding plates, and similar chemical vessels, a somewhat different glass is more suitable than true porcelaine. Dr. LEWIS says in his article and research that the cause of this cannot be an intrusion of cement, because the glass does not become heavier and, exposed only to as much heat as will blister the surface, should not change otherwise; on the contrary it seems that salt is exuded and decreases to maintain the glassy state. Cement should for this reason only effect the glass in shape, but it seems also to contribute to its color.

To Make Deliquescent Glass

§ 178 (a) To one part quartz is added 1 or 2 parts potash and [the whole] fused in a closed crucible with a bellows-blown flame, at first for 5 minutes without the bellows, and then for 15 minutes with strong blowing with the bellows, to form a clear glass which tastes strongly alkaline and dissolves in liquids. The crucible must be fairly large, because the alkali foams a lot and otherwise pushes itself out of the enclosure leaving the residue too rich in quartz.

(b) The glass dissolved in water is called Liquor silicum [water glass], and the earth of quartz can be precipitated with acids, but the saturation point should be approached with care, since if too much acid is added the earth dissolves again. Acid of vitriol is used to dissolve earth of quartz and the solution is evaporated until a film of proper alum separates, which is related to clay (§ 76). If another acid is used for precipitation it should be well sweetened before the acid of vitriol is added.

Note 1. If all the earth is precipitated similarly with acid and well washed, some of it can then be dissolved in acid of vitriol, but part of it never reacts. During fusion, the glass attacks some of the clay in the crucible, and it is this part which dissolves. Pure quartz fused with alkali in an iron crucible gives hardly a trace of alum with acid of vitriol, but quite a bit with flint since this contains besides quartz a certain amount of clay. While the flint and alkali fuse together, a strong frothing occurs and acid of fluorspar is driven off in the fumes; GLAUBER had noted that an acid like acidum salis was given off. Pure quartz or natural crystal contains acid of fluorspar, we know for certain, but the base which collects in the vapor is not yet identified. Quartz is dissolved by acid of fluorspar and I have gotten crystals to separate [from the solution] after 2 years, the largest of their kind.

(c) If alkali tartari is used as the potash we obtain liquor silicum, but it does not precipitate with acids, but remains in solution, forms a gel, and then becomes almost insoluble in water. This consists of calx of tartar, which is more plentiful in this case, when it has not been dissolved and filtered after ignition, than in the case of leached and calcined potash.

Note 2. The formation of a gel with alkali tartari can readily be avoided if the solution is properly diluted.

(d) Beautiful flint or gun-flint fused with 2 parts potash yields a glass which forms a gel when dissolved in acid of vitriol, and this attests to its lime content.

Note 3. Those who are little acquainted with chemistry maintain that it is impossible to decompose glass and reveal its salt; but this [analysis] is entirely possible and is accomplished by laying open the powder [glass] by boiling in mineral acid; if this does not succeed then fusion is carried out with equal or double the amount of pure lyesalt, so that a deliquescent glass is obtained, whose alkali neutralized with acid produces a neutral salt, which by its amount reveals both how much and what kind of alkali was used. See § 51, Note 3.

To Make Glass-Flux

§ 179. Colored glasses are made by the addition of metallic calces.

(a) Ruby-red glass is obtained with a few drops of gold solution, which is mixed with crystal glass and fused. This [process] fails quite often.

Note 1. If the author [Scheffer] would start with [1 part] crystal glass and add 1/4 part calcinated borax and 1/6 [part] calx of lead he would get a glass with an exquisite luster which is suitable for imitation precious stones. KUNKEL had to make ruby-flux in a large amount; among other examples is the calx which he used to color 24 pounds [of glass] for the Elector of Cologne. This preparation is still among the most difficult to make, since with the same materials in the same proportion it succeeds one time and fails another, although we attempt to keep all conditions alike. The most important condition is the control of the heat, for which there is still no dependable thermometer available. It is quite certain that it propduces a ruby-glass with the blow-pipe, and we can learn thereby what is produced with longer or shorter heating times. KUNKEL says that 1 part gold can produce 1280 parts of ruby-red glass. Purpura mineralis is commonly used, which was so named by its discoverer, Dr. CASSIUS; it is gold precipitated by tin solution (§ 143). We can accomplish the same result also with other additives, but they are to some degree always dephlogisticated gold. We often get colorless glass, but with the proper amount of heat we get the desired redness. (See Note to (h) [below])

(b) Garnet-flux is obtained from 2 ounces of crystal material, 5 oz. redlead, 35 grains brunsten, and 4 grains safflower. The glass gets very pliable.

Note 2. Crystal material 256 parts, vitrum antimonii 128, purpura mineralis 1 [part], and magnesia nigra [manganese oxide] 1 [part], gives a good imitation garnet color. We can leave out the purpura mineralis, but then it requires double the amount of magnesia and antimony glass.

(c) Topaz-flux is obtained from roasted antimonium and crystal material; as the [amount of the] former is increased the color increases also.

Note 3. Three parts of flint fused with 4 parts white lead and 2 parts chalk gets very close to topaz. If we want a darker color we can add a little calx of iron, well calcinated.

(d) Chrysolite-flux is obtained from 2 oz. crystal glass, 8 oz. redlead, and 12 grains crocus martis [probably iron oxide or sulfate].

(e) Green glass is obtained from calcinated Spanish Green [verdigris] fused with flint and potash.

(f) Emerald-flux is made with 2 oz. crystal glass, 4 oz. redlead, 48 grains pulverized Spanish Green, and 8 grains crocus martis. This becomes too soft to grind, so it is molded.

Note 4. Calx of copper produces a similar dephlogisticated red, green, and blue-green glass, and even becomes almost colorless after prolonged heating.

(g) Saphire-flux is made from 2 oz. crystal material, 6 oz. redlead, 2 scruples prepared safflower, and 4 grains magnesia nigra [manganese dioxide]. It becomes very soft.

Note 5. Calx of colbalt gives a strong and very persistent blue- ness, [to glass], which however turns to red in enough light, which can be of help in reducing the color. To imitate saphire, 1/100 [part] of calx of cobalt is used (precipitated out of some acid), with [1 part] hard crystal material.

(h) White glass is made from calx of tin 2 parts, to 3 to 4 parts of each of the other ingredients, namely potash and flint. We need here 1/2 to 3/4 hour of strong heat to fuse the calx of tin.

Note 6. What is called white glass here is called enamel in other places. In general white glass means that which reflects all kinds of light beams and involves complete opacity, but not what we often called colorless or transparent [glass], although the author [Scheffer] himself uses this meaning in referring to white glass, white water, etc.; but this serves to confuse and obscure [the meaning].

The transparency in both the wet way and dry way is a sign of com- plete solubility, and similarly often a solvent needs a certain amount of heat to keep it clear, but during cooling it gets cloudy, because it cannot retain all the solute that was formerly dissolved with the aid of greater heat, as occurs in fusion. Borax dissolves a certain amount of lime and after cooling leaves still a clear glass, but this also has its limit or saturation point, which when it is exceeded can still leave the glass clear, when it is still fused, but during cooling causes the excess first to form a cloud, and immediately thereafter plain whiteness; insoluble particles mixed with the clear [glass] produce a similar reaction of course. If we could fix the particles in the condition they have when fused, the mass would remain clear even after cooling. We can easily convince ourselves of this by research with a blowpipe; thus if a clear borax pearl is oversaturated with lime, and when it is still fused it is cast into likewarm water to congeal it instantly, we obtain a clear bead; it often breaks to pieces if the water is not slightly warm, but each piece is found to be completely transparent.

If the calx of tin is diminished in an enamel mixture we usually obtain a half-clear glass, which resembles opal completely; by adding a very small amount of brunsten, or other metallic calx, we get different colors on refraction. By varying the coloring and darkening materials we can imitate carneal, calcedony, and other semi-transparent stones; bone-ash, white arsenic, and white dephlogisticated antimony give somewhat the same effect as calx of tin. So-called Lapis philosophicus [philosopher's stone] is obtained from 80 parts of crystal material with 1 part of strongly calcinated iron, which is well pulverized after being well fused, mixed with gold leaf and fused again cautiously.

Enamel is tinged with different colors like opaque stones. In this way we obtain artificial Lazurstone, if dark blue flux is fused with it, which takes away its transparency; it can be colored with gold, calcinated borax, and oil of spike, but should then be exposed to some heat, so that the surface only becomes soft. Sea green flux made opaque re- sembles turquoise, etc.

Although the colors of precious stones, their clarity and luster can be imitated, the cost is still too high to bring the artificial ma- terial to the proper hardness and difficulty of fusion, which are the principle properties which make them infinitely less valuable than natural [stones]. A prolonged strong heat, which might reduce the shattering somewhat, also weakens the color and clarity, so that the state of the art up to now gives little hope for a complete imitation.

THIRD SECTION: CONCERNING WATER.

FIRST CHAPTER: Concerning Water By Itself.

§ 180. Water is a fire-quenching liquid, which by itself has neither color nor taste. This definition applies readily to the purest water that nature produces, although it [the water] is not always free from foreign substances; but it can be purified through distillation, but nevertheless still contains some earth, which can finally be separated by evaporation in a closed vessel, albeit imperfectly.

§ 181. Some people used to maintain that water can change into earth, and wanted to demonstrate that in distillation they always found some earth left behind. But I have, in a glass Pelican, which was well sealed and was held for two years in continual digestive heat, allowed the water to circulate back and forth without obtaining any more earth than showed itself in the beginning, and which came partly from dust contained in cracks in the vessel, and partly dissolved [from the vessel itself]. [Editorial comment by J.A.S.: I believe that this description of Scheffer's "Pelican Experiment" antedated that of Antoine Lavoisier. I have discussed this idea further in "The Pelican Experiment of H.T. Scheffer", published in the Journal of Chemical Education, Vol. 63, pp. 928-929 (November 1986)] The water and the earth are therefore two different bodies in nature, which are not readily interchanged, unless the former [water] is diminished, as it appears; so it must come from an entirely different cause, and must require that the latter substance [earth] increase in the same proportion as it decreases somewhere else.

<u>Note.</u> Whether water can be changed into earth is a question which has been much discussed in our time, but has still not been entirely settled. Mr. Lavoisier has very recently examined this [problem] and in the very same manner as our author [Scheffer], but not as long, but yet with heat that was close to boiling. After 101 days [over] a lamp flame, he has found the Pelican 17-4/10 grains lighter than at the beginning of the study, and at the same time he separated earth of weight 20-4/10 grains, a difference of 3 grains which did not come from the Pelican; but since the water on the opening of the vessel had not deposited more than 4-9/10 grains, other than the deposit gained when it was distilled in a new alembic, he believes that the 3 grains were taken from the new glass. If the earth so obtained is investigated properly, it can readily be determined if it could be derived from decomposition of the glass, especially if the composition of the glass used were known. We need to know in this instance that distilled water was used; how long and how strongly it circulated to make reaction possible; and in respect to the Pelican, whether the composition of its parts is known, the better to estimate the amount of earth to be expected. If ash is part of the glass, can a distinct earth be obtained, which must still always be mixed with flint and perhaps even glass dust, since when more soluble particles dissolve often others of firmer composition are dissolved

without decomposition. See more on this subject in my book "Physical Description of the Earth", § 155.

§ 182. It seems therefore that water usually contains some foreign material in it, earth at least; so some [authors] have taken this as a reason for giving it its own natural kingdom, the fourth kingdom, and divide it into separate classes such as vitriolic, alum-containing, lime-containing, etc. But they have not considered that these added materials are in any wise essential, because through their separation the water is not only not destroyed, but even becomes purer and better. When these materials which are obtained by distillation or evaporation are examined, we find that they all have their own different habitat, and it is consequently unnecessary to create a new kingdom for them, because all uncontaminated water always has the same properties. We could in the same way create an Air Kingdom, in consideration of the manifold particles which are carried up from the earth and collected there [in the air]. It is best therefore to leave water its place within the three kingdoms of nature, and since it neither increases, nor has life or motion, it consequently cannot be placed in either the vegetable or animal kingdoms, but in the mineral kingdom.

§ 183. Everything in nature, though it appears to be ever so dry, contains water, which is revealed by distillation, from which we always get some distasteful phlegm. Salts in particular are water-containing, from which they also get both their pelucidity and their crystal form, both of which disappear with the water. Metals however are so solid that they do not hold any water, and are consequently an exception.

Distillation of Water

§ 184. Through this operation water is converted into steam; at the same time it passes over and separates itself from any foreign substances.

(a) For this we use a copper vessel, since although it may be attacked, still nothing passes over of the metal; also we can eliminate a glass vessel, which can always break.

(b) The vessel is half filled with water, a glass still head is cemented on, and the whole placed on a blast furnace, which is gradually fired up and finally should be heated to that point where the water reaches a gentle boil, since if it is increased too much the steam blows out the water.

(c) A flask is placed loosely under the condenser pipe, which is first well washed with ordinary water, and then when anything passes over it rinses everything completely, after which it [the condenser] is cemented on tight with milk cement [mjölklister].

(d) The water should not be driven over completely, since if the residual impurities are burned up they carry over with them both a bad odor and taste.

(e) The distillate must be protected from dust and other contamination.

Note. Water undergoes noteworthy transformations when subjected to differing degrees of heat. At about 8 degrees above zero its specific gravity is the greatest, or what amounts to the same thing, its volume is the smallest [Bergman is here calling attention to the temperature of maximum density of water, now known to occur at about 4 degrees above zero Centigrade, not 8 degrees above, as B. says. J.A.S.] Below this temperature the volume increases, but not very noticeably before the freezing point is reached, and then most obviously freezing occurs. When the particles [molecules of water] pass from the liquid to the solid state they arrange themselves in the form of beams spreading out at 60 degree angles, from which we get the [hexagonal] shape of snowflakes, which form themselves variously accordingly as many orders of side-beams are added on. This behavior quite resembles crystallization. See "Physical Description of the Earth", §§ 117, 175.

The volume of ice is usually found to compare to the smallest volume of water as 9 is to 8, and is still further changed by different [degrees of] cold; if the water is free from acid of air, the above described behavior becomes as 22 is to 21.

On the other hand, if the heat is increased above 8 [degrees Centigrade], the volume increases more and more, so that at 100 degrees it is about 362/1000 greater, or a little more than an increase of 1/3. At this point [100 degrees] it changes to steam, or as the ancients put it, goes from liquidity to fluidity; they differentiated between a liquid, whose surface is always horizontal, and a fluid, whose particles, although they are readily mobile, yet possess collectively an uneven surface, such as is observed with smoke and vapors. Steam is elastic and at 100 degrees of heat expands to 14,000 times more space than the water, to which it once again condenses when cooled, as is noted during distillation. The boiling point changes according to the height pf the barometer; the greater the weight pressing on the surface of the water, the greater the heat required naturally to bring about boiling.

Second Chapter

Testing Water For Its Purity

§ 185 (a). Here we use silver solution (§ 94), or dissolved sugar of lead (§ 125 (a)), which is dripped into the water. If it does not become cloudy here, then it is pure and free from foreign substances. We never find wells, and seldom springs, in which the water does not change with sugar of lead, but sometimes lakes [have water that does not change with sugar of lead]. A very pure water becomes yellowish with blauholtz [word of German origin, meaning logwood].

(b) If the water seems to be made milky by the above named substances, and some precipitate forms, it is contaminated either with an acid, vitriol, lime, alum, or alkali salis communis, since lye salt is not found naturally in water, provided that it has not run through wood ashes (from a forest fire), or something else that was burned; alkali vegetabile is obtained only through fire, except in saltpeter which seems to have been formed by putrefaction or some similar process, but which is still unknown up to now.

If we wish to know more definitely what foreign substances are contaminating the water, we accomplish this in the manner described in the following.

Note. When analysis of water is being carried out, two questions should be asked, namely, **what does it contain**, and **how much of each contaminant is present.** The first question is answered by adding some reagent, and the second by careful evaporation and separation. The author [Scheffer] answers only the first question.

The substances which are found occuring naturally in water, as much as has been investigated up to now, are the following: free acid of air, and alkali minerale combined with it; lye salt, infrequently; alkali volatile, even less frequently in a few places; Glauber's salt; common salt; saltpeter; and very seldom, tartarus vitriolatus; gypsum; bitter- salt; alum, very seldom; lime and magnesia dissolved in acid of air, acid of salt, or acid of saltpeter; copper-, iron-, and zinc-vitriolate; and iron dissolved in acid of air.

To Detect Free Acid and Free Alkali

§ 186 (a) For this, tincture of Litmus and syrup of violets serve well; acid makes them [both] red, and the latter turns green in alkali (§§ 8, 31).

If acid prevails, it first must be neutralized with lye salt before further tests are undertaken, since otherwise it [acid] hinders the vegetable substance from showing its reaction with metals, which appears as ink, and becomes as clear as water by the action of syrup of violets, then somewhat yellowish; but as soon as it is reacted with a little gallapple it reacts with the iron and the difficulty reappears.

(b) Alkali other than minerale is not found in water; [this alkali] precipitates alum solution and silver solution in the form of a white powder [precipitate], is not darkened by syrup of violets, nor is anything precipitated if it [the solution] is first saturated with acidum salis and then lye-salt is added, which two latter circumstances occur with lime water.

Note. Sometimes we encounter some other free acid in water, other than acid of air, which when it is present in some abundance, betrays itself by its sharp acid taste; by tincture of Litmus which has been diluted to full blueness; by boiling, in such a case as I describe in KVAH 1774, whereby the acid expelled can be collected and measured. All water on earth contains acid of air, at least 1 or 2 cubic inches per kanna, but the so-called "luftige" [airy] is found to contain 10, 20, 30, or 40.

To Detect Common Salt [Chloride]

§ 187. This salt does not affect Litmus, syrup of violets, Bresilia, tea, gallapple, alkali, acid of vitriol, merc. sublim. corrosivus, but does affect solutio lunae, because the silver reacts with acidum salis and precipitates in a white powder [precipitate].

Note. Glauber's salt and saltpeter also occur in water, the former in certain health springs, and the latter in well water. They are detected most certainly by crystallization.

To Detect Earths

§ 188 (a). Lime water precipitates with acid of vitriol as gupsum; if silver solution is added to it [lime-water], the silver precipitates as a reddish brown powder; and if it is mixed with a few drops of acid of salt the earth of lime can be precipitated by lyesalt.

Such water also turns red with Bresilia.

> Note 1. Such a solution as the author [Scheffer] mentions here is quite like aqua calcis [lime water], does not occur in nature, nor in baths, and never in spring water whenever the lime is dissolved by acidum vitrioli, nitri, salis, or aereum, in particular by the last two.

(b) Alum-containing water is detected with alkali vegetabile, which precipitates earth of clay [from the alum], and if the clear solution is boiled we obtain Tartarus vitriolatus.

This water also reddens with Bresilia.

> Note 2. The author [Scheffer] does not recognize magnesia, which is found with acidum vitrioli in all bitter [spa] water, with acidum salis in ocean water and salt springs, with acidum nitri in different wells, and with acidum aereum in many health [spa] waters. It is precipitated by alkalis and burned lime, and is separated most easily from all others by reaction with acid of vitriol, since then we obtain bittersalt. See more in § 73.

To Detect Metals In Water

§ 189. No metals are found in water unless in the form of the vitriol[ate].

(a) Copper is detected by alkali volatile, which makes the water blue, and is darker, the greater the amount of metal present. If sufficient alk. vol. is present, so that the water plainly smells of it, all copper is dissolved, but if some other metal is present, it is precipitated.

Pure Fahlu vitriol does not get much bluer with alk. vol. because it contains mostly iron and zinc, but the color deepens when the solution stands a while.

Note 1. Cement-water is the name by which it is usually called; it contains copper-vitriol, and it is hardly found otherwise than in the vicinity of copper-pyrite, which by some process or other reacts in solution.

(b) Iron-containing water blackens with astringents like tea, gallapple, etc. Alk. fixum makes a greenish precipitate, but in air the precipitate turns yellow-red.

Note 2. Sulfur-pyrite [svafvelkis] occurs abundantly in nature, collected or dispersed in various minerals, often as imperceptible sparkles. These decompose with time to produce vitriol, and for many reasons are the origin of black springs, bog-iron-ore, and lake-ore, since ochre separates from the vitriol little by little, which [ochre] has collected and coagulated together in certain places.

(c) Zinc-containing water darkens and becomes brownish with gallapple, tea, etc.

Alkali fixum produces a white precipitate.

All three waters [copper, iron, and zinc] become blue with blauholtz, or brown Bresilia, as it is sometimes called, the iron solution precipitating something greenish. Scraped Bresilia is a poor reagent for this test although Hollanders use it by boiling the color out of it and then sprinkling with alum solution, whereby it reddens and looks fresh. It is best to avoid this deception and to use chips cut from Bresilia-wood itself.

Note 3. No metal is found in water unless it is dissolved by an acid, either acid of vitri- ol or acid of air. Blood-lye precipitates all metals, and the three named in the text [by Scheffer] have different colors; the first [copper] is brown, the second [iron] blue, and the third [zinc] white. The difficulty is that other metals precipitate with similar colors; consequently quicksilver, lead, bismuth, arsenic, zinc, and magnesium give white precipitates; gold, tin, and nickel yellowish; iron blue; copper red-brown; cobalt grayish-red; and silver dark brown. Antimony has a bluish color, but it is from iron contamination; [antimony] alone precipitates very slowly and white, for which excess acid should be present, because the solution otherwise reacts like pure water. Platina is in appearance difficult to distinguish from iron, but its precipitate appears to be whitish. Natural zinc waters are rare and get shiny by decomposition.

To Detect White Arsenic

§ 190. Arsenic waters are among the most difficult to recognize, partly because they usually are very weak, and partly because they have in some circumstances behavior similar to other substances; but it is very important to be able to recognize them, since they are most dangerous of all to health.

(a) Alum solution produces no changes in them [arsenic waters], nor does common salt; and alkali precipitates nothing, since arsenic is soluble by itself, without the aid of acidum vitrioli, which other metals need.

(b) Arsenic waters color Litmus red and syrup of vio- lets green, but then they must be strong. Borax, neutralized by mineral acid reacts similarly, but cannot be confused with arsenic.

(c) Positive tests are: to evaporate to dryness and ignite the residue in a crucible, at which an odor of garlic should be noticed, but these waters are almost never strong enough for any noticeable residue to be found.

(d) First add Merc. subl. corr. to arsenic water, and then alkali fixum, producing a white precipitate, which becomes grayish after one hour, although Merc. sublimatus is otherwise precipitated red by alk. fixum and white by [alk.] vol.

Sal ammoniac together with Mercury subl. gives a precipitate also with alk. fixum, but chalk-white, and when proper alk. fixum is used, [alk.] vol. is driven off, which is detected by its odor.

Quicksilver solution precipitates similarly gray with alk. vol., but such [solution] cannot be found in nature because acidum nitri is hardly ever found free in nature.

(e) If sulfur is boiled with ignited [unslaked] lime in water we get a clear yellowish solution which precipitates arsenic water; the sulfur combines with arsenic to form orpiment, precipitates as pale gold, and differs from sulfur in color, while the lime remains in solution.

Such liver of lime solution precipitates sugar of lead in large red flakes, but if sulfur and lime are boiled in arsenic water, the solution turns whitish and sugar of lead is precipitated black.

(f) The most dependable test is with solutio lunae, which precipitates arsenic as a red-brown powder. Initially we observe hardly any redness, but after it stands a short time [the redness] appears.

If common salt is also present we obtain for the most part only a white powder. The reason is that acid of saltpeter attracts arsenic more than [it attracts] silver, so that consequently the latter [silver] is released at least to the greater part, but then the arsenic has once more a closer attraction for silver than for the acid, and so it precipitates to the degree that acidum nitri is unable to retain it [in solution].

Lime also precipitates silver red-brown, so the water should be tested with syrup of violets, since if this [solution] turns green it indicates lime, but otherwise arsenic, which precipitates the silver.

Note. White arsenic is a metallic calx which is prepared artificially but which very seldom occurs by itself alone in nature; moreover it is so difficultly soluble that at 15 degrees of heat, 90 to 100 parts of water can scarcely dissolve 1 part [of arsenic]. Consequently without doubt arsenical waters occur very seldom, indeed we do not know for certain whether some, or maybe all, which are dangerous and deadly are necessarily arsenical waters. **NEUMANN** has with reagents and precipitation made careful investigations of arsenic solutions and found them to be precipitated by most metal solutions, although often a half day or a whole day was required before it [the precipitate] showed itself definitely; this investigation instructs us further how to detect the content, in case some arsenical water should occur.

§ 191. Lime-containing water is the best of all water in daily use, since others contain some acid which will retain the foreign substance in solution; but lime is soluble by itself and has this use, that it absorbs all acidity in the body, so that it cures arthritis, gout, and stones. Those who say it is dangerous should remind themselves that Selzer water, Carlsbad [water], etc., contain lime. Steffens pills are good for stones and have a lime-like quality. The stone itself, research shows, is a middle salt which contains alk. vol., and has other than ordinary properties such as disappearing in fire and leaving only a little ash behind; if the stone is old it grows a hard kernel inside itself, which is very fearsome because it cannot be dissolved easily.

The spring at the Stockholm deer-park is popularly praised because it contains both lime and iron; it must be drunk at the spring however, because the iron precipitates in free air. Lime has the property of precipitating whatever is dissolved by acid, but if sufficient air is lacking, as in springs or veins, it cannot react. We often find that if a strong metal solution is, poured into a flask and oleum tartari is added, a strong effervescence occurs and precipitation; but if it is immediately enclosed with a well-ground stopper the reaction stops and the solution remains clear as long as air is kept out; but as soon as the stopper is re- moved the fizzing begins again, along with precipitation. We should therefore not proclaim the raction of the Deer-park Spring to be caused by some fleeting substance which disappears in free air.

Note. The author [Scheffer] had no knowledge of the correct properties of this subtle martial material, whose iron is retained in solution by acid of air and is precipitated as soon as this acid escapes into free air. Crude vitriolic water is usually called "black spring [water]", and cannot be freed from all martial [iron] by boiling only once, much less by opening it to air alone.

What the author mentions as the use of lime-containing water cannot refer to what is usually called lime-water, because it never contains this so-called earth in the absence of acid of air, unless in the case where there is an excess of it, most likely unburned lime. This lime is separated [precipitated] by boiling, and tea kettles are caked with it; it occurs whether good or bad, but as a rule the lime causes obstruction; and in my experience here in Uppsala, splendid springs contain lime.

§ 192. Water that contains alum or some other earth dissolved in acid is very dangerous because both the base [alum] and the liquid stay behind in the body; not only humans but also other animals have an alkaline nature, and contain volatile alkali, which combines with acid to form a neutral salt and precipitate the earth, which then stays behind in the body.

A water which does not yield a precipitate with alk. fixum is never dangerous to drink except when it contains arsenic (§ 190).

Note. Very seldom do we encounter water which does not contain lime, and usually it is combined with acid of salt, but if this is decomposed by alkali vol. combined with acid of air, the latter then reacts with the lime by itself and causes it to behave like unburned lime or chalk (§ 78).

§ 193. Iron vitriol is among the most universal medicines: it is also an astringent, repels caterpillars, and in large doses serves as an emetic, e.g., up to 20 grains; 2-3 grains dissolved in 2-3 pounds of water confers properties [on the water] similar to crude mineral water. Better water is produced from the residue obtained when bloodstone [hematite] is distilled with sal ammoniac (§ 158). The flowers which pass over in this same operation are useful if they are dissolved in spiritus vini and the solution dribbled on sugar, which after the passage of time, 12 to 24 hours, is soluble in water. Some drops of the resulting solution are good for use as a water cure; the dissolved material is precipitated green by alkali vol. because of the phlogiston contained in the sal ammoniac.

<u>Note</u> The examination of Bitter-, Selzer-, Spa-, and Pyrmont- waters, and the method for imitating them, I have given in Vet. Ac. Handl., 1775. I intend also to publish various and complete articles on testing of water in general, which I do not include here for want of space.

We usually designate those waters as raw or hard, which do not soften and similarly benefit from boiling; they are slow to dissolve soft or hard soap unless they decompose them. The first reaction takes place naturally at the surface with lime dissolved by acid of air; when heat begins to penetrate the water, the solvent for the lime [acid of air] is driven away, and the earth precipitates in its finest form, then collecting into pea-sized globs, stopping up pores and excluding the water. This difficulty is relieved if the water is boiled first. That soap is disturbed from its combination is caused by all earths and middlesalts, the alkali in the soap being attracted to the acid in the middlesalt more strongly than it is to fats, and therefore the latter [fat] is released [precipitated]. Free or excess acid of air gives the same effect. Addition of a proportionate amount of alkali vegetabile can overcome this difficulty.

FOURTH SECTION: CONCERNING COMBUSTIBLE SUBSTANCES

FIRST CHAPTER: Concerning Sulfur

§ 194. Sulfur consists of phlogiston and acid of vitriol, which [combination] shows the following properties:

(I) It melts readily over a flame, and burns with a blue flame and gives off a stinging acidic odor.

Note 1. Sulfur is generally designated as an acid saturated with phlogiston, but except for the ordinary sulfur and urine-phosphorus we know of no other kind [of similar substance] unmistakeably up to now. Common sulfur melts at 118 degrees, and catches fire at about 312. If it is taken to the first melting point and then allowed to cool it separates [crystallizes] in rays, especially at the surface, and becomes electrically charged. If allowed to stand a few moments longer [molten] in the flame it thickens [becomes more viscous], and becomes very soft in a few days; if cast into water it is useful for making imprints; its color is now reddish.

(II) Melted together with alkali, it [liver of sulfur] dissolves metals (§ 198).

(III) It is dissolved in oils (§§ 202, 203).

(IV) Lead, copper, iron, and tin can be calcinated by it [sulfur] just as well as in other ways.

(V) It blends with all metals and semimetals except gold.

Note 2. The specific gravity of sulfur is about 1.800; drop- sulfur 2.00; its color is yellow, sometimes turning to green.

§ 195 (a). Sulfur is found in all ores with a few exceptions, such as hornerts, lead spar, bloodstone [haematite], and gray and white iron ores.

Ore is a natural mixture of sulfur and a metal; if the first [sulfur] is separated, what remains is no longer an ore, but rather a metallic earth or pure metal, according to the various methods of separation used. Consequently, because gold never lets itself be combined with sulfur, there is no entirely pure gold ore, unless it is pure [gold itself] or gold mixed with another metal such as in Adelfors guldkis [pyrite-gold], in that in the fused impure stone [skärsten] the gold combines with iron and sulfur; similarly it combines with Schemnitz golden fused glass [glaserts], in which silver also enters in.

Note 1. Usually each ore is a mineral named for the metal contained in it. In general metals occur in nature in three classifications, namely, pure, calcinated, and mineralized, and so we also have a like number of kinds of ore. Pure [ore] designates that which not only has enough phlogiston that it takes on metallic form, but also is free from sulfur; calcinated [ore] is that found deprived of the amount of combustibility [phlogiston] necessary for metallic sheen, and consequently having the appearance of either an earth or stone; mineralized is the name given to that ore which is intimately combined with or dissolved in sulfur. Sulfur is the only substance which mineralizes metals in nature, since if we designate the blending of metals when sulfur is present as mineralization it follows that no pure metal is produced, because hardly any [metal] is found which is entirely free from all other metals. Thus gold always contains silver or copper; silver contains copper or gold; platina contains iron; etc. When white arsenic is combined with another metal it always takes the reguline form, and when it occurs alone, without sulfur, it is nothing other than the pure metal. Acids sometimes disguise metals, as hornerts in [acid of] vitriol, etc., and therefore can with some cause be considered as mineralizing, but such compounds are up to now found in such small amounts that they are seldom used as ores.

Only gold does not combine with sulfur, but if some substance is present which readily combines with both [i.e., sulfur and gold], it can serve to unite them. Likewise zinc. Platina does not completely combine with sulfur without a similar helping substance, at least it must first be dissolved by pyrite, according to the research of Herr Prof. and Knight WALLERIUS.

(b) We can imitate nature and artificially produce certain mineralizations. If a metal is placed in a crucible set over a fire, and sulfur is added a little at a time until it is all well fused, we obtain on cooling a mass which somewhat resembles an ore. In this manner we get leadglance from lead, glaserts from silver, etc. Skårsten is indeed a blending of sulfur and metal, but does not always resemble an ore, but just as much as kupferglas resembles spårsten or trorsten, or liver-brown pyrite resembles kisskärsten.

Note 2. What our author [Scheffer] here calls kupferglas is not quite correct; it consists of petrified copper calx, without a mineralization of copper with sulfur alone, which gets this name improperly and is otherwise called gray copper ore.

(c) Pyrite generally consists of only sulfur and iron, but also contains a little copper; it is the richest of all [ores] in sulfur and could therefore be called a sulfur ore. Yellow copper ore is the second-richest sulfur-bearing ore, and after it lead ore, and finally iron ore.

Note 3. Pyrite [svafelkis] contains 30 to 40 percent iron; it is also simply called kis.

To Drive The Sulfur Out Of Pyrite

§ 196 (a). The pyrite [kis] is broken into small piec- es to pass through the neck of a clay retort (glass does not allow enough heat for this operation), which is filled one quarter full with it and no more, because part of the pyrite contains lime which swells in the heat and could break [the retort]. If broken too small and fine, pyrite depresses the heat and hinders its working. If a clay retort is not available, a glass one can be used if it is lined with linseed oil and unburned clay powder, which [paste] can be spread on and allowed to dry for 7-8 days, during which time the clay becomes cemented to the glass by the oily paste; then the mixture is rubbed thin with mufflar

(§ 171), but very thin and with drying in between [rubbings], continuing in this way until it is thick enough, as required.

(b) Then the retort is put on the grate of a blast furnace, a receiver half-filled with water is attached, and the operation is started up with gentle heat. Without the water the receiver might burst from the heat of the sulfur passing over.

(c) In the beginning a little phlegm passes over; then white fumes which condense to sulfur in the water; [sulfur] also drips from the neck of the retort when the heat becomes somewhat stronger. When white fumes seem to diminish, all joints are opened and the furnace is heated as high as possible, and this is continued until no more fumes are noticeable.

Note. If the air is allowed access, the sulfur will ignite. Such combustion is accompanied by very harsh fumes from volatile acid of sulfur; the longer this occurs the more combustible the acid it contains, which is consumed by more vigorous combustion. For silk and other sulfur-containing substances the combustion therefore should take as long as possible.

(d) The water in the receiver is now found to be yellowish and cloudy from fine particles of sulfur, which can be collected if the water is poured into a glass and allowed to settle. All the sulfur is weighed and its content is determined in this way. That which contains less than 15 percent is not worth working with. The sulfur which it contains is grayish and very impure but can be purified as described in the following paragraph. (§ 197).

(e) The residue in the retort usually consists of a grayish black powder and mostly some whole pieces of pyrite which lie on top and were heated least. They contain some sulfur yet, since there was released in the closed vessel only what was in excess to the skårsten. Iron vitriol can be prepared from this (§ 86).

To Sublime Sulfur

§ 197 (a). The sulfur is pounded into a powder and mixed with 1-1/2 to 2 parts of burned clay powdered fine, and poured into a clay retort, which is placed in a blast furnace and covered with a helmet [alembic]. Then when it is heated the sulfur rises in a fine powder which deposits in the hat and the top of the retort, and is called flowers of sulfur, but has no other properties than those of pure sulfur.

(b) This can be carried out more conveniently in a retort on a sand bath, when the dripping sulfur passing over is pulverized to flowers of sulfur. If the sulfur is still impure the sublimation is repeated until it becomes that which is desired.

§ 198. Liver of sulfur (hepar sulfuris) is the name given to sulfur combined with alkali: it got its name from its liver-brown color, it tastes and smells repulsive, which results from its high volatility if alk. veg. is contained in it, but not with [alk.] minerale. When fused it dissolves all metals, producing a substance which is soluble in water; and because gold can be taken up in water in this way, STAHL had reason to say that gold is potable, or more correctly abominable.

The sulfur can be separated from the alkali if the hepar [sulfuris] is dissolved in water and into this solution is dripped an acid, which then attracts the alkali and allows the sulfur to precipitate; this is best collected by filtration through heavy paper, but the product is very impure and should be sublimed if it is to be free from impurities. The filtrate is evaporated until a neutral salt is obtained, different from the alkali in hepar and the precipitating acid.

Note. Hepar is prepared in both the dry way and in solution. In the former case [dry way] similar amounts of both reactants are melted together; but in the latter case boiling in water is used and the amount of caustic alkali is doubled, since crystallized [sulfur?] does not dissolve. Hepar smells like rotten eggs; it dissolves carbon in both wet and dry ways; this solution precipitates metallic solutions, and its fumes reduce litharge to some extent. Dry liver of sulfur, added to metals heated to glowing in a crucible, dissolves them with fusion, from which we deduce that the sulfur is not consumed by long heating. Gold requires 16 times as much hepar as its own weight; silver, lead, and tin 8, but copper only 2; zinc is not attacked according to Mr. MALOUIN'S research in 1743.

Volatile hepar is prepared from 6 parts of burned lime, 2 [parts] salmiak, and 1 [part] sulfur, which are then distilled together. It is otherwise called Spiritus fumans sulfuratus Beguini, and is used in making cinnabar in via humida, if Merc. subl. corr., merc. dulcis, mercurius vitriolatus, Turpetum minerale, mercurius nitratus, or running quicksilver is added to it; but white arsenic produces realgar.

Hepar is decomposed by itself in free air because the alkali attracts acid of air. If hepar is precipitated by acid, and the precipitate is sublimed, we can determine how much sulfur is in it, if then a similar amount of hepar is gently calcinated until all its phlogiston is driven off, to give a determination of the approximate proportions of sulfur's amount in the compound. STAHL has found [the ratio], acid to phlogiston, 7 to 1; but Mining Counselor BRANDT, 17 to 1.

§ 199. Lime and sulfur do not fuse together readily in fire, but when they are boiled together in water, solution occurs, and much more lime is dissolved than when it is boiled by itself. The odor is more repugnant than that of the ordinary hepar. Alkali fixum precipitates the lime and combines with the sulfur to form ordinary hepar; in fire the sulfur burns up and leaves the lime behind.

Note. This is called hepar calcis and the preceding hepar salinum. If magnesia is contained in water with a little flowers of sul- fur 8-12 hours in a closed flask in boiling wa- ter, something hepatic is obtained which can be designated hepar magnesiae; by melting we also obtain a very soluble compound.

FOURTH SECTION, CHAPTER I, Sulfur

Burned lime and sulfur can be combined by fusing, if one part of the former [lime] is reacted together with 2 of the latter [sulfur], and finally one part lime is added little by little and the mixture held in fire until it hardens completely.

When liver of lime is burned in fire, this acid [acid of air?] always reacts with the lime.

The sulfur deposited from the hepar with acid is called magisterium sulfuris, milk of sulfur, etc. Liver of lime is precipitated by alk. fixum, borax, sedative salt, etc.

To Prepare Sulfur From Acid of Vitriol and Phlogiston

§ 200. This can occur in various ways:

(a) If essential oils are distilled with very concentrated acid of vitriol some sulfur is sublimed.

(b) If 6-8 parts of Sal Glauberi are ground together with one part coal dust and then melted down in a crucible on a blast furnace, and then this is poured out, we obtain a good hepar. The crucible should be large, because the mixture swells up very much.

We can also use Tartarus vitriolatus, with which only 1/16 coal dust is used, because alk. veg. reacts with less acid than does [alk.] minerale.

With coal dust the fusion should be carried out more slowly than with alkali and sulfur, and more sulfur is also burned up.

Note. We usually use a little free alkali to better conserve the sulfur. Impure hepar made in the manner just described is grayish from dissolved coal.

(c) Vitriol and coal dust are combined in fire into skårstenar [cut stones?].

§ 201. If sulfur is ground with iron filings and water in a mortar, and this mixture is then allowed to stand for one day in free air, the acid of vitriol releases phlogiston and attacks the iron; the oiliness again transfers to the water but does not mix with it. Brought to crystalllization, the solution produces vitriol, which when distilled per se leaves the acid alone.

Note. Sulfur, iron filings, and water when mixed begin to stink immediately like hepar. If the mass is left in a dish it quickly turns black, swells and heats up; a crust forms at the surface, breaks, and the heat drives out the stinking fumes, which finally catch fire and burn for several hours, if the mass weighs 3-4 pounds. LEMERY buried such a mixture to imitate an earthquake [volcanoe?].

Common sulfur is usually so acidic that it attacks metal mortars with pungent fumes, because it requires some phlogiston from the metal to release the acid; indeed flowers of sulfur which are allowed to lie in a damp room become so wet that they cannot be ignited without being dried beforehand. Undoubtedly this released acid first attacks the iron, and during this mutual reaction heat is produced, which disperses a lot of phlogiston.

FOURTH SECTION, CHAPTER I, Sulfur

To Prepare Balsamus Sulfuris Crassis

§ 202. Balsamus sulfuris is the name given to the solution of sulfur in oil. It is prepared as follows:

(a) A piece of sulfur is put into a glass vessel and pressed oil poured on it, e.g., olive oil, and [the vessel] is placed on a griddle [skedpot] over a gentle fire.

Note 1. Usually we use 1 part sulfur and [up to] 4 parts oil, and at least 2. If neatsfoot oil is used the solution is called Balsamus sulfuris Rulandi.

(b) When the oil boils it should dissolve the sulfur with a strong reaction, and when it stops no more can be dissolved unless the vessel is allowed to cool. The sulfur which is not dissolved is very black and viscous.

(c) This balsam is bright red, thick like syrup, strong smelling, repugnant in taste, with an odor of neatsfoot oil. It cannot be used.

Note 2. This balsam is usually called corpus pro balsamis after it is diluted with essential oils or other fatty oils, which [mixtures] when named are variously called balsamus sulfuris anisatus, amygdalatus, suc- cinatus, etc.

To Prepare Balsamus Sulfuris Liquidus

§ 203 (a) Drip sugar, almond oil, distilled oleum carvi [carroway oil], 2 parts each, are placed in a stirring pan [skedkolf] and dissolved over a low fire, until a drop placed on cold iron immediately hardens so that it does not run, at which time the pan is removed from the fire. For a more certain reaction we can use 3 parts sulfur, the excess settling out. The oleum carvi should be free from small carroway seeds but not from the long.

(b) Then 1 part oil of citron peel is poured in, mixed well and allowed to stand in low heat until it gets thick, after which it is kept in a flask.

(c) This balsam is very good for the chest. Almond oil is mollifying, citron oil purges and gives off an agreeable odor, cummin [carroway] oil comforts the stomach, although somewhat hot, but it is usually moderated by the other two. Peppermint oil is very mild, but very costly, and seldom free of imitations. Take 3 drops of it on a sugar cube; the taste is very strong but not unpleasant.

(d) Apothecarys prepare this balsam in this way: 3 parts alk. fixum, and 2-3 parts sulfur are fused to a thick mass, which is then removed from the heat and stirred until it hardens; it is then ground up fine and boiled with 4 parts of anise oil, juniper berry oil, or cummin oil, as before. In this way we obtain a balsam which is like those above, but which is much cheaper, although more fluid and volatile. It is usually called balsamus sulfuris liquidus and is [also] called anisatus, juniperatus, terebinthinatus, etc.; it is very unwholesome. Oil of anise is mild at first, but then exhibits a strong choking effect, in particular for hypochondriacs and hysterical persons; cummin oil and oil of juniper berries are very hot

The phlegm, and not the acid in sulfur prevents its combining with oils, and in order to absorb this the aforementioned alkali is used; and the acid in the sulfur con- tains more firmly fixed phlogiston, which does not go together so readily with the liquidity in essential oils. But concentrated oil of vitriol and oleum stillatitium, e.g. carvi, readily form a thick balsam, but a vapor forms very quickly and some of the acid separates, so that its use in medicine is uncertain. Acidum vitrioli combined with phlogiston works more consistently with oils.

(e) No thick balsam is formed with oil of yellow amber [bernstenolja], since it dissolves very little sulfur, and strong heat is also required. When the solution cools, long radiating crystals separate on the bottom; thus the oil acts in the same way towards sulfur as water acts towards salts.

Note 1. Oils, like water, dissolve more with heat than they can hold in solution, and therefore on cooling [crystals] separate which are called Rubinus sulfuris.

(f) All balsams from pressed oils settle down in water, but float after they are distilled.

Note 2. Usually 6 parts of oil are used to dissolve 1 part of sulfur with sufficient heat. The vessel is not entirely closed because it might be burst by the initial violent reaction.

The acid in sulfur seems to leave its union and is mixed with the water in the oils; at least this happens with time and with the help of heat. In distillatiopn of sulfur balsam we get no sulfur except volatile acid of sulfur and oil, just as when acidum vitrioli is used in place of sulfur.

One part of sulfur is dissolved in 72 [parts] spiritus vini, if both react with each other in vapor form, which GREVE DE LAURAGUAIS discovered in 1758; it [this solution] dissolves in water and is not nauseous in taste. Ether dissolves it also. Strong acid of vitriol, as it boils, melts sulfur and dissolves some of it.

§ 204. Sulfur forms various inflammable compounds.

(a) Fulminating powder [Knallpulver], or nitrum fulminans, consists of refined saltpeter 3 parts, alkali tartari 2 [parts], and flowers of sulfur 1 [part], well ground together in a glass mortar.

If a pinch of this is put on a black iron spoon and held over a candle and stirred well until it almost melts, it burns with a sudden spurt and the spoon is knocked out of the hand, and a report is produced like a gun shot. At the instant of melting, ignition occurs, so we should therefore stir well, so that nothing essential evaporates. The cause of this is the air in the saltpeter, which is freed from phlogiston in an instant and takes on its expansion.

This powder deliquesces strongly and loses its power when it becomes the least bit moist.

Note 1. Fulminating powder seems to get its power from hepar, which forms in the fusion, and releases its phlogiston so readily that all the nitrum can be decomposed in an instant, since hepar with two parts saltpeter gives a similar explosion. In an evacuated space it burns without violence and without explosion, and produces less elastic fluid [gas] than a similar amount of gunpowder, but a flame is hardly noticeable.

(b) Gunpowder [krut] contains 24 parts saltpeter, 3 sulfur, and 4 coal dust.

Note 2. Gunpowder is used for hunting, breaking up of stone and ore, and as an article of commerce; in fact the cost of war is chiefly based on it. In Sweden approximately 1200 hundred-weight of it is consumed in ore breaking, and 2000 for private use, which is only based on a 50-year estimate; for our castle's defense, and the needs of the navy and army, e.g., during a campaign, 126,600 hundred-weight.

The quality of this material, according to the research made by Mr. BAUMÉ, at Knight D'ARCY'S request, depends less on the proportions of the constituents than on the manner in which they are mixed. The degree of dryness which the material has in its granules is very important. The saltpeter and coal dust should be mixed first, or all together, after which the sulfur is covered with a layer of saltpeter, which hinders its combination with the coal dust. The material must be stirred about until it has reached sufficient dryness, since moisture is attracted by heat to the surface of the granules, by which [moisture] the saltpeter is separated and crystallizes. The size of the kernels does not affect the quality, nor does the hardness of the wood from which the carbon dust is made. Originally gunpowder was not granulated, and a similar volume was more powerful, but it attracted moisture more easily, did not pack easily into a musket and stick in place. Common salt must be well separated from saltpeter; also lyesalt, and all sand as well, which otherwise produce the most grievous accidents. Gunpowder can be prepared without sulfur, but the smallest amount of this substance [sulfur] doubles its strength; it requires more intense heat for igniton and becomes moist in air more quickly than usual. Gunpowder needs some degrees stronger heat for ignition than sulfur; it also burns in an evacuated space, although without explosion, but requires strong heat or otherwise the sulfur sublimes.

One hundredweight of Swedish gunpowder contains 75 percent saltpeter, 16 sulfur, and 9 coal dust; French [gunpowder] 75 saltp., 9-1/2 sulfur, and 15-1/2 coal dust; Portugese [Lissa] gunpowder 80 saltp., 12 sulfur, and 8 coal dust ; Dutch contains 71 saltpeter. Hunting powder, which has smooth kernels, is weaker and ignites more slowly. Mixing of materials occurs best with stirring; rolling is not suitable. The proportions in a sample of gunpowder can be determined readily enough. The saltpeter is separated by dissolving it in water, and its purity is found in the manner previously described (§ 185). The sulfur is gently burned out of the residue from the previous extraction, leaving the charcoal, which still contains 1/24 of the sulfur. Gunpowder is best preserved in paper which has been saturated with alum. Knight D'ARCY has determined its power in the best method used up to now.

§ 205. Pyrophorus is a black powder, like milkpowder, when it is not fully burned but yet reddish, which ignites spontaneously in free air and burns like tinder. It is prepared as follows:

(a) Sieved alum 3 parts, and 1 [part] of fine powder of white starch are mixed well, placed in an earthen crock while it swells and is then placed over low heat, at which the mixture begins to melt and should be well stirred with a knife so that it is evenly heated throughout. When the moisture evaporates the remainder adheres around the edges as a dry brown belt, which is scraped off, ground, sifted finely, and again put in a crock over low heat until it becomes entirely free of water, during which it is continuously stirred. This roasting requires 2-3 hours according to the quantity involved. When the powder gives off no more odor, and feels soft to the knife, it is removed from the heat. If the crock gets too strong a heat, too much phlogiston is driven off, and it glows, and becomes entirely useless for kindling material.

Note 1. This difficult roasting [process] can be entirely avoided if we take 4-5 parts alum added to 1 part coal dust. The less coal dust we use, the more yellow the pyrophorus becomes, but also the more slowly it kindles; when there is less than 1/16 used, we no longer get pyrophorus. Too much coal is also bad; the product here does not kindle. One part sodasalt, 1/4 sulfur, and 1/3 coal dust gives pyrophorus; but it still requires some moisture before it will kindle.

(b) The powder obtained [in (a)] is put in an iron grenade, which is closed with a well-fitting iron stopper, which for greater security is smeared over with clay and litharge, which form a glaze in fire. The trick is to apply just the right degree of heat, because too little [heat] will not cause the phlogiston and acid of vitriol to unite properly, whereas too much [heat] will open the pores of the grenade and the combustible material will escape; in both cases we get a powder which does not ignite. The grenade should glow dimly. It is most secure if we put it in a blast furnace, fill it up with coal, lay some kindling on top, and then close all openings. When everything burns up and becomes cool, the pyrophorus is ready. A hearth in a stove can also be used for treating the grenade, but if the ashes are too hot we will see and smell sulfur burning, which usually shows itself through the pores.

<u>Note 2.</u> I use ounce vials which are filled about 2/3 full with the material from which the pyrophorus is prepared, then put them in a crucible filled with sand to a depth such that only the neck sticks out, and the crucible is set on hot coals. When the sulfur flame begins to decrease in the mouth [of the vials], they are closed with dry fitted clay stoppers, and when the vials have cooled somewhat they are closed with a cork. If coal dust is used and no previous roasting, the material should be moistened slightly or at least well packed into the vials. Sugar in place of burned powder in the author's

method (a), gives from 30 parts of the mixture about 20 parts coal and 7 or 8 pyrophorus.

(c) When the material is fully roasted a ring is at- tached, made of wax melted together with turpentine, around the mouth of the vial, the stopper quickly removed from the grenade, the opening fitted in the wax ring, and so the pyrophorus is contained in the vial, which is then well sealed. When it becomes ignited by itself, only a white earth of alum remains, which previously had no other use than to hold pyrophorus in a dry and fixed form, because it would otherwise be a spiritus vitrioli fumans, or sulfur, which would escape during burning. Pyrophorus contains a regenerated sulfur, from which earth of alum when coming in contact with air attracts the acid with an avidity such as iron attracts sulfur moistened with water, which produces heat, and which because here it occurs without water, breaks into full flame. Vitriol cannot be used in place of alum, because earth of iron has too strong an attraction for the acid, so that phlogiston cannot unite with it. If it were sulfur pyrophorus, it could never be stored in open air.

Note 3. HOMBERG found by accident, when alum and human excrement were distilled togeth- er, that the residue ignited spontaneously in air. This gave us the first introduction to Pyrophorus in 1710. LEMERY the younger then discovered that flour, honey, sugar, etc., all vegetable and animal substances which contain phlogiston, could be used in place of excrement, in 1714. Finally Dr. SUVIGNY maintained that all salts which contain acid of vitriol could be used in place of alum, but his method was quite troublesome. That alum alone, as GOERITZ reported, could be roasted to pyrophorus, is not credible.

Various authors mention other kinds of pyrophorus, which I still have not yet had occasion to investigate. If orpiment [arsenic sulfide] and iron filings are mixed together and sublimed, and the sublimate is ground together with 10 parts luna nitrata, it ignites by itself on paper, and is called Phosphorus Menderi, or more properly pyrophorus, since phosphorus denotes the genuine [element] which only glows [phosphoresces]. HENCKEL also mentions a white arsenic, sublimed with black flux in a closed vessel, which produces regulus of arsenic in octagonal crystals, which when ground together with luna nitrata also ignite.

(d) Some authors take one pound of Roman alum with 16 [parts] egg yolk, which are calcinated together in an iron pan, then put into a glass or grenade, which is placed in a crucible with sand over a fire. An acrid smoke and odor is noticed immediately, the fire is taken away until this disappears along with the sulfur odor, and flame comes out through the opening. When the flame begins to decrease, [the vessel] is closed with a tight-fitting stopper and left to cool. We get no pyrophorus here if the opening is closed too soon, or if the flame goes out by itself. The whole operation requires 5-6 hours and seems not to be right because it is wasteful of phlogiston, but in the method described, with alum and egg yolk, we do obtain a regular pyrophorus.

We can also use alum, wheat flour, white or candied sugar, and sulfur together, but the powder [flour?] alone with alum is the best, because it contains so little water in it.

Note 4. The cause of the self-ignition of pyrophorus is not yet fully understood. Since acidum vitrioli must always be present, we have conceived the not unlikely theory that some free part of the acid is brought to a fixed position by roasting, whereupon it very greedily attracts the moisture in the air, thereby becoming heated, and ignites the nearby sulfur and coal. The sulfur present produces a hepatic odor, and it can also be precipitated with acid out of the solution of pyrophorus in water; that pyrophorus attracts moisture more strongly than alkali fixum causticum can be deduced from many circumsatnces, since a small amount held on a moist mirror makes the corresponding place dry immediately; a thread dropped into acidum salis fumans does not smoke over pyrophorus in a flask; etc.; that moisture favors ignition is perceived plainly from this, that it is ignited slowly and with difficulty unless it is laid on moist paper or breathed upon. All this notwithstanding, it still appears that the ignition may arise from another cause, since the water which is added to the pyrophorus gives not the least trace to the acid, or to the alkali either, although this is hardly noticable; but if any free and solidified acid of vitriol were present this should be detectable with tincture of litmus.

§ 206. Phosphorus is nothing else but a fatty substance combined with an ac- id, like sulfur; but the acid is of a different nature than vitriolic.

FOURTH SECTION, CHAPTER I, Sulfur

It is prepared, as KUNCKEL suggests, from fresh urine, which is evaporated over a low fire until a blackish material is left, then mixed with twice its amount of fine sand or bole [a kind of clay allied to bauxite], put into a retort which is attached to a receiver holding some water, and distilled. At first a little phlegm passes over, then a black, stinking oil, and finally with the highest heat phosphorus in the form of white fumes, which either condense in the receiver, or fall to the bottom like sand.

When the oven cools the bits [of phosphorus] are collected with an iron ladle, water added to them and the mixture held over heat until they melt like wax, but they are quickly congealed in cold water, and must be kept in a flask that is not only well stoppered but also sealed with wax and pitch, because phosphorus will otherwise fume away.

The first discoverer was BRAND, a bankrupt German merchant, who when he worked with urine to prepare Lapis Philosophorum [Philosopher's Stone] from it, found this substance instead which he showed to KUNCKEL, but without telling of what it consisted, but which KUNCKEL recognized, since he knew that BRAND worked much with urine; he [Kunkel] investigated it, to try to produce this sparkling [phosphorescent] substance, and he succeeded.

Note. Phosphorus of urine is a sulfur produced from acid of urine and phlogiston, so combustible in free air that it is always decomposed, and must therefore be stored under water. When we use it to draw or write on paper, it shines for all of a quarter or half an hour in a darkened room, without burning the paper in any way. But if a small piece, the size of a lentil, is put between folded paper and crushed to pieces, the paper is ignited. Mr. MARGGRAF has taught us to take one part of evaporated urine with 16 parts of crystallized sal microcosmicus, which are mixed with 16 parts sand and 2 parts coal dust, which are driven with strong heat in a stone retort to yield only 1 part of purified phosphorus; water is placed in the receiver so that it almost reaches the neck of the retort. Mr. SCHEELE has had the same intention, when he dissolved burned hartshorn in acid of saltpeter, precipitated with acid of vitriol, strained and then abstracted all the liquid, when finally acidum phosphori is left, which is taken up with a little warm water, filtered when it cools, evaporated to the consistancy of syrup, blended with coal dust to make it somewhat dry, and distilled in the above manner.

Phosphorus Balduini and Hombergianus were previousl;y discussed (§§ 77, 78).

On the other hand we obtain different kinds of phosphorus, natural and artificial. To the former belongs Flores mineralis, which in a certain degree of heat or enough severe grinding begins to appear; dazzling Schemnizer shines when it is scratched; similarly sugar, shining wood, various insects [glowworms], in particular larva which are found in the ocean, particulars of which are found in § 215 of "Physical Description of the Earth".

Of those which must be prepared artificially, we might first con- sider Phosphorus Bononiensis, which is prepared from heavy spar [barium] properly roasted on charcoal, which when then allowed to lie a few minutes in sunshine glows in the dark, like a hot coal; if the stone is ground to a fine powder beforehand, and then formed into stars or other shapes using gum tragacanth and water [as binder], the glowing [phosphorescence] appears more delightful. Mr. MARGGRAF has clearly described the preparation in Skomaker's VINC. CASCIOROLO in Bologna, as carried out by the Alchemists, and gave us the first reference. Mr. CANTON has produced a new phosphorus from 1/4 flowers of sulfur, mixed and roasted in a crucible with 1/4 white powder of calcinated oyster shells; then it is exposed to sunshine, as described above. Prof. BECCARI in 1744 produced a different material, which after preparation glowed for several minutes in the dark, when it had been exposed to sunshine beforehand; likewise tartarus, sugar, cotton, linen, sackcloth, etc., and especially paper; tinder, animal stones, Bezoard, etc., and in particular egg shell; likewise different kinds of stones and earths, especially diamond and precious stones, so-called. He has found also through a certain degree of heat, bone nerves, flesh of birds, ox bile, cheese, almonds, beans, coffee which is not brown roasted, all these glow also. DU FAY had previously made similar studies also, and comparable results were found by both [these men], i.e., that with the exception of a few metals, hard or dark materials have the best of such properties.

A peculiar circumsyance, which Prof. JOH. BAPT. BECCARIA of Turin has recently observed, is that Canton's phosphorus shining through colored glass, does so with its color unchanged.

SECOND CHAPTER

Spiritus Vini [Ethyl Alcohol]

§ 207. Spiritus [Vini] is miscible with water, a general property of three kinds of materials, namely alkalis, acids and inflammables.

(a) Alkali is obtained from both vegetables and animals and consists of alkali volatile dissolved in water or spiritus vini, which we have considered previously (§§ 48, 49).

(b) Acids obtained from minerals or liqueurs are used to help digestion. Vinegar is of the latter kind. They were described previously (§§ 11, 14, 17, 20, 24, 30).

Note 1. These first two foregoing spirits [spiritus] [alkalis and acids], get their name only from their use; they are not the kind of liquids which are usually called spirits.

(c) Spiritus vini is a very volatile oil, combined with water during fermentation. It is made from vegetables, which undergo fermentation, since if [unfermented] wort, or decoction of malt, is distilled, we get only a tasteless phlegm. Corn, grapes, all kinds of fruit, etc., which can ferment, give spiritus after this operation. Wine is obtained from sugar or honey dissolved in water through fermentation, and that which is obtained from the others is called mead; from these wines spiritus can be separated by distillation. Here belong also all brandies, which contain more water than pure spiritus vini.

Note 2. Effervescence or Fermentation is the name given to the internal reaction in a mass by means of which either spiritous, acidic, or volatile alkaline materials are produced, as has been noted previously. Vegetables go through almost all of these alterations, one after another, and this operation can be divided into 3 stages. The first [stage] is completed when the mass becomes spiritous; the second when it becomes acidic; and the third occasions a total destruction of the organic structure, and also produces a stench, and at the proper time a volatile alkali. Animal substances give hardly a trace of the first [spiritous] stage; spiritus can be readily made from milk, but this originates naturally from the so-called saccharum lactis [sugar of milk], and consequently milk is not entirely properly classified with animal substances. These two first stages are accompanied by heat, but concerning the last [stage] it is not yet established. The internal reaction begins slowly, little by little, and increases to the highest point at which particles are first seen in the fermentation; the second stage begins before the first stage is completed, and this is signalled when Litmus juice is reddened by the fresh brew. Yeast or fermentum is the name given to that which is added to bring the mass to fermentation. Acid of air does not only hinder the putrefaction, but can also restore that which has already begun to stink and fall to pieces; it [acid of air] is produced abundantly , particularly in the first stage.

Distillation of Spiritus Vini

§ 208 (a). A fermented liquor, such as Rhine wine, which produces most spiritus, is placed in a vessel which is sealed with flour paste to a receiving vessel, after which it is distilled with mild heat.

(b) Since spiritus is lighter than water, it goes over first, in long narrow streams, but as soon as round drops of moisture [water] appear, the operation is completed, because spiritus has passed over and water has begun to come over.

(c) The residue contains most of the water and an earth from the wine, producing a thick, brown, repugnant broth.

Dephlegming or Rectifying Spiritus

§ 209. The spiritus obtained in the foregoing operation is no stronger than ordinary brandy, and still contains a certain amount of water, and from sour wines an acid, which ought to be separated.

(a) A little ash, lime, and decrepitated salt are put into a vessel, and the spiritus poured on it, and the operation repeated as before (§ 208). Common salt prevents the fatty essence from uniting itself with alkali or spiritus, or otherwise a soapy substance forms if anything alkaline is produced more than the acid can neutralize, in which case the soap unites with spiritus, whereupon it passes over more slowly and any empyreumatic oil follows from the ash.

If it is distilled many times without [new] addition the spiritus becomes then quite water-free.

Note. Addition of excess alkali decomposes some of the spiritus, especially with the help of heat (§ 210, Note). An essential oil comes over readily at length and produces an odor of spoiled brandy; this is kept in the residue by charred bread, calcinated alum, or Sal Glauberi.

(b) This operation is repeated often until the spiritus boils away without leaving any water behind.

(c) It still contains some water nevertheless, which is separated when the spiritus is poured on alkali fixum in a flask. If there is much phlegm, the salt dissolves in it and it remains on the bottom under the spiritus; but if only a small portion [of phlegm] remains it is alkaline, like mush. The flask is allowed to stand a day with shaking now and then. Then the spiritus is poured on fresh alkali and the process repeated in the manner described until the alkali stays like dry sand.

(d) The spiritus, now freed from phlegm, contains some alkali and also phlogiston, from which it is yellowish, and should therefore be redistilled, whereupon it becomes entirely white and pure, and is called alkohol vini.

(e) One kind of weak spiritus poured on lyesalt, and believed to be rectified through distillation, is called spiritus vini tartarisatus, but although alkali has attraction for water it is not so great but that heat is able to drive it off and the spiritus becomes weak again.

(f) If warm spiritus is poured in a certain amount on hot lyesalt it becomes yellow-brown or almost red, and is called Tinctura tartari.

§ 210. Alkohol, or spiritus vini rectificatissimus is considered good if it goes off on gunpowder and finally ignites it in a flash, without effervescence; but a better test is to pour it on dried lyesalt and allow it to stand well enclosed for some hours in mild heat, with occasional shaking, since if the salt remains as dry as dust, the spiritus is good, but if it gets ever so little moistened, it is a sign that water is present. All the water cannot be separated without destruction [of alkohol], since a portion becomes essentially combined with a fine oil, but by what means we do not know; some believe that some alkaline bond is formed, but it seems it could not burn so well if this were so.

All spiritus contains oil and gets its strength from this, and not from some salt, as some believe, because then it would not be fit for paintwork.

Note. Alkohol vini has specific gravity 0.820. It is ignited without prior heating, and burns with a white flame, which is blue on the edges, with no soot or smoke, but produces acid of air.

It is clear that acid of vinegar unites the fine oil with water, because according to MANGOLD'S research, spiritus rectificatissimus extracted often with alkali tartari is decomposed, and finally the alkali itself is changed into Terra foliata tartari.

To test spiritus vini we have developed different methods, most of them uncertain. If a lot of gunpowder is put into spiritus it does not become damp enough to prevent it from ignition, which still often occurs when a smaller portion is put into a large amount of the same kind of spiritus, as was used for the previous test. If in a cylindrical box, which is held under running water until it reaches the same temperature, spiritus is burned on a layer of ice of a certain height, in a considerable proportion of the tests the water produced extinguishes the flame. Spiritus rectificatissimus does not produce 1/2 its volume of satisfactory brandy, doubled 1/4, etc. The aerometer [hydrometer?] is quickest, but if it is to give dependable results, correction must be made for different temperatures.

§ 211 (a). Spiritus vini rectificatissimus is miscible with water in all proportions.

Note 1. The volume is always diminished when this mixture is made, which Mr. NOLLET be- came aware of through ingenious research; the water attracts spiritus into itself. The least change occurs when 1 part by volume of spiritus is mixed with 2 [parts] of water; and the great- est [change of volume] when equal parts of each are taken; the decrease in this latter case is about 1/34 of both volumes together before mix- ing.

Ordinary brandy is about 65 parts per 100 Of water, and 35 [parts per 100] spiritus; but the French [brandy] has 44 of the former and 56 of the latter.

(b) [Spiritus] does not dissolve salt.

<u>Note 2.</u> Spiritus rectificatissimus, 288 parts, dissolves fully dried salts with boiling as follows: nitrum 4; nitrum quadr. 15; nitrum flammans 108; sal digestivus 5; sal ammoniac 24; calx nitrata 288; calx salita 288; argentum nitratum 84; cuprum nitratum 48; ferrum nitratum 4; mercurius subl. corr. 204; cuprum salitum 48; and ferrum salitum 36; all these according to Mr. MACQUER'S research which was restricted to 3 mineral acids combined with alkalis, calx, silver, quicksilver, copper and iron. In addition to the aforementioned, terra foliata tartari, plumbum nitratum, sal seditivus, sal succini, sugar, flores benzoin, etc., [are soluble in spiritus vini]. Any [salts] which contain acid of vitriol do not dissolve [in spiritus vini], probaby because it [acid of vitriol] combines more clearly and more rapidly with bases than the other [acids].

Green flame is obtained with sedative salt, cuprum nitratum, cuprum salitum, and a few others.

Spiritus often precipitates salts from saturated solutions by taking the water from them [the salts]. <u>Offa Helmontii</u> is such a precipitate with mild volatile alkali, but caustic [alkali] does not succeed. Mercurius nitratus is similarly precipitated, the yellow [precipitate] which is prepared by boiling, and the white which is prepared without heat.

(c) It [spiritus] does not dissolve gums, but [will dissolve] essential oils readily, and resins, and also sulfur and gold with special processing.

It is used therefore for extracting certain parts of plants, an operation which is called infusio sprituosa, and if water is used it is called infusio aquosa, in the same manner as tea is prepared. The French name the product of the first operation an essence, and when it is distilled, an eau. It is called a decoction when a plant is boiled in water; and an extract when the decoction is boiled down to form a syrup; the former is the same as an infusio aquosa, and spoils quickly.

Note 3. Tincture is the name generally given to transparent, thin extracts with some pleasant coloring [added]; they may be made with whatever liquids we please; Essentiae [essences] are somewhat thicker and less highly colored; and elixiria still thicker, dark, and hardly transparent.

Spiritus vini dissolves something from oak wood, which when mixed with vitriol gives a beautiful blue color; [this is] a test by which pure French brandy can be recognized, but to no purpose. [French brandy must have been stored in oaken casks; hence B.'s test. J.A.S.] The gummy and gelatinous precipitate dissolves in water.

(d) Very little heat is required for the extraction. In fact the lighter and more volatile the substance is, the more easily it is extracted, and vice versa. In this way are prepared pressed vegetable oils, and animal fats, except the latter are much lighter than water and require strong heat to bring them out. The reason is that they expand less with heat than water.

Note 4. Alkohol vini apparently begins to evaporate at 32 degrees of heat, and boils at 74.

(e) Spiritus has blood-coagulating and astringent power, heats and inebriates both violently and quickly, so that it also injures the body.

THIRD CHAPTER.

Concerning Ether

§ 212. Spiritus aethereus Frobenii is nothing other than oleum vini rectificatissimum and is not miscible with water as spiritus vini is; free from all spiritus, it is heavier than water, but floats on that [water] which has some alkali dissolved in it, is turned yellow by phlogiston and readily takes up such contamination, but can become entirely clear and colorless again through distillation.

Liquor mineralis Hoffmanni, or the so-called Hoffman's drops, consist of the above mentioned oleum vini and spiritus vini combined in such proportion that the mixture becomes miscible with water.

Note. Ether and oleum vini are two different substances which are confused by many authors [including, apparently, Scheffer], although they have unmistakeably different properties. Ether is like an intermediary substance between spiritus vini and its oil, since 1 part of it is soluble in 10 parts of water, and burns with a strong flame which produces little soot, and leaves hardly a trace of carbon; whereas spiritus mixes with water in all proportions, but its oil not at all; spiritus burns quietly without soot or residue, the oil with strong blazing, a whitish and smoky flame, and a carbon-like residue. Liquor mineralis anodynus Hoffmanni has its natural power from ether, (consisting of about equal amounts of spiritus rectificatissimus and ether, which dissolves a very small amount of

oleum vini), so I have thought that it should be treated in a separate chapter. What the author [Scheffer], and many with him, calls spiritus aeth. Frobeniii is a mixture of ether and oil, contaminated with acid of vitriol. The pure [ether] is so much more delicate. If it is dropped on a piece of sugar and put in a glass of water, a vapor is given off which becomes noticeable immediately by its odor and inflammability, which burns on the surface of water as long as it is present.

To Prepare Hoffman's Drops

§ 213 (a). Spiritus rectificatissimus is poured into a large retort and then a similar amount of oleum vitrioli concentratissimum is added, little by little, since much heat is produced from it. A receiver is cemented on immediatly with fresh ox-bladder, which best prevents escape of spiritus; this should be done with the greatest possible speed, otherwise a lot of the liquor mineralis will excape.

Note 1. We choose a retort such that about 3/4 of its globe can be empty. When in yield of both mostly ether is produced, very free of phlegm,, and more of the acid is held by the spiritus than otherwise, the heat may be increased. Now the mixture should be stirred as vigorously as possible and brought to mild boiling. Some of it reacts immediately, but with acid of vitriol hardly any ether is obtained without boiling.

(b) Then it is distilled over a low flame, and as soon as the retort has become warm, liquor mineralis passes over in long streams in the neck of the retort and receiver. When drops of moisture are observed among the streams, all the liquor mineralis has distilled over, so the receiver is changed immediately, and the distillate is stored in a well corked flask so that it does not evaporate.

Note 2. Liquor mineralis, which is prepared with 3 to 4, and often more, times more spiritus than acid, produces little other than spiritus vini, somewhat freed of phlegm and mixed with volatile acid of sulfur.

(c) As soon as a new receiver is attached the heat is increased a little, and then spiritus aethereus Frobenii distills over like oil drops in water. When white fumes arise and volatile acid of sulfur is observed by odor, the receiver is changed again. The distilled spiritus is separated in a separatory funnel from the water and stored tightly enclosed.

Note 3. If all that which passes over, before the white fumes begin to appear, is distilled in a retort with some alkali tartari dissolved in it, over mild heat until about half has passed over, we obtain an almost pure ether which is very volatile, ignites easily and burns, as described earlier (§ 212, Note); it has on standing a very agreeable odor, about like that which rennet has, but stronger. If it is rectified in the same manner one or more times, it becomes so volatile that a drop of it evaporates before it has fallen to the floor, and is ignited at a great distance by a candle flame; in this case it also completely dissolves Refina elastica [Hevea rubber?] which is obtained from America, and through its peculiar elasticity has attracted much notice. Its evaporation is always accompanied by cold, so that ice is conveniently produced with ether in the hottest summer; a glass tube with a globe is filled with water, wrapped with linen cord, dipped in ether and swung about to increase evaporation; the water freezes after several dippings, and indeed we can in this manner cool the quicksilver in a thermometer down to 40 degrees below zero. Ether exhibits a special solvent power for all fats. Its specific gravity is about 0.732. Its taste is hot and harsh at first, but after some time becomes more agreeable, something like acid of citron. It was named ether by FROBENIUS in 1730, is also called Spiritus aethereus Frobenii, Naphtha vitrioli, etc.

(d) After the last receiver is changed we get nothing but heavy fumes of sulfur.

Note 4. This result shows that the author [Scheffer] did not make this change hastily, since when a real oil usually passes over, which freed of acid and ether is volatile enough by itself, it resembles ether neither in odor, lightness, or power. It is called with reason oleum vini, and other names such as oleum dulce, oleum dulce vitrioli, etc.

(e) Caput mortuum is a black earth of oleum vitrioli and a fatty material from oleum vini. If it is distilled again we obtain a phlegm and very little acid of vitriol mixed with water. The water comes from decomposed spiritus, the rest from oleum vitrioli, producing with phlogiston itself caput mortuum, which is a proper sulfur, which burns with a blue flame and leaves a red earth behind, not unlike terra tartari.

Note 5. The residue seems to be oleum vini refined by acid of vitriol, which produces on calcination both lime and flint.

(f) The cause of the entire reaction is the concentrated acid of vitriol, which during distillation attracts the water from the oil in spiritus and destroys it. During this separation everything is completed; both the unchanged spiritus and the oil distill together; and liquor mineralis collects similarly. After this comes oleum vini and the phlegm, which formerly made up part of the spiritus, but now are separated from each other, which clearly shows that some third substance is needed to unite them in spiritus; and this would seem to be some analog of alkali. During distillation some acid passes over, both in the liquor mineralis and the oleum vini, which is easily detected with litmus juice.

Note 6. Whether ether is generated during the reaction by spiritus, which by degrees loses more and more of its own best combined water, or whether it ought to be regarded as oleum vini attacked and changed by acid of vitriol, which [oleum vini] was originally in the spiritus, has not yet been completely worked out. At least it is known that if the first proposal is right the reaction must not only consist of a separation of water without any phlogiston also being added, since ether nitrosus betrays itself somewhat in its taste and color, even though carefully purified; ether acetosus by its odor; etc.

(g) Liquor mineralis, when obtained pure, which seldom occurs, is very useful in medicine. The small amount of acid of vitriol which it contains causes no harm unless used too quickly, because the drops are heated as much as desired, and tempered by the cooling power of the acid. Acidum vitrioli is used for cooling a fever and to go into Mixtura simplex, which is prepared from spiritus tartari, acidum vitrioli, and spiritus teriacalis.

Spiritus aeth. Frobenii can be transformed into a liquid if it is diluted with a little spiritus, and in fact it has also lately been used by itself.

Note 7. We have traced ether to BASILIUS VALENTINUS and still older authors; but VALERIUS CORDUS who died in 1544, first described the preparation clearly. BOHNIUS, BOYLE, and NEWTON, with a few others prepared it also, but FROBENIUS particularly aroused the interest of chemists in this material, which we still do not know fully, nor can we bring it to complete purity.

Dr. NAVIER in 1742 discovered how to prepare it with acidum nitri; Marquis DE COURTENANYAUX with acidum salis, but in the form of Spiritus fumans Libavii, in 1762; and Count DE LAURAGUAIS with vinegar, distilled from verdigris, in 1759.

The so-called Liquor mineralis came first into use in about 1710, and was prepared by a pharmacist in Halle by the name of MARTMEYER, who called it Panacea vitrioli. He consulted at length with F. HOFFMAN concerning its improvement, which he carried out, and called this preparation Liquor mineralis anodynus. Aqua Robelii is also a mixture of spiritus vini and acid of vitriol, but which has not been subjected to distillation.

§ 214 (a). If a gold solution is poured into spiritus aethereus, it [ether] attracts most of the gold and becomes dark yellow or brown, whereupon the aqua regia becomes quite pale, like aqua fortis. Then the mixture is diluted with water, and the aqua regia separates better from the oil, by which means the latter sinks to the bottom and can be drawn off by itself. The water attracts the excess gold like aqua regia and becomes somewhat yellow.

Note 1. When ether is poured into gold solution it stays the same and attracts the gold gradually without precipitating it; through extraction it gets stronger and the ether floats up; as soon as the whole becomes quiet, the gold stays where it is and does not precipitate. With platina solution something similar occurs, except that the ether does not attract as much of the metal, and little by little it precipitates. This reaction must be carried out in a tightly closed vessel to prevent volatilizing [of the ether].

(b) The oil, containing the gold, is mixed with finely ground [powdered] sugar and a little cinnabar, and produces a powdered gold of a bright red color, which has a very astringent taste and sticks to the tongue. It is dissolved by deliquescence and consequently can be used only in liquid form, without addition of sugar and cinnabar. It is a good antispasmodic.

(c) Essential oils dissolve gold out of hepar sulfuris, in particular that made with alk. minerale, when the phlogiston sulfuris is calcinated away, after which the salt becomes most insoluble; also the gold dissolves when spiritus Frobenii takes it out of aqua regia.

Note 2. As essential oils also attract gold out of aqua regia, and then it is dissolved in spiritus vini with the color retained, we have fancied that this method produces proper tincture of gold; but Mr. LEWIS notes that the gold separates without changing the color of the tincture, because it arises from destruction of the acid by the oil.

FOURTH CHAPTER

Concerning Oils and Fats.

§ 215. Oils are thick fluids which in summer at least burn in fire. All of nature's kingdoms have some oily substances or fats, but particularly the vegetable and animal kingdoms. [Scheffer was probably unaware of mineral petroleum.]

(a) No plant or tree is found which does not contain oil, and consequently can burn. They differ in many ways. Pine, fir, and turpentine trees have such an excess of this [oil] that it is exuded and congeals to a gum. Mastic is nothing else than fir- or pine-gum, which falls on an ant hill and is attacked by the acid which they have [formic acid]. In this category belong native balsams, which are usually obtained from India and warm countries, including Copayba [copra?] from Gilead and Mecca, and other balsams from Peru, Tolu, Palmae, etc. Camphora is a strong vegetable fatty substance which catches fire readily and therefore is much used in fireworks. It is sublimed without leaving residue, and indeed must be preserved well enclosed, since otherwise it vaporizes after some time and disappears completely. It makes an excellent cordial. One kind is believed to be produced from the cinnamon tree or its roots; another is exuded from a large tree in China and India.

Note 1. Camphor is obtained from a tree called Laurus camphora, from which it is separated in Japan by means of hot water, and is then allowed to harden in straw. To remove dirt and straw, it is further refined in Holland and England, and indeed a similar installation is now set up in Stockholm. The process consists of sublimation in a special vessel, in the presence of chalk or lime slaked in air. One kind of essential oils, such as Rosmarin, Lager, Thymus, etc., should in time give off some camphor, and similarly water vaporized many times over menthol [mentha], threaded or single ply [piperitus eller enträd]. Camphor resembles essential oils, is completely and wholly volatile, has a strong odor but no sharp taste. Vegetable juices which flow out naturally of their own accord and turn hard are in general of four kinds. One kind dissolves completely and entirely in water, and is called a gum [Gummata]; another is not attacked by water, but dissolves completely in spiritus vini, and gets the name of Resin or rosin, and seems to be an essential oil hardened by acid; still another kind is partly soluble in water and the remainder in spiritus [vini], and is mixed and called, after the excess of Resin, Resinae gummata or gummata resinosa; finally there is a kind which is never attacked in any way by water or spiritus, which appears to be a dried fatty oil which has hardened, and I have at least one sample of hempseed oil which has become half solidified.

(b) All animals have tallow or lard, and their flesh and blood also contains fat.

<u>Spermaceti</u> or valrat is nothing else than oil from the brain of a whale, melted free like tallow.

<u>Note.</u> Tallow, lard, marrow, butter fat and egg fat, are all fats obtained from the animal

kingdom, but similar in their nature to fatty vegetable oils; they yield on distillation burned-smelling oil and acid, but never the least bit of alkali volatile; their nature is somewhat acidic; spiritus vini can dissolve but little of them, if they have been burned or hardened; when they harden they can be separated from all spoiled material by spiritus vini.

§ 216. In regard to the way in which oils are ob- tained, it is usually either by (a) mechanical compression; or (b) by rendering; each kind has its own odor and taste distinguishable from that of another oil, especially olive oil or linseed oil; or (c) by distillation. Thus we obtain pressed oils, rendered oils, and distilled oils [olea expressa, olea cocta, and olea destillata]. The latter are of two kinds: essential [oils], which are distilled with water or another liquid; or empyreumatic [oils], which are distilled by themselves by suitable heat.

Note. Oils are classified according to their nature as: (a) essential, which dissolve in spiritus vini and are volatilized at 100 degrees of heat; and (b) fatty (unguinosa), which do not dissolve in spiritus, nor are they volatilized by heating to the temperature of boiling water, but can still by repeated distillations at higher temperatures eventually obtain both properties [solubility and volatility], and are called empyreumatic [vidbrände]. The first kind [essential] have approximately the same odor, which distinguishes them [from other oils], and as a consequence of this quality they also have a distinguishing taste; the second kind [empyreumatic] are bereft of either taste or odor, but when we add a third kind they obtain both properties through burning, and lose more and more of their repulsiveness through many rectifications. The essential oils contain a modified acid, which brings about their solubility in spiritus vini, and which can be separated by repeated distillation, whereupon, for the same reason mentioned, their solubility is diminished; the fatty [oils] reveal no acid, but decompose under distillation and thus become soluble.

Of Essential Oils

§ 217. Essential oils are thin, clear, and have the same odor, taste, and reaction as vegetable [oils] from which they are distilled. Some of them can be produced with no addition, as for example oleum terebinthinae, but most of them are distilled in the way shown in the following example (§ 218). Those which are distilled from materials grown in the tropics are so dense that they sink in water. Essential oils are still volatile after they lose their strength with time, but do not then deserve the name "spiritus", because they cannot be mixed with water.

<u>Note.</u> Essential oils contain an acid which decomposes more and more with time and exhibits much acridity. In the same way the most volatile part usually dissipates and what remains is of decreased odor, volatility, etc. If such an aged oil is distilled, with only the heat of boiling water, that small portion which passes over is still unchanged, and the remainder must be driven over with stronger heat and becomes empyreumatic

[vidbränd]. Through repeated distillation this empyreumatic [oil] can be removed and the oil regains all its former good properties, such as fineness, volatility, and ability to pass over at boiling temperature, but not its odor. This deficiency cannot be helped in any way other than by extraction of fresh material, from which it is drawn out. The delicate substance, which usually causes the odor of vegetables, was called Spiritus Rector by BOERHAAVE; it is infinitely delicate, volatile, miscible with water, oil, and spiritus vini, and also inflammable, which shows itself when we approach Dictamnus albus [tinder?] at night with a [lighted?] taper. This subtlest of oils is likely to be the vital element in essential oils, which [vitality] is lost when this goes away. Vegetables which have a strong odor, without yielding any [considerable] proportion of oil, contain spiritus rector almost alone, and Lifjor, Jasmine, Rosae, rubrae, etc., are examples.

To Distill Cinnamon Oil

§ 218 (a). Cinnamon is put into a retort, which is three-quarters filled with water, some common salt is added, and the whole is allowed to digest for 12 to 14 days. Heat helps the water to dissolve the oil from the earthy material, and the salt serves double duty, since it partly prevents rotting during digestion, and partly it helps the oil to separate.

Note 1. During digestion much of the spiritus rector is dissipated and so less of this oil is obtained than on distillation (§ 217, Note). Therefore this is not the best method, and experience shows that we get more oil by other methods. Soaking hard substances in cold water cannot hurt them if they are not left too long. If common salt is added, the water boils higher than 100 degrees, and so more extract can be obtained in this way, but poorer oil, since when the specific gravity is much greater it cannot be so harmful unless much more is used.

No more water than necessary should be used, since it always takes away much of the spiritus rector to the detriment of the oil.

(b) A receiver is then sealed on and distillation carried out over low heat until no more water is left than will produce a mist in the retort. If the operation continues longer the oil becomes empyreumatic.

(c) That [oil] which passes over is separated from water in a separatory funnel.

Note 2. The water is very fragrant from spiritus rector, which at first, notwithstanding all speed, is partly separated even before the boiling point is reached, so it is entirely possible for the water to be used again for a new distillation of material, but at present it is used in a variety of other ways. The murkiness it has comes from its fine division and the half-dissolved oil in the spiritus rector, which is still collected after this when it clarifies again.

(d) Cinnamon gives very little oil and moreover is used in Holland in distilling from it one part, yet without salt, which would spoil the taste, and this is mixed with good cinnamon and sold.

That which has almost no taste has undergone such deterioration.

Note 3. SLARE has observed a salt, half crystallized, in 20-year-old oil of cinnamon; GAUBIUS has found something similar in Pomeranian oil.

(e) Cinnamon oil is treated and adulterated by addition of linseed oil, oil of turpentine, and other oils which are lacking in distinct taste themselves. Fresh linseed oil is harmless, although it deteriorates, but turpentine is not particularly wholesome.

Note 4. Essential oils are subject to three kinds of adulteration in particular, namely with some fatty oil, with a cheaper essential oil, and with spiritus vini. The first is de- tected when a drop is put in spiritus vini and does not dissolve completely; if some of it is rubbed on paper, and well warmed over flame, it still leaves a grease spot, though a portion is distilled, etc.; the second kind of deception is more difficult to detect, but since commonly turpentine oil is used for such adulteration, it can hardly be concealed if a rag is dipped into the mixture and allowed to dry for an hour in warm free air, when the genuine oil vaporizes first and what remains reveals itself by its odor; the third kind becomes apparent if it is added to warm water, which attracts the spiritus and precipitates the finely dispersed oil and separates it from the mixture. Moreover essential oils are differentiated by color: some are blue, such as oleum chamomillae, millifolii, sesli montani, etc.; some are yellow, such as oleum Majoranae, Saturejae, Thymi, rutae, menthae , etc.; others are as clear as water, which color is still with

the older varieties. Most of them are volatile, but partly congealed by cold, such as oleum anisi, carvi, foeniculi, macis, etc., whereas others are perpetually solid, when not melted, such as oleum baccharum lauri, fol. laurocerasi, rad. enulae, anisi, rosarum, etc.

To Distill Oil Of Turpentine.

§ 219 (a). Turpentine is a resin from a tree which is called Terebinthus in Latin, and comes mostly from Mediterranean regions. Two kinds are found: the so-called common turpentine is viscous like mush, and yields little oil of spike; but the Venetian is thinner and yields more.

(b) Turpentine is put into a retort with 6-8 times as much water, a receiver is cemented on, and the mixture is distilled gently, so as not to boil over, until what remains begins to thicken, after which it is cooled and the oil separated from the water in a separatory funnel; it is pure, fine, somewhat murky the first day, but becomes quite clear afterwards, and is called oleum tereb. aethereum or Spike oil. The residue is colophonium, a sticky oil, which is left at the end and may not be dry.

This [turpentine] is very good for removing stains, especially grease stains, both from silk stuffs and cloth. A piece of the same kind [of cloth] (other kinds give new stains) is dipped in the oil [of turpentine] and rubbed over the stain until it disappears. Common [undistilled] turpentine oil is too thick to completely sink in without leaving a stain itself afterwards.

(c) Common oil of turpentine, which is called <u>oleum terebinthae</u>, is distilled itself until the residue becomes quite dry; then it is cooled, but when it is warmed again it smells like resin. This oil is a proper essential oil, but not as fine as the former [oleum tereb. aethereum, § 219 (b), above], because something viscous comes over during distillation, and although the oil seems quite clear, still after a while it betrays itself when it begins to thicken and turn yellow in free air, and even in a small flask or large vessel with a little in it and sealed with an ordinary cork. Distillation is not continued longer than to dry the residue, otherwise the oil gets spoiled by burning. If the operation is carried out in a retort with a cap, the oil becomes purer. Colophonium, which we get here, is solidified.

(d) Oil of spike and turpentine burn strongly without a wick, and are known to be crude and tar-like whereas the other (b) is fatty.

When oleum tereb. per se is distilled the first distillate has an acidic, urine-stone, and sudorific sourness. Then follows an oil which heats, heals wounds, and is good for congealing blood. The residue, or colophonium, when pulverized, is good for treating bone aches.

<u>Note.</u> The proper oil of spike is distilled from flores Lavandulae spicae in Provence and Languedoc, where this plant grows in abundance,

but that which is sold commercially is commonly only oil of turpentine furnished with a little proper oil of spike.

The turpentine tree is called Pistacia terebinthus in Latin, and grows around the Mediterranean Sea. Spermaceti is dissolved with gentle digestion in oil of turpentine, and then is placed in a glistening ice bath as cold as possible for some minutes; copal [copra?] is also dissolved in oil of turpentine with boiling over a water bath to get a beautiful yellow-gold product.

To Distill Essential Oil Per Descensum

§ 220 (a). In a large glass [vessel] place a piece of linen, which droops down 1/2 to 1 inch [inside the vessel], fasten it [linen] tightly around the glass with wood, which is [then] wrapped tightly around with iron wire at the brim, so that the piece [of linen] has a flat bottom on which are laid cloves or other spices [filling up] even with the upper edge. On this are laid 2 or 3 sheets of paper, and then a thin sheet of iron on which sand is poured to a depth of 2 inches.

(b) When this is ready some small hot coals are put on the sand, by which heat is maintained, increased or decreased as necessary. When the sheet [of iron] gets warm, water is driven out first and then comes the oil; to prevent overheating the sheet is not allowed to get hotter than one can stand [to feel], and when no more [oil] comes off at this heat, heating is stopped.

(c) In this way the oil from the spices can be driven out without burning. Clove-oil seems thick and fatty, does not smell bad, but is potent enough.

Note. This procedure is useful when we need a little bit of pure essential oil, and do not have access to a Laboratory.

Small amounts of essential oils can also be obtained by simply pressing them out, as from citron peel; this kind is better and less likely to spoil than that prepared with the help of heat.

§ 221. Uses of essential oils:

(a) Elaeosaccharum. Three parts candied sugar and one part essential oil of cinnamon, moschati [musk?], or other oil, are ground together well in a glass mortar, and preserved in a flask which is well sealed, until it needs stirring, when instead it is put into rectified wine. This medicament is in convenient form and does a good job.

Note 1. This refers to the so-called Huil de girofle [oil of cloves], which is beginning to be imported here and is in great demand; it is nothing else but elaeosaccharum of oil of cloves dissolved in a certain amount of water; an agreeable but very warming liqueur.

(b) [Essential oil] combined with alkali volatile in distillation produces sales volatiles oleosi, which is mixed with wine and drunk.

(c) Essential oils which feel greasy can be used to test the strength of brandy, since the stronger it [the brandy] is the better it dissolves [the oil].

(d) Some essential oil is used by Nurembergers together with three other substances to represent the 4 elements [earth, air, fire, and water] in the following way: powdered blood-stone is put in a flask to signify earth; on it is poured deliquescent alk. veg. to signify water; then spiritus vini rectificatissimus to denote air; and over all, oleum ligni sassafras [oil of sassafras wood], which is red, to signify fire. These substances have specific gravities such that they can be decanted in the same order as they are added, because they are immiscible with each other.

Note 2. A more likely method for this representation could be one where the lowest substance [earth] is black, and for this we could use ground slate [skifermjol]; the second, or water, is signified well by clear and colorless lye [solution]; the third, or air, could be light blue, and for this we could use oleum terebinthae colored by copper; the fourth, or fire, can be represented by spiritus vini rectificatissimus colored red with radix alkannae or something similar. Oleum ligni sassafras does not float on spiritus vini unless it is mixed with oil of cinnamon or cloves; and other spices from India sink in water itself; most of them float on water, but very few on spiritus vini, such as oleum florum Lavandulae tenuifoliae; wormwood oil suspends in water usually, which indicates that it has about the same density.

FOURTH SECTION, CHAPTER IV, Oils and Fats

Concerning Fatty Oils

§ 222. To press out oils from fruits and seeds is more a mechanical than a chemical operation. This is how olive oil, almond oil, etc., are produced. We often use small mills or stamping machines to crush the material out. This is so with linseed and turnip seed. They dissolve sulfur (§ 202), and amber also, from which is obtained a very glutinous and sticky product, especially if calx of lead or minium is added, good among other things for covering glass which must re- sist high temperatures.

Note. Fatty oils are produced partly by pressing of crushed material encased in cloth between warm metal plates, whereupon the mass is usually either sprinkled with seething hot water, or exposed to steam for the same effect; or partly by boiling in water, when the oil is found at the surface after cooling. In general the oil is poorer and more likely to spoil the more heat is used in its preparation. Of these oils, part of them are always congealed unless melted by heat, such as oil of cacao, Lagerbär [barberry], etc. Some dry up very strongly and are therefore called siccative, such as linseed [oil], oil of hemp, filbert [oil], etc., whereas others cannot be brought to do this, such as olive oil, almond oil, etc. The former [drying oils] are therefore naturally used for paints, and the latter [oils] for greasing, annointing, rubbing on, etc. They dissolve different fatty substances, among them amber, which product can be prepared as follows: amber powder wet with linseed oil is burned black in a clay vessel, then after cooling it is ground

fine and spread with ordinary linseed oil, then cooled until it becomes as soluble as can be; finally it is mixed with some oil of turpentine.

To Make Soap

§ 223 (a). Two lod of alkali fixum and 3 lod of burned calx [unslaked lime] are put in an iron pan with 1 stop of water [1 stop = 1/2 kanna = 1.3 liters], which is then evaporated one-half over mild heat, then removed, allowed to stand until it settles thickly to the bottom, when the clear [liquid] is carefully decanted. It should then be well filtered, but because of its causticity it should not remain on the filter. One stop of water is again poured on the residue in the pan, boiled as before, and filtered after the coarsest part settles.

In the soap boiler with the alkali is put lye from wood ashes, which has not yet been boiled.

Note 1. We can also filter the lye first, if doubled heavy paper is laid on a sieve, since then nothing heavy can force a hole in the weakened paper.

(b) Then both lye solutions are boiled together and into this is poured 6 lod of olive oil, turnip oil, or hemp oil, whereupon the mixture gets very milky, and care must be taken not to overcook it, although strong heat is very necessary. The boiling is continued with heavy stirring with a knife, so that nothing sticks or burns, until it becomes like thick gruel or pudding.

(c) When it has thickened, a little litmus is added which makes the yellowish soap green through its blueness.

(d) When the soap is reasonably thick it is removed from the fire to cool. The longer it stands the harder it gets, and better in winter time than in summer time. It still does not get as hard as it does when 4 lod of white wax is used instead of oil.

Note 2. I have made such a soap from white wax; it weighs about three times more than the wax, becomes solid and gives off an agreeable odor, almost like almonds.

(e) Good soap should not, after dissolving in water, feel greasy to the hands; it raises bubbles when beaten; it goes thickly through a filter and does not show any traces of the [original] oil in the water.

To Make Tvål [Hard Soap] From [Soft] Soap

§ 224 (a). Soap is dissolved in a little water and boiled for one hour, after which sour wine or vinegar is poured in until we note that it turns [coagulates?] and the hard soap [tvål] gets much thicker than the rest, which is as thin as water. If we take out some of this crust with a knife we can see that it is quite different from the thin [watery liquid], so then the whole pot is removed from the fire to harden.

Note 1. Generally when hard soap [tvål] is made directly using lye, it is brought to sufficient strength to float a fresh egg, and then divided; one half is diluted with water until the egg sinks, and this is gently boiled and mixed with a similar amount of oil, stirred until a change appears, and then three times as much strong lye is added, which balances the oil, after which it is stirred about and boiled until it reaches the proper consistency. Some handfulls of common salt are added, enough to cause coagulation, so that the tvål may better separate itself from the rest of the lye. Mixing in of dissolved vitriol of iron makes Venetian stained [soap].

(b) If to the soap from white wax we add Bourgogne wine, we get an excellent tvål, which makes the hands white and soft.

(c) The reason for using this procedure is that the hard soap [tvål] contains less alkali than [ordinary] soap. Then when the acid begins to react with this excess, most of the remainder becomes tvål, which is harder. We should still take care that not too much acid is present, which would remove all the alkali and no tvål would be produced.

Note 2. A fatty substance made soluble in water and spiritus vini by means of some salt [alkali] is generally called soap or tvål. Salts which react with fats are properly either alkalis or acids [Remember that B. considered acids and alkalis as "salts", in addition to the so-called "neutral" salts]; the former are the best known and are produced with skill very easily; but the latter, in which acid is used, are still prepared almost only by nature, or if desired, by the Acad. Sci. in Dijon, who for their established prize award, found the means to prepare it.

Fatty oils and animal fats are used with vegetable or mineral alkali for all the usual tvål and soap; but alkali volatile causticum seems to combine just as easily; because I have ground together with it unsalted butter, and after a few minutes obtained a white mush, which for the most part was soluble in spiritus vini. Similarly olive oil. Essential oils do not appear to have any propensity to combine with alkali fixum. Starkey's Soap is nothing other than a durable combination of oil of turpentine and lye-salt; it is hardly more than a mechanical mixture. Other essential oils seem to behave similarly. Burnt oils are combined with alkalia fixa caustica; at least we know this for certain, although diversely, that they dissolve in alkali vol. causticum sparingly and transitorily, since exposure brings in acid of air and thereby destroys the dissolved compound; but if a little tvål is dissolved in spiritus vini, which is then saturated with oil without heating, and this solution is poured into 4 times as much caustic alk. vol., it becomes milky immediately and produces a kind of liquid soap. If oleum succini is used, the preparation gets the name of Eau de Luce. Oleum animale appears through long stirring with alk. fixum causticum to give something which resembles soap. GEOFFROY suggests making medicinal tvål for internal use without boiling. Lye is prepared from two parts of the best soda, 1 part unslaked lime, and 12 times as much water, which is boiled together, filtered and boiled until its spec. grav., compared to water, is as 11 is to 8. One part of this lye is mixed with 2 parts of good olive oil, allowed to stand at mod- erate heat and stirred daily, and in 7-8 days we obtain a solid tvål.

For fine kinds of tvål we use almond oil or olive oil, refined potash, or best soda; it is customary also for them to be colored and perfumed in various ways.

For coarse soap, train oil, tall [pine] oil, turnip seed oil, and similar fats are used, and likewise ashes [for the potash].

A tvål should not grow moist in air or be sharply alkaline, dissolves in both water and spiritus vini with no oil drops floating up. If tvål solution in spiritus vini is poured into water, a precipitation and decomposition takes place, since a large part of the precipitate is not further soluble in water. Tvål is decomposed by acid and fatty material separates from the alkali, which then combines with the acid; the separated oil is then soluble in spiritus vini.

Lime water by itself, without [other] alkali, actually makes a kind of soap. I have put unsalted butter in a flask of lime water and let it stand well sealed for several days; the butter gets changed in appear- ance after a while and becomes partly soluble in spiritus vini. It is mentioned above (§ 168, Note 3) that milk is curdled by unslaked lime; that which is precipitated from the whey by acid is caseosum [casine?] with a little butteriness, which now is soluble in spiritus vini. The whey itself is an excellent soap which quickly makes the hands very white.

Note 2 [3]. Ignition of oils by acid of saltpeter is a singular phenomenon which BORRICHIUS first mentioned in 1671; he used oil of turpentine. They worked a long time in vain to repeat it. Finally in 1701 DE TOURNEFORT ignited sassafras oil, HOMBERG in 1701 thick oil of turpentine, ROUVIERE in 1706 oleum empyreum Guayaci;

HOFFMAN and GEOFFROI found that fuming acid of saltpeter mixed with strong acid of vitriol worked somewhat better than the former alone; and finally ROUELLE, the elder, announced in a 1747 publication that he had learned how to ignite all essential oils, and also those oils which became dry [drying oils], such as linseed oil, hempseed oil, and nut-oils; but alomond- and olive-oil, etc., which dry slowly, could not be ignited by acidum nitri alone, unless strong oil of vitriol were added, which probably dephlegmed both acidum nitri and the oils. When we wish to repeat this experiment, 1 or 2 lod of oil is poured into a teacup or large wine glass, a similar amount of acid is put in another glass to the base of which is fastened a long handle; from the latter [glass] about 1/2 or 2/3 is measured into the former [glass], whereupon heat and motion is aroused, but seldom burning; as soon as the last of the acid is added, which should not be delayed over 3 or 4 seconds, burning then breaks out with great vigor, splashing about wildly.

Camphor dissolves in acidum nitri clearly and without reaction; this solution is usually called oleum camphori, can be precipitated with water, and if the precipitate is well washed it is found to be pure unchanged camphor. One part concentrated acid of vitriol dissolves 8 parts camphor with digestion.

Note 3 [4]. Lead, metallic and calcinated, dissolves in fatty oils to form a plaster; also regulus of arsenic.

Concerning Burned Oils

§ 225. Burned oils are obtained when they are distilled per se, without water or other liquid, are brown and vis- cous, smell burned, and have neither the odor, taste, nor properties of the [oil] from which they are distilled. All are good to spread on reformar [puzzling word; could signify some kind of skin growth or infection, since Scheffer is often concerned with medicinal compounds] which are removed [fördrifvas] by this means.

To Distill Wax-Oil

§ 226 (a). If wax per se is distilled in a retort it becomes thick on passing over and hardens in air like butter.

(b) If distilled again in a vessel with an alembic and prepared sand to prevent the viscous part from going over, we get a volatile oil, which all becomes thinner the more often it is distilled before the oil rises, and acidic vapor comes over.

(c) This oil is softening, good for sores, cuts, etc.

§ 227. With the burned oils are classified also those which are obtained from amber, which were mentioned previously (§ 68 (b)); that which is driven out of staghorn (§ 47 (b)), and tartar (§ 66 (d)). From other single and mixed materials, when they are distilled without supplement, we get similar burned oils, also when they are brought down in strong heat (§ 220).

Note. *Oleum animale Dippelii* is a peculiar oil, which differs from lard, tall oil, and marrow, in that when found in certain places collected in the body, without entering into the composition of the flesh, fin, horn, hair, etc., dissolved by saline material to a kind of soapy consistency, known by the name of gelatin. When a gelatin or something similar is distilled we obtain oleum animale, very acrid, burned, unpleasant, and mixed with alkali volatile. By repeated recti- fication it can, after a while, be brought to a very great fineness, clarity, and volatility, so that it resembles ether in its properties, but appears to differ from other animal and vegetable oils by its volatile alkaline nature, whereas the others contain acid. The most common and best material from which to distill oleum animale is staghorn. We should, with each distillation, not drive over more than the most volatile [part], and perform each [distillation] in a new or carefully cleaned vessel, and so obtain in 3 or 4 distillations as fine an oleum animale as would otherwise require 50 or 60. It should be kept tightly closed in very small flasks, partly so that it does not evaporate, and partly so that it does not get dark and thick, both of which properties are brought about by exposure to air.

FIFTH SECTION: CONCERNING METALS.

FIRST CHAPTER: Concerning Noble Metals.

§ 228. Noble Metals are so-called because they cannot be calcinated per se, and when that [calcination] is brought about [indirectly] by additon [of other substances] they are easily reduced again without addition of phlogiston.

Note. Because certain metals resist water, air, and fire, without alteration, when others are more or less attacked and calcinated, a classification has been made between the former and the latter, under the names noble and ignoble. Noble [metals] can nevertheless lose a part of their phlogiston, seldom during melting, but readily when they are first dissolved in acids and then precipitated with alkalis. That they are now readily calcinated is shown by their appearance, increased weight, their behavior with acids which is quite different from before, like that from combination with glassy materials which are colored by it [the metal]. But it does not require more phlogiston for reduction than that which it can get from a glowing container.

Another useful classification of metals is that into whole and semi-netals; the former are malleable to a certain extent under a hammer; examples are gold, platina, silver, quicksilver, lead, copper, iron, and tin; but the latter shatter under the hammer; zinc is still an intermediate [between semi-metals and whole metals], since it is quite brittle, but not so [brittle] that it is easily pulverized.

FIFTH SECTION, CHAPTER I, Noble Metals

Metals are the densest bodies in nature; they are about 6 times heavier than water, but the heaviest exceed this by about 19-1/2 times. All have their own particular luster and are opaque to the highest degree.

Concerning Gold

§ 229. Gold has the following properties:

I. It is the heaviest of all metals, even [heavier than] quicksilver and lead, as 19636 is to 14019 and 11345.

Note 1. The specific gravity of gold is reported variously, which can be attributed to the differing accuracy of instruments, different purity of the gold, different temperatures of the water, etc., as Dr. LEWIS very carefully reports. From even the more trustworthy accounts the least is still 19,300 and the greatest 19,640. Notwithstanding such considerable density, NEWTON believed that, of the volume of gold, at least 1/2 was void or pores because it did not interfere with the magnetic flux, and gold leaf held against the sun let light through. If this be so, we can get some idea of the magnitude of the porosity of cork, feathers, etc.

II. It is the most malleable and most ductile [of metals]. With 2 lod [1 lod = 1 oz., approx.] of gold we can gild a cylinder of silver which weighs 96 lod, and when this cylinder is spun out into such a fine filament that 2 yards of it weigh no more than 1 grain or 1/240 lod, so then 98 yards are overgilded with 1 grain of gold, and the best microscope cannot detect the least showing of silver bar [underneath the gilding]. 98 yards contain 2352 verktum [1 verktum = 1-1/2 inches], each of which can be divided in 300 without a microscopic trace showing. Consequently for complete coverage to the eye, 1/705,600 grain or 1/169,344,000 lod of gold is sufficient. [If 1/705,600 grain = 1/169,344,000 lod, then 1 lod = 240.14 grain. J.A.S.]

Note 2. Goldsmithing shows still further the remarkable extensibility of this most prec- ious of metals. When a piece of the above-men- tioned filament is dissolved in very dilute nit- ric acid, we obtain the gilding itself in the form of an incomparably thin cylindrical shell. 10 [Swedish] acres or 140,000 square yards can be covered by 2 lod [of gold].

When gold is hammered quite a bit it becomes hard and cracks after an hour under continuous hammering, but if it is heated in well glowing coals it becomes malleable again. All whole metals, except quicksilver, lead, and tin, behave in a similar manner, but require fusion in [hot] coals; or with precious [metals] it is sufficient to heat to glowing in a muffle [furnace].

There is another property in close connection with malleability, namely toughness or tenacity. It is valued next to density, when a filament of known diameter is wanted without breaking. MUSSCHENBROEK has found that a gold filament of 1/10 Rhenish tums diameter requires 500 pounds. [tensile strength?]

III. It is flexible and without clang [bell-like quality]

Note 3. Gold is harder than tin, but more flexible than silver, and should therefore be mixed with another metal at manufacture to provide greater wearability.

IV. Next to iron it is the most difficult to fuse and most resistant to fire of all metals. Gold melts when a good burning lens is used, and gives off fumes which leave a red stain on a fragment of crucible or cover, but a green glass on a coal, which comes from coal ash.

Note 4. If a pyrometric observation can be made with surety of the degree on a thermometer, then gold requires a heat of about 705 degrees for melting. While it is fused the surface has a blue-green color. KUNCKEL has kept gold fused almost one month, and BOYLE two [months], without noticeable alteration. HOMBERG reported in 1702 to have vitrified gold with a burning lens, but its purity was not determined beforehand, and with the same or stronger lens he has then been able to investigate it.

Well fused gold, which was cooled slowly, shows facets like leaves on the surface.

V. [Gold] does not blacken when roasted, like silver and other metals, and cannot be calcinated by itself, but only with reagents; [the calx] can be reduced by fusion without addition of phlogiston.

Note 5. Strong electric shocks bring gold leaf to a dark purple-colored powder, which produces a ruby-colored glass.

VI. [Gold] is dissolved only by aqua regia. It is then precipitated with alk. fix. and the solution becomes green, indicating the presence of substances which produce such a color with acidum salis.

Note 6. See § 80, Note; § 143, Note.

The properties enumerated, together with the intense color have given gold first place among metals. Its value compared to silver in Europe is now 14-1/2, highest 16 to 1, but in China only 9 or 10 to 1.

The beautiful color of pure gold does not change; when surface dirt attaches to it, it is washed off best with spiritus vini.

§ 230. Gold alloys with all whole metals and semi-metals in fusion, but nickel is still to be investigated.

Quicksilver has a greater attraction for gold than some other metals, and therefore amalgamation occurs most readily [with gold]. If too much quicksilver is used, so that the amalgam gets too thin, the excess can be passed through linen, or better yet through chamois.

Spaniards produce all their gold in Peru by quicksilver, because lack of wood prevents the use of smelting [of ore].

When a girdle maker [Gördelmakare] wants to gild brass or other metals, whose surface tarnishes in air, he first amalgamates the surface; that is with the help of strong nitric acid it [the surface] is amalgamated or covered with quicksilver, since otherwise it will not fix gold amalgam.

Note. Nickel combines well with gold in smelting, but makes it much whiter after this and more brittle.

The combining of gold with quicksilver is improved by heat. If the cold amalgam remains fluid, a part of the quicksilver can be removed from it through a chamois, after which the remainder becomes a white buttery mass, which contains hardly more than 1 part quicksilver to 2 parts gold, hardens in cold and softens in heat, and is suitable for making impressions. In hurried heating this amalgam swells and effervesces, but gentle heat drives away the quicksilver and a very fine powdered gold is left behind. When 6 parts quicksilver dissolves 1 part gold it forms crystals. Amalgamated gold feels hot to the touch.

When silver, copper, or brass is to be gilded, the vessel is first scoured; heated; amalgamated; and covered over with amalgam of gold; the quicksilver is evaporated over fire; it is lightened or made shiney with hot wax, mixed together with saltpeter, sal ammoniac, iron vitriol, and verdigris, mixed into a powder, moistened, spread on, held over fire until it fumes, and quenched in urine; finally it is held or placed in boiling water, in which 1 part tartar and 2 parts common salt are dis- solved. In the hot wax [mixture] it is customary for many useless things to be added, and also in the holding water. Iron can be gilded similarly if a copper plating is put on it first.

Removing the gilding on silver again is easily done with aqua regia, [or] with sal ammoniac worked into a dough with nitric acid, spread on, heated to fuming, dipped in water and scratched off.

Gilding on copper will come loose if borax solution is spread on to make sulfur adhere which is spread on, then heated to glowing and quenched in water.

Gilding can be scraped from filament, when the adhesive loosens in hot water; all which is removed is burned and then ground with quicksilver.

To Purify Gold By Cementation

§ 231. Before casting with antimony became well known, the ancient cementation process [for purifying gold] was used, which was so-called because of its resemblance to masonry, since the ingredients were piled on like bricks in a wall. The operation was set up as follows:

(a) Powdered brick, 1 part, is mixed with powder of sal ammoniac, sal gemmae, and sal marinus, the same amount of each, moistened with urine and placed in cement boxes, in- terlayered with laminated gold [sheets], so that the upper most [layer] and lowest [layer] are both of cement. Finally a lid is put on and smeared with clay. If a little calcinated vitriol is added, acid of salt is set free more readily.

(b) The cement box is then put in dull glowing heat for 16-24 hours, during which time all the other adhering metals or impurities are calcinated so that, as brittle material, it can be broken off with a hammer and washed away. It appears then that the gold has not reacted; the powder described is called caementum regale; it has little or no use now because antimony and aqua regia serve this purpose better.

Note. Saltepeter or common salt with a similar amount of calcinated vitriol and double [the amount] of brick meal give a cement which reacts with all metals except gold and platina. The vitriol serves to set free the acid, and the brick meal to prevent fusion. The greatest care must be taken that no saltpeter or common salt are contaminated with each other, since then we get aqua regia.

To Cast [Purify] Gold By Antimonium Crudum

§ 232. The purpose of this operation is to separate foreign metals, silver and copper particularly, whereas gold remains undisturbed for the most part.

(a) For this we choose antimonium crudum [antimony sulfide, see § 232 (k)] with broad and straight radiating shafts, which contains less lead than that which consists of fine hooks. If you can get lead-free, it is the best.

(b) The crucible is made ready as follows: it is moistened, sprinkled over with borax powder and heated to glowing in a forge, so that the walls are glazed inside and reinforced, so that no antimonium can escape, which occurs readily otherwise.

(c) Into the glazed and glowing crucible is first put 3 parts antimonium crudum, and immediately thereafter 1 part gold, which should not be under 16 carat. Immediately thereafter it should be blasted with vigor on the bellows, so that smelting may occur so much more swiftly and less antimonium be fumed away; if it does not become fluid quickly enough, [the blasting] is doubled in vigor. We succeed readily when it is smelted promptly, since then the black layer is cast off, the mass becomes clear and out of it arises [something] very hard and glistening. The material should now be poured out into a casting mold which has been warmed and smeared with wax, to prevent sticking, then it is tapped gently with a light hammer or key, by which means the regulus is well gathered together.

(d) When the casting hardens it is turned out. The regulus consists of gold and the reguline part of antimony, which lies underneath and should be separated from the overlying black plasma, which contains a little gold, antimonium crudum, and the foreign metals which are now separated from the gold, sulfurated.

(e) The plasma is melted down in the same crucible, which immediately after pouring out is put into ashes, so the solid is kept glowing. When the material is fully melted it is poured out, as before, the regulus is separated, the impure plasma remelted, and this is repeated until no more regulus is obtained, and the plasma is then put aside.

(f) After this, all of the reguli are melted together except the first (d), with 3 parts antim. crudum in a newly glazed crucible; then the plasma is melted as before as long as it produces regulus, and finally it is added to that in (e). The first of these reguli is added to (d).

(g) The other reguli from (f) are melted together in a newly glazed crucible with 3 parts antim. crudum, and the plasma are melted together, as long as regulus is obtained, if we are working on some larger quantity of gold; but if it is pure, it does not need to be cast more than once, because the plasma contains very little gold and can therefore be added to that from before.

(h) Then all the reguli remaining from the first of each casting with fresh antimonium are melted together again for greater surety with 3 parts ant. crudum as before; the first which is now collected we can be quite sure contains only gold and regulus of antimony. The other is melted together with three times as much ant. crudum by weight as the first gold, it is cast as before, the first reguli preserved, and the rest melted with new antimonium, etc. But for the most part we now get in the first castings so little reguli that we must pay attention if we want to get any more.

(i) All impure reguli, now having been freed from all impurities, are scraped clean from the plasma which usually adheres to them. That which is scraped off is added to the rest, which are melted together without addition, and if we work on the pure quantity of gold, the regulus is so small that it can be remelted with a blow pipe; if it gets too large to be worth the trouble, the whole casting process should be gone through as with gold; it should not be mixed with the pure reguli (h).

(k) The basis of the entire process is the following. Ant. crudum consists of regulus of antimony and sulfur, and often lead. Now sulfur has attraction for all metals except gold, and among the others the least for regulus of antimony, and therefore this regulus has a great propensity to be united with gold. Consequently, during melting the sulfur goes together with the foreign contaminating metals, but the regulus of antimony combines with the gold. This does not occur completely in one melting, and therefore many castings must take place. The plasma consists of sulfur, copper, silver, or lead, and also some antimony and gold, because the foreign metals are not enough to saturate the sulfur and not leave some excess which attracts a part of the regulus of antimony, with which then follows a little gold.

Note. Few take as much trouble as our author [Scheffer], with repeated castings in particular, when 1/2 to 1 ducat is purified. We usually melt [gold] with twice as much ant. crudum, the plasma is remelted with 2 parts ant. crudum, and likewise the other plasmas. Then all the reguli are recast with double the amount of antimony crudum, and there we let it rest. If the gold is under 16 carat it should be mixed with a similar amount of sulfur in the first smelting. The crucible is often fitted inside another, to prevent the gold running out in case the inner one should break. Gold is purified by antimony crudum with the least expense, and also the most perfectly, so BAS. VALENTINUS was the first to report. Antimony was called by the ancients balneum regis or solis, and also solius solis.

To Blow Away Antimonium

§ 233. When the gold is free from all other metals, antimony must still be removed, which is easily done, because it fumes away in smelting. Great heat is needed for this, greater than can be reached in an assay oven; and [the heat] ought to be increased quickly when it is required, which can be obtained in the following manner.

(a) A muffle or large crucible is placed in ashes so that the blast plays on a closed end. Into this we put a splinter, and when it glows we put in the reguli (§ 232, (h)), or regulus if they have previously been melted together for convenience. They become fluid immediately and begin to smoke; to provide some air we blow very gently on the surface with a hand bellows, whereupon it will harden at last. A burning coal is then placed in the muffle mouth, which, when the material begins to move, is taken away again, and the hand bellows used as before. This kind of cooling and heating is continued until the burning stops by itself.

[(b) was missing here.] Now most of the antimony is gone, and the remainder requires fairly strong heat to drive it away; therefore a burning coal is put in the muffle mouth and the blasting is begun again with particular vigor. When the material melts and smokes the burning coal is removed, cooled with the hand bellows, but put in again, when it will harden again. We continue thus until few fumes are detected, when a little borax and saltpeter is put in, from which the gold becomes bright.

(c) When no fumes are noticed, although strongly blown with the hand bellows, and no dark spots float on the molten material or lie to one side without becoming very bright, the blowing is stopped, and when the gold solidifies it is dropped from a splinter into water, and gets from this a higher color.

(d) If we wish to be still more certain of the purity of the gold, it is melted on a splinter in a muffle furnace with a bellows, and with a burning coal in the door, and after this a little borax is laid on. If any antimony is present a black spot shows up which plays off and on on the surface, and in this case the blowing is continued until the spot becomes bright, when it is dropped as before into water. The gold is now certainly 24 carat.

(e) To test the plasma, it should be pounded up finely, put on a glowing splinter in the muffle used before, and the antimony blown away; the over blown mass then shows, [on examination] in a crucible test, either quivering, cupelling, or dissolving in acids until its make-up is determined. The above described procedure has as its purpose the recovery of that portion of gold that was contained in it.

<u>Note.</u> Those who find blasting off too difficult put 3 times as much pure and dry saltpeter in a wide crucible, fasten a crucible cover on tightly with a small hole in the bottom, place this in ashes and bring it cautiously to melting with a bellows. Niter attacks regulus of antimony, consumes its phlogiston through detonation, whereby it becomes unsuitable for combining with

gold without cupelling. If the gold is not quite malleable after the first treatment, it is melted in a similar manner, but usually only with double the amount of saltpeter.

By casting in the manner described, gold can be freed from all metals except platinum. Gold plated silver seldom relinquishes silver-free gold, except when it is dissolved and [silver] precipitated as silver chloride. In general it is very difficult to separate the last traces of impurities; the more they are diminished the more strongly they adhere [to the gold], and what is left is shielded.

To Test Gold For Its Fineness

§ 234. One pound [mark] of gold is classified into 24 carats which is therefore called its fineness; 24 carat [gold] is entirely pure, 21 carat contains 3/24 of its weight of another metal, etc. Now to determine the carat of a given sample of gold requires 3 operations:

(1) to reduce approximately the fineness or the amount of pure gold in a certain mass with 3-1/2 times as much silver, by which aqua fortis becomes able to separate most of the silver without precipitating the gold.

(2) To part the gold, or separate it from the silver.

(3) To find the content of the aqua fortis, which will tell us the amount of silver which the aqua fortis was able to take out, without leaving any trace in the gold.

Note. One mark of gold is equal to 4384 ass at the mint, but according to the goldsmith's system it contains 4424 [ass?], or 1/2 skålpund victualer's weight. One carat is divided into 12 grains [grän], and each grain in 4 parts; consequently 1 mark [=] 288 grains, and 1 grain = 16 äss. In France 1 carat is divided into 32 parts, and in England into 16. [According to Uno Boklund, in the old Swedish system of weights and measures, 1 ass (victualer's weight) was equal to 48.042 milligrams. Then if 4424 ass = 1/2 skålpund, this makes 1/2 skålpund = 212.54 g. 1/2 pound (advoirdupois) = 226.80 g. Then one skålpund is approximately equal to one pound advoirdupois, and one mark of gold about equal to one-half pound of gold. Yet he also says that 1 mark = 288 grains, and if 1 grain = 0.062 grams, in the old Swedish system, then 1 mark = 17.86 grams. Trying to make sense of these terms, as B. describes them, is difficult. For example, is 1 ass = 1 äss? J.A.S.] The first operation which is mentioned in the text [Scheffer's] is called quartering, because the gold usually makes up 1/4 [part] to [3/4 part] of the silver; we can still use the same name even though the silver should consist of somewhat more than 3 parts [to 1 part gold]. In France only 2 parts silver are taken, and Messers HELLOT, MACQUER, and TILLET, who by order of the [French] government made many investigations concerning testing of gold and silver, main- tain that if the smith's mixture [bläcket smides] is well thinned and properly parted there is not the least bit of silver left behind.

In Sweden only 3 kinds of gold are allowed to be worked. Ducat-gold should contain 23 carats 5 grains of gold, and 7 gr. silver to the mark; Crown-gold should contain 18 carats 4 gr. gold, and 5 carats 8 gr. silver; and Pistol- gold 20 carats 4 gr. gold and 3 carats 8 gr. copper. With the two latter kinds 2 gr. leeway is allowed because the goldsmith cannot demand as accurate a composition as they do at the mint. A mixture of equal amounts of gold and silver is called "cheap custom" [billon plägar]. The Chinese never alloy gold with anything but silver, and classify it with 100 points, so that gold of 100 points corresponds to 24 carat gold. In France gold should be 20 carats, which is worked into trinkets, as long as the piece being worked is not over 1 mark, and no leeway; but larger works must be 22 carats with 1/4 carat leeway. Louis d'ors gold is 22 carats with 2 carats fine ingredients and 3/8 carat leeway.

To Quarter Gold

§ 235 (a). The gold which is to be used is stroked on touchstone, so that its purity is determined.

Note 1. The proper touchstone is a fine black Trapp [basalt]; this has a proper hardness and color, and can be purified without harm from impurities with acids, which do not react readily with the black marble. In this stroking of gold, the color is matched with test pieces of known composition, and a customary degree of color can be determined fairly closely. These [test pieces] should be prepared for each 1/2 carat, at least from 15 to 23-1/2 carats, and in 5 different ranges, namely: pure gold with addition of 2/3 silver and 1/3 copper; with 1/3 silver and 2/3 copper; with 1/2 silver and 1/2 copper; with silver alone; and finally with copper alone. It

is understood that the [pure] materials must be completely pure, and the [mixtures] of the most accurate compositions.

(b) Then the gold is beaten out until it has a smooth finish, using a burnishing hammer on a polished anvil, until it is very thin; it is cut into small squares and 2 half marks are weighed out for duplicate tests, for the sake of certainty.

(c) With each 1/2 mark, [the composition of] which is determined approximately with the touchstone, is weighed out 3-1/2 times as much native silver, with which the gold is found to hang together best during separation, and most of the silver is dissolved. If this were to give the gold a value of 20 carats, 20 is multiplied by 3-1/2, which gives 70 carats, but as 24 carat or 1 mark native silver only contains 15 lod 16 grain pure, so it lacks on 70 carats, 5-5/6 grains, whence the whole addition of silver ought to be 70 c. 5-5/6 gr., if no correction is made for the silver which the gold might have contained originally; let that be 3 carats, so it is 70 c. 5-5/6 gr. - 3 c. = 67 c. 5-5/6 gr. Half of this is 33 c. 8-11/12 gr. for each of the marks weighed out; but as the fraction of a grain in the result is negligible, so for 8-11/12 grain we may substitute 8-3/4, or better 9; in accurate calculations the fraction must still be considered, otherwise a mistake can more easily enter into the result. Finally we weigh out with each 1/2 mark, together with its addition of silver, 3 times as much lead.

Note 2. It is usual, on account of the copper which also is present, to take 8 times as much lead.

(d) In the assay oven, fired up as high as possible, are placed two inverted baths, at first hindmost, then transferred to to the front of the oven, and then set up- right when no darkening is observed on the bottom of the muffle. On each [bath], well heated and moved out to the middle of the [muffle] bottom, are first put the lead which was weighed out, the muffle top is closed, but opened again as soon as it begins to change; the baths are taken out to cool, then inserted again immediately and the silver which was weighed out is put in; when this mixes with the lead the gold is added, and when this mixes in the muffle top is closed.

(e) The reaction seems to be the same as when silver alone was used (§ 247), yet more heat can be used without loss of gold, and less suffocation.

When it gets through the glistening [stage] so that the bead becomes high and round, and produces large flowers, the baths are returned to stronger heat and the erect [bead] has only a thin skin left, which shimmers with rainbow colors and is called lead glance; then the baths should be placed innermost [in the oven], since if they are put in the strongest heat they usually become entirely mirror clear after glistening, which indicates the presence of some grains of lead, and we can consequently gain 17-18 pure mater- ial, because the bead weighs more than the gold and silver together before the reaction.

In the strongest heat the bead should remain molten for the length of an hour, otherwise it retains lead blemishes which become brittle on hammering out. Then the lid is closed down so that the bead may cool off a little, and it remains standing until it turns to silver glance, and the raised surface subsides and buckles, when the baths are tak- en out a little and finally removed, the bead loosened, cleaned and weighed; it should be the same if the test goes properly. The added weight of the silver is subtracted from that of the bead and the gold content, which is obtained after separation from the rest, the silver content of the gold in the composition is recalculated, which is accounted for when the gold is separated and when the separate accounting is subtracted from the value of the gold; but not if the gold is over a ducat´s content, or such that some work can be done on it, or has already been done and used unmelted, since then the silver not in the gold is evaluated, just as the copper is not calculated in determining the silver.

In the manner just described, the gold is freed from base metals, and recalculated so as to eliminate the silver. It is noteworthy that if gold with 1/24 copper is taken, and three times as much lead is added, during reaction all the lead is separated, but almost all the copper remains behind; if the added silver is somewhat more than the copper, the last named metal is separated, but as soon as the silver becomes equal to or more than the gold, both copper and lead are completely separated in the bath.

To Part Gold

§ 236 (a). First the bead is beaten out very thin and smooth so that it is not cracked, since when the gold breaks down the proper weight cannot be obtained. Beating out is done with a polished hammer on a polished anvil; the bead is first beaten on the edge and then flattened somewhat, then on the edge again, after which the piece is heated to glowing in the test oven; meanwhile the same thing is done with another bead, and so forth in turn. The first bead is taken from the oven and beaten thinner with the smooth edge, but as soon as it becomes hard it should be heated again. Then it is beaten out to about the same width and size, smoothed with the flat of the hammer and reheated, and this is continued with the peen [hammer] to a thinness which can be rolled with the fingers, heated again after rolling to free it from all dirt, and finally allowed to cool by itself.

(b) The rolls are dropped carefully with tongs into a retort, aqua fortis poured on which already contains some dissolved silver, a like amount of distilled water added, to stand altogether about one inch above the rolls; weak aqua fortis must necessarily be used in the beginning or the gold will be broken apart. The retort is held over a candle, when the silver reacts with effervescence, but as soon as it gets active the retort is removed from the flame until the reaction subsides somewhat. When no more can dissolve, it is boiled with large drops, half of it is held back in a flask, a like amount of silver solution added in its place and the heating continued as before; when no more dissolves it is collected all together and the unmixed silver solution poured in; when this is done some of the mixture is held back and plain aqua fortis is added in its place; when this ceases to react, half of it is held back and plain aqua fortis poured in again; when this finishes its reaction the whole collection is held back and unmixed aqua fortis poured in in its place, when usually no fizzing with large bubbles occurs without boiling, which still should be continued for about an hour so that all that possibly can may dissolve, after which the liquid is removed and the gold has been parted.

(c) Then the rolls are washed twice with distilled water, which has been heated in a pure tin basin, after which similar water is boiled with it in the retort, which [process] is repeated as long as the decantate shows the least cloudiness with acid of salt or common salt.

(d) The rolls are carefully placed on a dish, since they are quite brittle, heated to glowing in a test oven, when they come together noticeably and become firmer and more manageable. When the dish is as hot as the oven, it is removed; the rolls are quite green at first but become properly gold colored on cooling.

(e) Finally the rolls are weighed, but the proper caratage is not yet known, unless the content of the aqua fortis is determined (§ 237).

To Test the Aqua Fortis For Its Content

§ 237. Plain aqua fortis is never strong enough to separate all the silver from the gold, without leaving some behind, more or less according to its strength. To determine how much the used aqua fortis has left behind we follow this procedure.

(a) Two half marks of 24 carat gold are weighed out, combined with 3-1/2 times as much silver, then with 5 times as much lead, heated on a bath, parted, washed and weighed (§§ 235, 236); then the excess in weight over the gold weighed gives the [amount] held behind, which is usually 1/2 to 1 grain.

(b) If the amount left behind is found greater, the aqua fortis should not be used in testing gold, unless in case of necessity.

Note. Messers HELLOT, MACQUER, and TILLET melted together 12 grains of purest gold calx with 24 gr. fine silver, and heated it in a cupel [capellerte] with 2 drachmas of lead. The parted bead gave gold of 23-30/32 carat, from which they concluded that no silver could have been left behind in the roll (cf. § 234, Note 1). In France the test weight makes 6 grains of pre- scribed weight 1 mark or 24 carats, which is the usual amount used for a test.

All of the foregoing tests are concerned with getting gold-free silver. This is investigated easily if the solution occurs with plain aqua fortis, since if some gold is in it, it precipitates in the form of black powder, which gets gold-colored under heating to glowing. Now to obtain pure silver it is dissolved in fresh aqua fortis, filtered, precipitated with copper, which is not gold-colored, washed carefully and fused with the help of borax. The silver solution can also be precipitated with common salt, the precipitate washed, and then reduced with double its weight of alkali minerale, or otherwise with quicksilver (§ 106, Note), to obtain pure silver. If we want to know the gold content of the silver more exactly, it can provisionally be done by means of the blackness of the powder collected from a certain amount, and weighing it after heating to redness; if then the silver in the reference [sample] is mixed with so much pure gold that the proportion of the former [silver] to the latter [gold] in the whole is as 3-1/2 to 1, the same mixture is cupelled with 5 times as much lead, parted, and the amount retained by the aqua fortis determined, the gold content [of the silver] is [thus] obtained.

Dry parting can also separate silver from gold. The silver is first fused, poured through a bundle of twigs into water, and so is brought to a fine graininess; the size [of grains] which are to be used in the next operation after this is still to be determined. The granulated [silver] is weighed out, 1/8 of it saved back, and the rest mixed with 1/8 or 1/6 powdered sulfur, after which it is smoothed out or compacted, moistened, and packed into a graphite crucible, covered, put into a blast furnace and cautiously fused. After one hour, 1/3 is taken from the part saved back and added; after 2 [hours] again 1/3; and after 3 [hours] the last 1/3. Each time it is stirred with a stick or a new tobacco pipe, and this repeated every 1/2 hour while the 3 crucibles remain in the flame. They should not be taken out unless large silver beads are seen on the surface which stay longer the more silver drops down with the gold. The mass can either be poured into a warm greased mold or allowed to harden in the crucible; in the first instance the plasma fastens very firmly to the regulus. If 100 parts of silver are brought, through the first operation, to 10 parts, all has gone well. If the plasma is found to contain gold it is melted and for each pound is added 1 lod [half ounce] of iron, but preferably silver; In the former instance the mass contained melts in 1/2 to 1/4 hour, in the latter 1-1/2. By repetition of the preceding procedure described, in which 1/2 lod of sulfur is taken for each pound of silver, the gold can be brought to 24 carat. The plasma, which is an artificial glass, is freed from sulfur on a lead forge with muffle; the sulfur is burned off by itself, so that no silver is lost. The silver is finally obtained alone and should not contain [any] gold; if iron was used it covers the residual silver and makes it less pure.

To separate gold and silver in the wet way with aqua fortis at least 3/4 granulated silver is required in most cases, and strong plain aqua fortis is poured on quickly, and when it becomes saturated it is poured off and new solution added, which operation is continued until the silver is all dissolved. The gold powder [left undissolved] is moistened well and boiled in pure water, until it produces no sediment with common salt, and this is then finally fused with borax. The gold cannot be made higher than 23 carat in this way, and highest 10 grain. 1/4 to 1/2 grain of gold per pound cannot pay the cost of separation.

Gold is freed from a lot of copper most easily if litharge is mineralized with some sulfur, and then is cast on the fused gilded copper, until as much lead is produced as the weight of copper. It is then precipitated with coal dust and stirred with iron-tin, at which the gold follows the lead to the bottom.

To Make 24-Carat Gold With Aqua Regia

§ 238 (a). One part gold, mixed with 3-1/2 to 4 times as much silver, is beaten out thin, and it doesn't matter if it breaks; then pure strong aqua fortis is immediately poured on it, and this is repeated as many times as necessary until all the silver is dissolved. Then the gold powder is washed until the wash water gives no precipitate with common salt.

(b) The gold powder is then put into a retort with 4 times as much aqua regia, and a solution formed over a flame; the impurities fall to the bottom in the form of a white powder, and the solution is diluted so that [the precipitate] may sink better. The gold solution is filtered through heavy paper into a retort and the filter washed well so that no gold is lost.

(c) Then a receiver is cemented on with flour paste and the liquid distilled over to dryness. The residue is scraped out, put into a wet crucible sprinkled with borax, somewhat thicker on the bottom, borax strewn evenly above, a crucible cover cemented on, this arrangement placed [over a flame] 5 minutes without [blowing] and 10 [minutes] with vigorous blowing on the bellows, taken out, cooled, and emptied out, when a gold bead is [found] wrapped around by a brown, somewhat gold-colored glass.

<u>Note.</u> In general the gold becomes purer than it was before by means of the liquid which dissolves it without attacking the contaminants. In this case [however] it is most advisable to precipitate [the gold] with 10-12 times as much iron vitriol as the weight of gold, since the precipitate is freed in this manner from both copper and platina, if they are present, both of which are dissolved in aqua regia; and in the distillation they are left behind mixed with the gold.

To Assay Cast Iron For Gold

§ 239 (a). Cast iron is broken into small pieces , placed in an open crucible with a little sulfur over and under it, some borax at the bottom also, so that the pieces [of iron] do not stick to the crucible, [the crucible] is placed on a forge with the crucible cover on top, which when the crucible bottom becomes red hot is taken away and blowing is begun. When the material is fused the bellows is used vigorously, a little more sulfur is cast on it until it becomes clear, which is brought to completion with a little borax, after which the crucible is quickly removed.

All metals can be transformed in this manner into skårsten [pure stone?] except zinc, and gold when the latter is not contaminated by other metals.

(b) The skårsten is roasted, then fused with 2 parts lead glass, 2 parts tartar flux, 1/2 [part] borax, and 1/2 [part] glass powder for each part of the skårsten. The fluxes are mixed with each other in a crucible, so that the skårsten is put in [first] and the lead glass last, after which [the mixture] is fused in an open crucible in the usual manner.

(c) The lead-regulus obtained takes the gold to itself, if any is present; after heating on a sand bath a golden bead of silver is obtained, which is then parted (§ 236). The silver comes from the lead glass.

<u>Note 1.</u> <u>To assay gold ore</u>, I use almost the same process on a small scale as the one used at Adelfors [smelter] on a large scale; at least both depend on the same basis. Two lod [approx. 27 g.] of pulverized gold ore are mixed with 1-1/2 [lod] litharge and 13 lod of glass powder, put in a crucible, sprinkled with common salt,

covered with a clay lid, and fused on a forge using a bellows. Then a nail is put in and the crucible covered again; when this is consumed another [nail] is put in, and we can with continuous additions of new nails so let each [nail] undergo reaction with sulfur, since when this is complete, the iron is not further attacked. The iron precipitates the lead, which takes the gold along with it, and when this occurs the crucible is removed, and on opening it the regulus is found under the pyrite. The lead carrier is driven off on a sand bath, when a gold kernel bleached by silver is obtained; these two metals are finally separated by the usual method.

To assay gold sand is an easy process when the gold kernel is mixed with stone particles, by washing and amalgamation; if the gold is firmly attached, fusion must be employed or much will be lost. The sand is heated to glowing and doused in water 2 or 3 times to produce a fine dust, which is fused with double its amount of lead kernels and a like amount of very diluted black flux, when a gilded carrier lead is obtained. We can also, probably to more advantage, grind the stone powder with litharge, fuse and precipitate it either with coal dust or iron. Dr. LEWIS has discussed this very well, along with other cases concerning gold.

Note 2. [In the following paragraphs, Bergman outlines his basis for quantitative analysis. He has often been called the "father of quantitative analysis".]

The art of assaying, in regard to its subject matter, has nothing to do with investigating the purity of metals themselves, or the ores from which they came. Here we find two main classes into which the metals are divided. The methods for performing the assay are also two-fold, namely dry methods and wet methods. The former [dry] are generally used, although still very imperfectly, partly in regard to wanting a procedure for obtaining the proper degree of heat; and partly in regard to the fluxes which react more or less readily with the metals themselves. For ores, which are free from mixing with stoney material, we need use only coal dust, but this case is very rare. Precious metal ores are usually treated on a dish covered with lead (§ 252), but when the content is small [this method] is unsatisfactory because so little is contained in the vessel, and [the method] works best using carrier lead in a crucible (Note 1). Ores of non-precious metals are assayed in a crucible, except for quicksilver and zinc which should be distilled.

Wet methods have had little use up to now, although they will probably receive more use. When we dissolve a certain amount of metal, precipitate it, wash and dry the precipitate in moderate heat, we get, after many confirming tests by rigorous methods, the conclusion that the weight of the former [metal] agrees with the weight of the latter [precipitate] without having to reduce or smelt it. It is assumed that the substance and procedure should be the same each time, which can be done more easily than by dry methods. It is necessary to consider very carefully how to discover how much precipitate a certain amount of metal gives, with alk. causticum, alk. crystallisatum, bloodlye, etc., but once this has been determined, the greatest difficulty will have been overcome. More difficulty seems to be encountered in finding the weight of corresponding metal from the weight of the metallic calx, because calcination has so many different gradations, and it is very difficult to determine which of them is encountered, and we must use here heat of indeterminate strength.

Concerning Platina

[The following account of the discovery of the element platinum in Sweden, by H.T. Scheffer and Torbern Bergman, is some of the evidence for giving these two men at least part credit for the discovery of platinum. At first, Scheffer called the element "white gold", and Bergman referred to it as "platina". Later Bergman changed his name for the element to "platinum" to correspond with his Latin names for the other metallic elements, with the neuter ending "-um". J.A.S.]

§ 240. From Mr. Assessor RUDENSCHÖLD in June 1750, I received 100 grains [6.2 g] of dark sand, with the information that it came from the West Indies. It consisted of dark kernels of sand, particles of iron-colored iron ore, which were attracted by a magnet, a few gold kernels, and 40 grains [2.6 g] of silvery-white, flat, triangular scales. In Nov. 1752 I got a small portion of the same sand from Mining Counselor BRANDT, which he also had received from Mr. RUDENSCHÖLD. By study of the white scales, I have found them to be a new precious metal, which can be called White gold.

Note. Messers SCHEFFER and LEWIS have at almost the same time discovered this new metal; the former [Scheffer] published his research in

1752, and the latter [Lewis] in 1754. And although Mr. SCHEFFER had only a small amount to investigate, it is still noteworthy that he discovered from it its most essential properties. Mr. LEWIS, who had sufficient material, has given a complete and very enlightening history, in reference to which Mr. SCHEFFER made remarks in 1757, concerning the proper density of the new metal. Mr. MARGGRAF published his research in 1757, which also came out in print 2 years after.

§ 241. Its properties are as follows:

(1) Its specific gravity is greater than that of gold, for which hydrostatic observation on mixed melts can be used, in particular with tin. When the specific gravity of one metal, such as tin, is $G = 7.180$, and that of another metal, such as platina, is $g = 17.000$; then the specific gravity of the mixture of P parts tin and p parts platina, according to Mr. LEWIS, is: spec. gr. $= (GP + gp)/(P + p) = 12.090$, if $P = p$. If mixtures were made in this way, this [formula] would be correct; but it allows neither accurate weighing nor for other [weights] to be used [other than $P = p$]. If the volume of the mixture is $(P/G) + (p/g)$, and its weight is $P + p$, then the correct specific gravity is given by: spec. gr. $= \{(P+p)Gg\}/\{Pg+pG\} = 10.090$ in the present case, but according to real hydrostatic weighing is equal to 10.827. If this last value were taken for the specific gravity of the mixture, it follows that [spec. gr. of] platina is 22.003. The roughness of the scales of platina seems to cause air bubbles to attach themselves

to the surface and reduce the specific gravity, which Mr. LEWIS by weighing found to be 17.000; but when when [platina is] fused with the help of another metal it gets a smoother surface, and thus by chance gives the correct result. Iron varies in its weight according to the amount of phlogiston [it contains]; copper, after different amounts of heat, becomes porous within or without, hence mixtures with it are not suitable; but with gold very dependable data are obtained without difficulty for determining the specific gravity of platina.

Note 1. To find the formula for the specific gravity of a mixture from one or the other of the ingredients is not difficult, since if:

$$\{(P+p)Gg\}/\{Pg + pG\} = a$$

we get:

$$g = \{Gap\}/\{PG + Gp - Pa\}$$

and if $P = p$, we get:

$$g = \{Ga\}/\{2G - a\}$$

By weighing of the purest platina, Mr. LEWIS found its specific gravity to be 18.213. He observed also that the mixture of platina with silver by mutual despoilation diminishes its weight, whereas that with iron seems to be increased; what can occur in this case undoubtedly takes place with other [metals], tough in different proportions. One part platina with 47 [parts] gold gives the former the largest specific gravity of all the mixtures made up to now, but still not greater than 18.711. From these and other investigations it is reasonable to conclude that the specific gravity of one or the other of the ingredients cannot be derived with certainty from these mixtures.

(2) It is as malleable as some iron can be.

Note 2. The ductility of the platina scales is quite varied; some break immediately under the hammer, and Mr. LEWIS has found that the brittleness is increased mostly by heating. After smelting with a burning lens, Mr. MACQUER found the ductility at least as great as that of silver; it hardens under the hammer like silver, but recovers its ductility through heating and gentle cooling.

(3) It is harder than gold, similar to forged iron.

Note 3. Its hardness seems to be greater than that of copper, but less than that of iron.

(4) It does not melt by itself, unless by chance between [hot] coals, but for such research we need pounds of platina.

Note 4. If we could build ovens with bellows on all sides in which the crucible and the iron bars which support it melt down completely, then the platina kernels wopuld only weld together, and could be easily separated again with a hammer. Among [hot] coals it cannot be held longer either. In an investigation in 1758, Mr. MACQUER and Mr. BAUME´, with a burning lens well supported, 22 tums in diameter and of 28 tums focal length, melted iron bars in an instant, Hessian crucibles in 3-4 seconds, and flint in 1/2 minute. Platina was placed in the focus of this [lens], [platina] which had become dark in fire, and it immediately got a dazzling lustre, sparkled and melted completely after 1-1/2 minutes.

(5) The [platina] scales darken under mild heatinbg, but in stronger heat they become white and glistening. Other precious metals are not darkened in heat. A platina kernel which at first is attracted by the magnet, loses this porperty in heat.

<u>Note 5.</u> Platina which has been fused gleams like silver. Otherwise it is like gold in its resistance to heat, air, and water. If it gets white by strong heat, it darkens again with mild heating. If through heat or acidum salis it loses the property of being attracted by the mag- net, it recovers it again by strong electrical shock.

(6) It dissolves only in aqua regia.

<u>Note 6.</u> See § 144.

§ 242 (a). Platina mixes in fusion with all metals, although cobalt and nickel have not yet been investigated.

<u>Note 1.</u> Both cobalt and nickel also go together easily with platina in heat.

(b) It mixes with silver most slowly, since three parts by weight of it [silver] are required to 1 part platina, if the mixture is to melt with a blowpipe, and it becomes white but brittle. It does not serve well for casting.

<u>Note 2.</u> Thirty [parts] of silver fused with 1 [part] platina gives a very white mixture, but still reveals to the eye some dullness. If the platina and silver are smelted ever so dil- ute, on casting the platina still goes more to the bottom, and in case the mold is narrower, the

material hardens quickly. They can separate more or less completely by enough heat as to melt only the silver, which then comes out the victor; by parting; by cementing and amalgamation.

(c) Quicksilver does not combine with platina if the latter is not [first] melted together with gold or lead; but in the first instance [with gold] platina can be rubbed in by long grinding; and in the latter case lead is separated and the quicksilver is left with the white gold [platina].

Note 3. Amalgamation [of platina] has succeeded for Dr. LEWIS through many hours grinding in an iron mortar with salt, water, and some drops of acid of salt.

(d) Copper fuses with it [platina] as easily as copper fuses by itself and [the mixture] becomes quite malleable, and resembles gold [in this respect]. The fusion is promoted by strong heat, as in refining, it is fanned until it sparkles, and when taken up [the mixture] is found to contain both metals; the mixture is less malleable then, just as with highly refined copper. In this mixture, platina loses its resistance to rust and flame.

Note 4. This mixture [of platina with copper] is whitish, loses 1/46 of its weight during fusion, is broken into pieces only with difficulty, shows granular fractures, takes a beautiful polish, which is not changed in 10 years, as Dr. LEWIS found. Increasing the copper makes it more malleable, but diminishes its hardness.

(e) Twenty-four parts platina with 1 part white arsenic fuses as easily as copper and iron together with the same semi-metal [arsenic], but the mixture is gray at fractures, like silver and arsenic together, and similarly very brittle. The combination needs no flux, as do copper and iron when arsenicated, but as soon as the arsenic reacts with it the heated platina fuses together immediately.

<u>Note 5.</u> With larger amounts of platina, such as one or more lod, reaction does not occur as long or as quickly as when the test is made with very small portions.

§ 243 (a). The purification of platina by means of refining [afdrifning] cannot be done. The small amount of lead makes it very brittle. If as much lead is added as is usually used in refining, it cannot be made to glisten in the strongest heat, unless congealed somewhat previously, the kernels become dark overall and rough underneath, are white underneath, and as brittle as when it contains 2-3 percent lead.

<u>Note 1.</u> Mr. MACQUER and Mr. BAUME´ have held the cupell in constant heat for 50 hours, which heat, especially during the last 24 [hours], was very strong. After this, the mater- ial was found to be flat, weighed somewhat less than the platina put in, was soluble in aqua reg- ia without residue of lead, and was malleable.

(b) Similarly, if platina is fused with antimonium crudum the [state of the] regulus is not completely reached, since at the end the mixture cannot be volatilized with the strongest heat.

(c) If mixed with gold, it cannot then be separated in heat, since both [metals] are incombustible and neither is attacked by sulfur. Both resist aqua fortis and acidum salis in wet methods, but both dissolve in aqua regia, both are precipitated by quicksilver; one procedure is to precipitate the gold with iron vitriol, which does not precipitate platina; the precipitate is amalgamated with quicksilver.

Note 2. How gold and platina can be separated was mentioned previously in § 144, Note.

§ 244. From what has been related, it is shown that platina is a new and precious metal, most nearly like gold. If some non-precious metal is found mixed with it, this is accidental and platina is distinguished from it basically.

Because this metal is dense, without color, very hard, and unchanged in air, it must be suitable above all others for telescope mirrors; so it needs only a method to put the white gold together in the right form, or to find the proper additive for casting with gloss and resistance in air, since its malleability is not suitable for this purpose.

There is no advantage in mixing gold with platina, because of the cost of the mixture.

Note. A few chemists will not consider platina as anything other than a mixture of gold and iron; but to obtain a specific gravity of 17.000, 10 parts gold must be used with 1 [part] iron, and then the color does not become the least bit white; most [chemists] regard it with reason as a distinct metal.

For telescope mirrors, copper seems to be the most serviceable additive (§ 242, Note 4).

If we had some less costly method for separating gold and platina, some method other than those known up to now, there would be no more serviceable additive [than platina] to give gold hardness with durability. One-fifth platina does not make gold as pale as 1/12 silver, according to Mr. MACQUER'S research, and consequently an amount sufficient to produce the same hardness as the ordinary addition of copper will not alter the beautiful color of the gold, nor make it as pale as silver does, nor red like copper, just as though the whole mass were precious.

Platina could properly be called 'small silver', and it was also named 'Platina del Pinto' and 'Juan blanco' [white John]. Don A. ULLOA mentioned it first in [his account of] his American journey in 1748, as a naturally occurring material. WOOD sent some of it to England in the beginning of 1749 with a report; and sometime later we in Europe obtained some small amounts, although all such exportation is strictly forbidden by the Spanish government, to the great detriment of metallurgy and the arts dependent on it.

Concerning Silver

§ 245. Silver has the following properties:

(1) It is among the metals next after lead in density, and its specific gravity is between 10535 and 11345.

Note 1. Compared to pure water, its specific gravity is as 11,090 to 1,000. [Modern value:

10.5 at 20 C.] Hammered [silver] is somewhat denser than cast [silver].

(2) It is the most ductile metal after gold.

Note 2. One grain [of silver] can be spread out to a length of 3 aln [1 aln = 1 ell = 45 in.], with a breadth of 2 tum [1 tum = 1 inch], or a surface area of 288 square tums. [3 x 45 x 2 = 270]. Silver leaf and filament are also a noteworthy test of its ductility. It seems to harden during hammering more than gold, but becomes malleable again with heating.

Its tensile strength is less than that of iron, but greater than brass, since a thread 1/10 tum in diameter breaks at a weight of 270 pounds.

Its hardness is greater than that of gold, but less than copper.

It is elastic, and consequently ringing [as a bell].

(3) It melts more easily than copper.

Note 3. It melts long after glowing [heat], later than antimony regulus, but before gold, at about 540 degrees [modern value, 962 degrees Centigrade]. It solidifies during slow cooling with raised ridges on the surface. Polished silver does not melt under a burning lens, which brings platina to a liquid state; the reason is its smooth white surface, which reflects all rays.

(4) It is not calcinated in fire when no contaminant is present, unless it [fire] is very constant, .

Note 4. KUNCKEL reports that he held silver in a glass oven for over a month, with no loss more than a few grains. Mr. TILLET has demonstrated in 1760 that 16 lod of silver, held in fusion for 2 hours with strong heat, lost about 1/24; when he placed a cupel over the one which held the silver, he found at the end that the upper cupel was sprinkled over with silver kernels so fine that he could find them only with a magnifying glass. Silver therefore can undergo a kind of sublimation with the help of fire alone, or more properly volatilization.

(5) It is soluble in acid of saltpeter and concentrated acid of vitriol.

Note 5. See §§ 80, 81, 94, 106, 123, and 139.

§ 246. Silver can be fused together with all metals except nickel and cobalt. A copper-silver mixture is used for minting [coins], and for silverware.

Generally, when ores yield silver mixed with copper, a method for separating them, used in various places, is called Segring; it is carried out by addition of lead, by which the entire mixture becomes more fusible than without the lead, which combines more closely with the silver than with the copper, and flows out at a temperature lower than that required to fluidize the entire mass. But tin cannot be removed from the copper, although many times more readily fused, since tin dissolves the copper.

Silver is amalgamated easily with quicksilver when it is pure, but consequently when it is contaminated with copper a little acid of saltpeter is used.

Note 1. For amalgamation silver should be in the form of leaf or filings, or precipitated with copper. Ten parts quicksilver is combined with 1 part silver in heat without rubbing, and then becomes fluidized. The amalgam is easily made to crystallize, but no more than 8 parts quicksilver, since otherwise if more is used the crystals liquify. In the Tree of Diana, no more than about 4 parts quicksilver is used.

The amalgam is used for silverplating of silverware.

Note 2. To blend metals of various manufacture so that the mixture reaches a desired quality, we can insert all the different cases in a single equation, which is found when we seek the content of the fusion mixture from the given amount of the components and their strength.

Let ´a´ be the content of the better kind [of metal], and its gross weight ´d´; the content of the poorer kind ´b´, and its weight ´e´; and the content of the mixture sought ´x´; then we get:

$$ad + be = dx + ex$$

and:

$$x = (ad + be)/(d + e)$$

If now the question is to find the amount of purer kind of additive obtained from a poorer kind of manufactured goods, we can use the same kind of equation, since when we determine ´a´ of the best kind, and ´d´ its gross weight, and ´b´ of the poorest kind, and its gross weight ´e´, the desired quantity ´x´ is obtained. Consequently:

$$e = (ad - xd)/(x - b)$$

Similarly, when it is required to determine the content of the certain amount of better quality metal and more costly, because 'a' corresponds to the highest content, and 'd' to its desired weight, 'b' to the lowest content and 'e' to its known gross weight, and 'x' to the desired content. We therefore derive from the first equation the value

$$d = (xe - be)/(a - x)$$

which is the formula for the property. One example designed to show each of these is enough.

From the first formula, the rule follows that:

$$(d + e) : (ad + be) :: 1 : x.$$

Let a = 10 carat (k.) 6 gr., b = 8 k. 10 gr., d = 10 marks (m.) 4 lod, and e = 6 m., so then:

16-1/4 m. : 7 k. 7-1/2 gr. + 53 k. :: 1 m. : 3 k. 8-30/39 gr.

The second formula gives this rule:

$$(x - b) : (a - x) :: d : e$$

so then, a = 16 lod, b = 12 l. 9 gr., d = 16 lod, and x = 13 l. 4-1/2 gr., and we get the following set-up:

13-1/2 gr. : 50-1/2 gr. :: 16 lod : 59-23/27 lod.

e = 59 l. 15-1/3 gr.

By a third formula it appears that:

$$(a - x) : (x - b) :: e : d$$

and consequently, if a = 15 l. 16 gr., b = 11 l. 5 gr., e = 16 l. or 1 mark, and x = 13 l. 4-1/2 gr. , or the desired control-content is:

47-1/2 gr. : 35-1/2 gr. :: 16 l. : 11-91/95 l.

d = 11 l. 17-1/4 gr.

On this we base the usual art of assaying, but we presume that the volume of the mixture is similar to that of the ingredients jointly, which is often not the case. If there are more than 2 kinds of different ingredients which we wish to blend, we can include also all possible [materials] in a general equation of an indeterminate number of terms; if we let the weight of another item be ´g´ and its content ´f´, etc., so then:

$$(ad + be + fg + etc.) = (dx + ex + gx + etc.)$$

To Test Silver For Its Fineness

§ 247 (a). One standard mark of silver consists of 16 lod; when these [lod] are all silver it is called ´16 standard´, or entirely pure, whereas if only 15 [lod] of the mark are silver and the other lod copper, or some other non-precious metal, it gets the name ´15 standard´, etc. This [system] is used for native [silver] which is usually not quite free from other metals such as gold, lead, copper, etc., but gives the amount [of the other metal] per mark as 1, 1-1/2, up to 2 grains. a silver which contains 1 lod of gold per mark shows itself on the cupel as 16 standard, and is so calculated until the question of how much gold it contains is raised.

Note. One mark = 288 grains, and 1 lod = 18 gr. = 288 ass. Consequently the same weight can be used for both gold and silver tests (§ 234, Note). In France they use: 1 mark = 12 deniers, and each denier = 24 grains, so 1 mark = 288 grains : 1 mark test weight makes 36 actual grains.

(b) Silver, which is to be assayed, is beaten out very thin with a polished hammer on a polished anvil and cut into small pieces. 2 half marks of these [pieces] are weighed out on a grain balance, in duplicate for the sake of certainty. Each of these [samples] is folded in on itself in triangular pieces of paper, which are called cut blanks [Skernitsel].

Note 2. Folding [the sample] in paper can be avoided if we use beaten-out lead [sheets]; paper gives charcoal, which sometimes is unde- sirable.

(c) To adjust the weight of lead [blyschweren], we should probably know the standard of the silver, which, if we are unaccustomed to estimate it by appearance, must be determined by testing; when test pins, from 1 to 16 standard, are rubbed [on a test block], the color [can be] matched to that of the silver which is being assayed.

(d) When, one way or another, the standard is known, the weight of lead is obtained from the following Table.

TABLE (BLYSCHWEREN FROM SILVER STANDARD). (SCHEFFER)
[m. = marks; l. = lod].

Std. of Ag	Pb-Ag, (m-l)	Pb to ea.pt. Cu	Togeth in (m-l)	Accur. sum, Ag + Cu	Grains Pb/mark
16	5-0	---	---	5-0	5-3-5
15	4-11	28	1-12	6-7	6-5-6
14	4-6	27-1/2	3-7	7-13	8-8-9
13	4-1	27	5-1	9-2	9-8-9
12	3-12	26-1/2	6-10	10-6	10-10-13
11	3-7	26	8-2	11-9	12-10-13
10	3-2	25-1/2	9-9	12-11	13-10-13
9	2-18	25	10-15	13-12	14-14-16
8	2-8	24-1/2	12-4	14-12	15-14-16
7	2-3	24	13-8	15-11	16-14-16
6	1-14	23-1/2	14-11	16-9	17-17-18
5	1-9	23	15-13	17-6	17-17-18
4	1-4	22-1/2	16-14	18-2	18-17-18
3	0-15	22	17-14	18-13	19-18-19
2	0-10	21-1/2	18-13	19-7	19-18-19
1	0-5	21	19-11	20-	20-20-21
Cu	0-0	20-1/2	20-8	20-8	21-20-21

[For example], if now a 12 standard silver is tested, we need first 5 lod for each lod silver, which together make 60 [lod] or 3 m.- 12 l. [of lead], and [this amount] is found in the second column, corresponding to 12 in the first [column]. Then we need 26-1/2 lod lead for each of the 4 [lod] copper which is present, which together makes 106 lod = 6 m.- 10 l.; the former is obtained from the 3rd [column], and the latter from the 4th column. The 5th column shows an accurate sum of the lead [required] for [both] silver and copper, but in which sum it is not required to calculate the [total] number of lod, so the 6 lod are ignored and only 10 marks taken [i.e., the amount of silver is rounded off at 10 m.]. If the number of lod is over 8 [i.e., 1/2 mark] we should take the next higher number of marks, which therefore is adopted; this gives the 6th column. The last [column] shows the grains of lead weight [blyschweren] because we seldom know the standard beforehand.

The lead weighed out is beaten cylindrical and the corners rounded, so as not to damage the cupel.

Note 3. According to research by the previously mentioned French Academy, the [French] government has, through a regulation, established the amount of lead which should be used in testing silver. I have reduced their method in calculating fineness for inspection, and I will hereafter quote the result of this rule which orders us to take 4 parts lead for each part silver of 15 lods and 6 grains [silver] content per mark; 6

for that which holds 14 lods 12 grains; 8 for 13 l. 6 gr. content; 10 for 12 standard; 11 for 10 l. 12 gr. content; 12 for 9 l. 6 gr. content; 13 for 8 standard; 14 for 6 l. 12 gr. content; 15 for 5 l. 6 gr. content; 16 for 4 standard; 17 for 2 l. 12 gr. content; and 18 for that whose standard is 1 l. 6 gr. According to their research, the amount of lead requires more silver in the cupel than a well-suited amount, and still not in proportion to its excess, since 30 to 40 parts lead to 1 [part] silver yields greater reduction than 16 to 20; this seems to be certainly an excess which works the strongest.

(e) When the test oven is heated up so that the muffle glows white, along with half the base, it is ready; this can be hastened by placing a large burning coal in the muffle. On each of the two well heated cupels prepared on the base is put one [piece] of the weighed out lead, the muffle hatch is then closed until the lead darkens and begins to move; it is then opened, the cupels slowly taken out to cool, after which they are immediately inserted again and cut blanks with folded-in silver laid each on its cupel at the innermost corner, partly to better observe when the silver goes into the lead, and partly so that the union may best occur, because the silver blooms spread out from the corner. As soon as accomodation occurs and the paper burns up the muffle hatch is closed, and when all the silver combines with the lead the cupels are moved out enough so that each gets both heat and cold. It is then noticed that the more copper is mixed with the silver the less heat is needed both for testing and for shining. Because strong draft makes the test go very quickly, neither cupel draws strongly enough for all the silver to soak into [the cupel] with the lead, or becomes like wet litharge; and in both cases the test is worthless because of loss [of silver].

Note 4. Strong draft is hindered by compression of the charcoal in the oven, and small coals laid in above the cupels. It is the same either for high or low heat; the former [high heat] is shown by the light red color of the cupels, and the latter by their darkening. Little or no draft with high heat makes the kernels almost spherical and dark brown, without congealing, but with little heat the cupels are quite dark and the kernels dark red, and about to congeal quickly. So at times it should be aided by regulating the heat and the draft.

(f) The indications that the proper heat has been applied are: (1) the kernels appear very white; (2) they lie upright; (3) they do not volatilize; (4) the application of heat takes a long time; (5) the cupels are very hot.

Note 5. That silver can be volatilized was previously mentioned (§ 245, Note 4), and with the application of strong heat and vigorous draft; that some [silver] can follow the lead fumes is so much more believable when [we remember that] at Sala and other silverworks we find smokestack scourers [rökhvalfskurarne] over the large smelter ovens [to recover the volatilized silver].

(g) Coolness is reached if: (1) the cupels are not able to draw the wet litharge into themselves, unless there is an oilyness around the kernels, and they almost run in in stronger heat, so this always results in loss of silver. Also the cupels become so clogged with wet litharge that the kernels harden at shining and can hardly be absorbed. Wet litharge is produced if the cupels absorb a little at a time, as soon as the kernel begins to flow and becomes like grease when the wet litharge is not far off.

(2) If several dark colors appear on the kernels, after which they also harden completely in case absorption does not occur in time. They solidify together and finally are brought together again in strongest heat, and some silver is lost.

Note 6. The wet litharge glazes the cupels, not only around the kernels, at least on some sides, and always causes a little loss of silver, although in most cases in the test this loss is prevented.

(h) The proper degree of heat is reached when: (1) the kernels appear somewhat yellowish; (2) when they flow a little, then have time to become small; (3) dry litharge grows like grass on the edges of the cupel; (4) the cupels shine around the kernels.

Note 7. A suitable procedure for measuring the heat in the test oven with accuracy will undoubtedly bring the test for purity [of silver] to completion. Mr. TILLET used a square of irontin with an arm at the proper angle; the tin was placed beyond the middle of the muffle bottom, and the arm was held horizontally outside on some hard support. At the end of the arm was a groove for a thermometer ball filled with iron filings. This was a sort of pyrometer which imparted to the thermometer a heat which existed in a certain proportion throughout the oven. From the relative degree [of heat] which is indicated, we can consequently have reliable information [about the temperature of the oven]. In his [Tillet's] oven, 120 degrees on the **REAUMUR** scale, or 125 on the Swedish scale, is that [temperature] at which lead should be placed on the cupel; the oven provided with [hot] coals increases its heat by itself, and at about 130 or 140 degrees Swedish [the lead] gets shiney. The heat in his oven can be brought to considerable height whenever needed by its installation over masonry pipes, which by means of certain opening shutters give a draft of different amounts as needed.

(i) When the lead so changes that large silver blooms begin to form at the edges, a stronger heat is applied to the light red base until shiney, when the kernel becomes spherical and silver blooms break out afresh, until so little lead remains that that which is laid on top likewise becomes colored, which is called leadshine; then the lead layer is quickly poured off and the silver remains very clear, and is called silvershine. This is accomplished with just enough heat, but if too cool the silvershine is not produced fresh unless some lead is attached to the kernel, which makes the test entirely unsatisfactory.

The same thing can occur with great heat, with which the usual silvershine does not appear, unless as soon as the kernel forms leadshine, and no particular fresh shimmer with color thereby, it stays clear and molten until it hardens again; in this case it is not advisable to take it out early, because then with heat the shimmer shows on the kernel much higher and gleaming, [the kernel] attaches firmly to the cupel and loses some in content.

When the kernel shimmers it is taken out by degrees so as not to spatter, and is loosened as when it lies on top of the cupel.

Note 8. The French Academy has assured us that a 16-standard silver, which is tested with lead, loses some of its weight each time [it is tested], and by repetition finally completely disappears; that 36 gr. or 1/2 quintin [drachma] with 4 quintin 8 gr. lead gives a kernel, about 11 deniers 21-8/16 gr., which by our reckoning makes 15 lod 15-3/8 gr; the litharge of the cupels is reduced 8 times in succession with borax and 3 times as much black flux, the lead added is driven off, whereby the chief kernel gains 5 times, and leaves 4 impure kernels, the latter being so small that they must be seen with a microscope, which together weighed 12 d[rachma] 10/16 gr., when the lead content is driven off, or only 1/16 gr. more than the first kernel with the silver, which is obtained by the first reduction and testing, namely 3-1/16 gr.; whereby the lightness is detected, that the lead is not changed to silver, since when the adhering litharge-skin is scraped off some of the excess disappers, without any discrepancy being noted, until at least a part is found in the last litharge, and occasionally part is dissipated; that entirely pure silver tested with lead usually retains on the under side a litharge-film of approximately 1/2 gr. weight, which cannot be separated by boiling in concentrated acetic acid, but by strong fusion or better by heating to glowing on a new cupel; that on testing with lead about 1/16 fumes away, and the remainder gains 1/16 by changing into litharge, but with bismuth the gain in weight is 1/6; that a cupel of bone-ash can absorb as much litharge as its own weight, after heating to white heat; that the cupels should only be made from finest bone-ash and be entirely smooth, since a rougher surface steals more of the silver, and the part staying in the pits easily holds small portions, from which the chief kernel's diminution is to be distinguished. [All this was assured to us, the readers, by the French Academy, in case you have forgotten.]

(k) That the test is proceeding properly is shown by: (1) the kernel does not have lead stains, nor is it lead colored; (2) the kernel is somewhat cracked on top and porous underneath; (3) it loosens easily from the cupel without tearing anything away with it; (4) it is not very high nor glittering; (5) that no silver lies on the side of the cupel, which has not combined with the lead; (6) it did not spatter, though there is always some loss; (7) no film of wet litharge forms on the cupel; (8) the kernel is cleaned off and weighs the same as it would if the same silver were taken away, and not as if an excess and a deficient test of the silver ingot were being made, which when copper is mixed in, is always of different content.

The lead content of the silver, which should be knowm, is subtracted at the end.

Note 9. The reason the test proceeds as it does is as follows: Lead by its ease of melt- ing makes silver molten at a lower degree of heat than is otherwise needed; but the lead at the surface, which the air mixes with the kernel, soon loses its combustibility and the calx then melts to produce a film of litharge which is absorbed by the cupel. By and large it [the cupel] is used to filter away part of the litharge. The blooms which appear are partly from lead calx, partly from an intermittent reaction which is stirred up by the heat. But lead is calcinated and vitrified easily not only by itself, but also takes other non-precious metals along with it. Finally, when the lead film gets very thin, there appears, as is usually the case with thin bodies, rainbow colors, and when the film is completely absorbed, the silver becomes clear and hardens quickly; thus the additives are removed with the help of its fusibility.

(1) As a control-check the silver is folded into a thin lead planchet, which is weighed first and then subtracted. The test proceeds with small yellowish blooms, which shimmer quite a while. A suffocating reaction is produced by this, since the oven is packed full of coals, and a small coal is laid on top of each cupel and the door is tightly closed. The coals are removed at the end because a strong draft is needed for a proper burn-off. When the burn-off is completed both doors are closed on the muffle, so that only a small opening is left, and the coal put into the oven fumes away, during which process both

fresh heat and strong draft is produced in the oven, important for burn-off.

Note 10. Silver produced in Sweden should contain 13 lod 4-1/2 grains fine [silver] per mark, with 2-1/4 grains leeway; 13-standard also bears a stamp, with double stamps for coinage. Our riksdalers weigh 608-8/9 ass and contain 14 lod 2 gr. fine [silver] per mark; a so-called double piece weighs 146-2/5 ass, and a single 73-1/5, and contains 7 l. 2 gr. fine; a white penny [styfver] 25-109/171 ass, and contains per mark only 3 lod 2 gr. In France wrought silver should contain 11 deniers 12 gr.

To Refine Silver With Saltpeter

§ 248 (a). Cupelled silver is fused in a crucible with a little pure saltpeter, is then removed from the fire, whereupon the slag is found to be all green from copper.

(b) The silver is fused anew with saltpeter until its slag is found to be completely pure, when we can be certain that it is free from copper.

Note. A covered crucible with a hole in the bottom is enough to hold coals, and allow air for detonation. If there is much copper in the silver more saltpeter is required and with this it always loses some of its fineness.

To Test Copper For Silver

§ 249 (a). The copper is beaten out somewhat thin, cut into pieces and sulfurated like cast iron (§ 239).

(b) The piece [skärsten] is then fused in an open covered crucible with 2 parts red lead and 1/2 [part] coal dust, and so we obtain the silver in the lead melt [verkblyet], if any is found in the copper.

(c) If the piece is roasted the copper ash returns to the metallic state and is preserved in the lead, since the sulfur has stronger attraction for the copper than does the lead, but silver just the opposite. An iron piece can be roasted without risk, because it is easily made into slag, but copper keeps itself free from the lead in this case.

To Test Tin For Silver

§ 250 (a). The tin is calcinated, then mixed with 12 parts of lead kernels, 8 parts borax, and 3 parts washed flint, in an open crucible, over which another is overturned containing coal dust, blown to ashes mildly for 5 minutes, and vigorously for 15 minutes; it is then found that when a fine iron sliver is probed into it that a lump of glass attaches itself to it; blowing is continued until this [lump] becomes clear and smooth on the sliver. On the bottom is found the lead melt, which is tested.

(b) In this manner we remove silver from tin which contains it, but not as well as with arsenical lead glass (§ 266).

§ 251. Removal in slag occurs only when we test some material for gold or silver, and is performed on a planchet with the help of lead kernels, which have the strongest attraction for the precious metals mentioned, collecting them and retaining them, but facilitating the calcination and removal in slag of all non-precious metals.

If we have no burned cupels ready, an unburned one is put into the muffle on edge while the test oven is being heated up, to burn it the same.

Slagging [Of An Ore]

§ 252 (a). The ore is finely ground, sifted through gauze, [a sample] weighed out and at the same time, in turn, about the same amount of lead kernels.

(b) On a burned cupel we lay half of the lead kernels, and on this the [ground] ore, and on top the second half of the lead kernels. Then the mass is gently stirred with a sharp pointed knife to mix it a bit, being careful that no part of the ore is without lead, either underneath, on the sides or on top, since it sticks to the cupel. The overlying lead seeps down through the ore while it is melting and takes the precious [metal] along with it.

(c) With the cupel tongs the cupel is placed at the front [of the oven], is turned frequently so that it gets evenly heated and spattering is avoided; then the oven door is closed for a little while to hold the heat in, the cupel is turned and can be moved further in without risk. If any cold gets to the hottest part [of the oven], it breaks the cupel so that the material is spalattered around.

If the cupel is burned during heating up of the oven and is thus glowing [hot] when the lead kernels and ore are laid on it, mixed in a manner similar to that described above, [the test charge] is folded in a square slip of paper, moved to a second cupel placed close to the first, and this should be done quickly before the paper catches fire, or otherwise [the charge] might be spilled.

(d) Different kinds resist [differently], so after warming up the cupel, it is introduced into the strongest heat, and then a small well heated coal is laid in the mouth of the muffle and the door is closed; the coal should not crackle, since when coal-dust is cast onto the material and reduces it, it will be slagged.

(e) When all is fused the cupel is moved farther in, the door is closed so that only a small opening remains in which a coal is laid, whereupon the charge begins to change and the sulfur fumes away, which does not occur in the absence of air.

Before the sulfur is removed no slag can form, other than in the form of pure stone [skärsten]. Zinc- and antimony-ore are treated differently in the following manner.

(f) During the process the degree of heat must be well regulated. When the fuming stops, it is hot enough, and the cooler it is the more it fumes until the material hardens, when the fuming also ceases. So the cooler it is when the material is completely fused the better. When it cools enough it then becomes slightly black when hardening, and slowly swells up and rolls up into the regulus; in such a case the cupel must be immediately moved to a higher heat. When it is placed just right the material lies inside [the cupel], the slag collects around it, a bit of shimmer forms on the regulus and it fumes moderately. Sulfur never takes gold or silver along with it, if no antimony is present, and the draft in the muffle must be carefully controlled, since if it becomes so strong that lead rises up in the fumes along with sulfur and regulus of antimony, then the silver also follows; therefore the antimony should fume away with the least amount of draft possible, which is adjusted with a coal in front of the cupel in the muffle, which breaks off the draft so that air does not produce too rapid cooling of the material.

(g) At such a degree of heat that volatile minerals fume away, the material is held, at which slag should form quickly; if it is slow in forming, the cupel must immediately be moved further in. If unmelted lumps are noted on the edge, they are moved in with a heated stirring rod; that which sticks to stirring rod is broken off and laid back on the cupel.

(h) When the [amount of] regulus appears somewhat less than a white penney, it weighs about 1 Centner and is right for most common cupels; it should then be poured out, to which end it is first moved to the hottest part to make the material quite fluid, and meanwhile a fold in the [metal?] test sheet is chalked and heated. Then the cupel is taken out, shaken so that all comes out, poured out in a pile over the fold until all runs out. The chalk prevents the material from sticking to the sheet and warming keeps the lead from spattering.

(i) Now and then it is found that all is not fused when it is poured out, in particular with a high melting ore. Also often lumps seem slow [to melt] so that the regulus does not get changed, since if it gets richer it cannot hold together unless it is beaten into fine crumbs, which go through a sieve with the powdered slag, and the richer it is the greater the loss. When it is poured out, if lumps still remain on the cupel, they are immediately mixed anew with 2-4 Centner of lead kernels [folded] in paper, melted and reacted in the previously described manner. When the material melts, the outcast regulus is added and reacted with the rest to its proper degree, and is finally poured out.

(k) When all is cooled, the regulus is separated, which we call work-lead here, and the collected crums, which can be spread, are ground with the slag, and the whole sifted through a sieve; what is obtained is added to the whole.

(l) The slag should be entirely smooth and shiney, and if it is the least bit metallic it must be resmelted with work-lead, to which some [additional] lead kernels can be added.

FIFTH SECTION, CHAPTER I, Noble Metals

To Test The Work-Lead On The Cupel

§ 253. The regulus should be tested for its proper quality. The difficulty resulting when the fusion is incomplete has already been mentioned (§ 252 (i)), but it doesn't get purer when more of this occurs, since the more lead used in the reaction the more the cupel is robbed.

On the other hand, precautions which should be taken with the reaction were described previously in detail (§247).

At the final weighing it is best to have reacted silver kernels for 2, 3, 4, and more Centner of lead kernels as counterpoises, from which can be determined very quickly how much lead was used in the kernels, since by weighing we can otherwise determine the troublesome fraction and [thus] allow the least error to occur in the weight.

Note. When we want to react a considerable amount of work-lead, it is poured on a bone-ash hearth on a stove top in an iron ring, 2 to 3 quarters wide [1 quarter = 6 inches] and a hand's breadth high. Ashes from tree leaves, well leached, can also be used, either alone or in a mixture. The ash is moistened, put into the [iron] ring, packed down with an iron ball the size of a fist, starting from the midpoint, so that the hearth becomes firm, smooth and packed from the edge 1 inch deep. Over the edge of the forge [extending] one inch is placed a round iron plate, resting on pieces of fire-resistant brick. A hand's breadth from the ring is built a round oven of stone bricks with an opening in it, which is filled with cold coal and carefully ignited;

then the lead is added for the test, and when this is ready, the silver [is added]. During reaction it is heated evenly, and in front of the oven is put a coal, which now and then can be taken out and the material cooled with a hand bellows, which is not good for the cupel. Wet litharge [våtglete] is also necessary to prevent the fuming away of silver. On pure wood ash the burning-off occurs almost pure. After the burning-off, when the silver is well solidified, water is poured on it, the mass is loosened, brushed clean, and for more certainty it should finally be refined with saltpeter.

To Slag Gilded Pyrite From Adelfors

§ 254 (a). To one part [ore] is added 8 [parts] of lead kernels and 2 [parts] lead glass, to make the fusion so much easier. Since the ore does not contain any blende or antimony it can be fused right away with strong heat, after which the sulfur should fume away on cooling.

(b) The work-lead is driven off on the cupel and the lead is separated.

To Slag Pure Leadglance

§ 255. To one Centner [of ore] is added 8 [Centner] of lead kernels, after which the above described procedure is repeated. A pale golden glass is produced.

To Slag Leadglance Mixed With Blende and Pyrite

§ 256. To 1 part [leadglance] is added 8 of lead kernels. The cupel is first placed at the outside of the muffle to fume away the sulfur from the blende and pyrite, after which it is moved into the heat, where the material fuses very readily. We know that otherwise the blende vaporizes around the edges along with flowers [of sulfur], which hinders the fusion like sulfur when it is fused at the same time as the lead and goes to slag. As long as the zinc contains sulfur it is not volatile, so therefore the sulfur must first be burned away at low heat, and then the zinc reduced by phlogiston from the lead; following this the heat is increased immediately, and the zinc-like parts go into the lead glass, without fusing, making the slag thick, and taking part of the silver into it, which then cannot be taken up by fresh pure lead or by new fusion.

To Slag Gray Copper Ore

§ 257 (a). To 1 part [ore] is added 20 of lead kernels.

(b) As soon as the material is well fused and the sulfur fumes away, it is poured out without regard to the a- mount of regulus, which should be tested on a cupel larger than usual.

(c) The reason for this procedure is that rich copper ores, some of which contain arsenic or iron, are reduced by the lead which at the same time is continually calcinated and vitrified. If now much lead reacts, the remainder becomes insufficient to react with all the copper on the cupel, unless silver makes coldglance and the material consists of more copper than silver. Also if the lead is so pure that it reduces [copper] and in the strongest heat takes the copper with it into the cupel, much silver is lost since the copper stays very firmly in the cupel.

If arsenic is present, as in Fahlerts, Fahlkupferets, etc. [ores from Fahlun?], it is fused and slagged very easily and needs no casting out before the regulus becomes just large enough, because arsenic slags the copper. Gray copperglass is very easily fused and gives a green glass.

To Slag Yellow Copper Ore

§ 258. To 1 part [ore] is added 12 of lead kernels, and this is treated in the usual manner, because sulfur is not contained in any large amount.

To Slag Iron Ore

§ 259. To 1 part [ore] are added 4 of leadglass, the whole well mixed, and then we add 8 [parts] of lead kernels and the mixture is treated in the usual man- ner. Because the ore is difficult to fuse, lead glass is added, which is more easily mixed with it than lead kernels. If it happens that the lead should remove the phlogiston from the iron, [wood] splinters must be added with some cooling so that it does not burn away. The glass becomes black.

To Slag Pyrite, Containing Some Blende

§ 260. To 1 part [pyrite] is added 8 of lead kernels and the whole treated as in § 256, but so that all the volatiles do not escape before the ore is fused, but then the remainder is fumed away, while the foreign materials are converted to slag by the lead, which does not occur until the sulfur is driven off.

To Slag Antimony Ore

§ 261. To 1 part [ore] is added 12 of lead kernels. It should be put in a red-hot dish, since in low heat the regulus goes to lead glass, and makes it so dense that the test is never completely slagged.

To Slag Arsenical Pyrite

§ 262. To 1 part [pyrite] is added 8 of lead kernels. It should melt before any gets fumed away, otherwise part of the silver fumes away before it reacts with the lead. Arsenic rises up in strong heat, before it is fused, and produces an odor of garlic.

FIFTH SECTION, CHAPTER I, Noble Metals

To Slag Pure Blende

§ 263. To 1 part [ore] is added 8 of lead kernels, and we see again on the other hand what was observed in §§ 256 and 260.

To Slag Azurestone

§ 264. This usually contains up to 1 lod silver per centner. To 1 part [ore] is added 12 of lead kernels.

To Slag Iron

§ 265 (a). It should first be calcinated, before slagging can be carried out easily. Machine turnings, which are cut into fine bits, are placed on a pan in the test oven; when they are warm, the surface phlogiston of the iron burns away like coal dust, and meanwhile they should be continually stirred with a stirring rod, or they will stick together. When burning is complete, [the mixture] is ground, and whatever will do so is calcinated anew. Finally it is left collected together in the muffle for 1-2 hours, and then it is ready. This iron ash is violet and more increased in weight than other metallic calces, namely 36 marks and more per centner, whereas lead only gains 4 lod per mark. [We can note here that Scheffer, and therefore also Bergman, although they were phlogistonists, were fully aware that metals gained weight when they were calcinated. It was this gain in weight on calcination that led Lavoisier to propose that something, i.e., oxygen, must be added in calcinattion, rather than something, i.e, phlogiston, being lost. J.A.S.]

Calcinattion can also occur in other ways: 4 centner of iron filings and 1-1/2 to 2 centner of sulfur in powdered form are mixed well in a tea cup, moistened with as much water as is needed to make the mixture just hang together, which mixture begins to fizz in 24 hours, as when wet gunpowder is ignited, and then forms a lump which is ground down, sifted and roasted free from sulfur. A similar mixture can be buried to imitate an earthquake [volcanoe], and has such a strong attraction for the air that no matter how great a weight is put on it, it is thrown off and the charge forces its way out.

Note. See § 201, Note.

(b) To 1 part crocus martis [iron sulfate] is added 8 [parts] of lead kernels and 4 of lead glass. At first the degree of heat should be regulated well, since if it gets too warm the material swells up from the phlogiston of the iron which was not completely driven out in the calcination, and when it cools again the mass solidifies. On weighing it is noted that the augmentation of the calx is not accounted for unless 1/2 Centn. is added for the reaction, weighs 50 + 36/2 mark, without taking into account more than 4 Centner lead kernels and 2 lead glass.

(c) Slagging also occurs without calcination, but takes more time, and the material is apt to boil over.

To Slag Tin

§ 266 (a). To convert it [tin] from metallic form into slag takes a long time, so it is best to lay the tin on a roasting pan and on top of it to place 2 to 4 times as much lead, put it in the mouth of a muffle and close down the lid to hold in the heat strongly. During fusion the lead sinks down and takes the tin with it. Some rosin added after fusion promotes mixing. Then it is all stirred together until it forms a powder. Calx of tin increases in weight like lead. [See § 265, (a).]

(b) Tin calx cannot be fused or slagged, either with lead kernels or lead glass, but only lumps itself together with them. Some believe that if copper equal [in amount] to tin calx is put together with 20 parts of lead kernels, the copper will attract the tin to itself and promote its slagging, just as lead after a fashion combines with silver, and this operation takes a long time and uses much lead. But it does not succeed.

With 32 parts lead to [one of] tin [the whole] is slagged completely, but the silver content is not determined with certainty.

(c) To accomplish this [slagging of tin], white arsenic, finely powdered, is mixed with some redlead, [the mixture] poured on a red-hot dish, which is taken out quickly, since the charcoal of the paper reduces it, immediately placed in the strongest heat, a coal laid in the muffle mouth and the lid closed again, to deprive the arsenic of air until fuming. The mass begins to swell strongly, but as soon as it subsides it is quickly poured out on a test dish and ground fine after it cools; this is arsenic-flux.

Take 4 parts of this flux, 12 of lead kernels to 1 of tin ash, which last-named is mixed well with the flux. The dish is put in [the oven] so that the arsenic gets driven off and then we continue with the usual procedure.

To Separate Silver From Tin On A Cupel

§ 267. To 1 part is added 100 parts of lead, which is driven off, which procedure requires a larger oven than usual, but the silver content is much more certainly determined than by calcination, during which a large amount is lost [bortsmörjes].

To Separate Silver From Bismuth On A Cupel

§ 268. This semi-metal [bismuth] slags itself and reacts alone on a cupel like lead, and therefore [needs] nothing but the cupel. Zinc does not have this property. Bismuth sometimes unites with silver, and sometimes not.

That which soaks into the cupel can be obtained in pure form again with reducing agents, but free from silver. It leaves a yellow color behind on the cupel.

Bismuth is slagged in the usual manner.

Note. Gold and silver are used for coinage because of their durability and beauty, as well as for precious vessels and various small trinkets. Platina could also without doubt serve a similar purpose if it were available. The gold smelter at Adelfors has produced about 20 marks per year, but our silver smelters, which formerly were so conspicuous, now seldom attain over 3000 standard pounds.

SECOND CHAPTER

Concerning Ignoble Whole Metals.

§ 269. Ignoble metals allow themselves to be converted to powder in heat, and so lose their phlogiston and allow it to escape, but still increase in weight, so that a calcinated metal always weighs more than before, but the increase is different for different metals. [See § 265 (a).]

Note. When metals are calcinated in the dry method heat is required, which increases the volatility of the inflammable material [phlogiston], but the presence of air is also required, which strongly attracts the same [phlogiston]; both circumstances must be combined in order for any separation of consequence to occur.

Mr. LAVOISIER maintains that instead of phlogiston being separated, there is combined with the metallic calx aer fixus, or more properly, acid of air; thus tin and lead can be partly calcinated in a hermetically sealed vessel, but only in proportion to the size of the vessel. Air is always mixed with [contains] acid of air, so a larger vessel must contain more of such mixture. The vessel, he has found, weighs the same before and after calcination; likewise air streams in when it [the vessel] is opened, more in a larger than in a smaller [vessel], wherefore after this communication [with the air] the vessel becomes precisely as much heavier as the enclosed metal gained during calcination. Mr. de MORVEAU urges that since phlogiston is lighter than air, its presence makes the weight less, (just as iron, by using a wide bend in it, can be made to float on water); likewise decomposition of its calx is the reason for its gain in weight.

[I can hardly pass over this section without again calling attention to the fact that both Scheffer and Bergman were very close to understanding, and thus discovering, the role of oxygen in the process of ´calcination´. It was in 1775-1776 that Bergman received Carl Scheele´s book "Air And Fire" from Scheele with the request from Scheele to write a Foreword for the book. In this book Scheele revealed his discovery of oxygen, but he too did not yet fully understand its role in the oxidation process. Bergman at the same time had his own book on "Elective Attractions" on the press, and at the last moment added a section describing how he had repeated Scheele´s experiment in heating mercury oxide and producing oxygen gas. I have described this incident in my biography of Bergman. It is one of the tragedies of history that Bergman was unable to put these two ideas together, as Lavoisier did, and come up with the discovery of oxygen. Eight years after his publication of "Scheffer´s Lectures", Bergman in Volume III of his "Opuscula" (1783) included an eleven page article on aer vitalis, column 35 in his Table of Attractions, in which he detailed some of the properties of this new element, six years before Lavoisier published his "Traité élémentaire de Chemie" in 1789. These last words of Bergman on this subject, published just a year before his death, are worthy of more study. J.A.S.]

FIFTH SECTION, CHAPTER II, Ignoble Metals

On Quicksilver

§ 270. Quicksilver is: (I) a semi-metal which is found in cinnabar, and also pure, when it is called Mercurius nativus or virgineus.

Note 1. When quicksilver solidifies through cold it can be hammered much like lead or tin. Metals have different heats of melting, Pla- tina most, since it can only be brought to it by a burning glass; and Quicksilver least, because it needs so little heat for melting that the atmosphere alone in only a few instances can be deprived of enough degrees, by 2[00] to 300 below the freezing point [Hg melts at -39 C]. Mr. PALLAS has in two cases in Siberia found the cold so intense that quicksilver solidified. So we have here the two extremities of fusibility. When running quicksilver now is classified with the semi-metals because it cannot be hammered, it is like saying that molten tin cannot be hammered and consequently belongs to the semi-metals.

Note 2. The specific gravity is reckoned to be 13.590, but quicksilver which BOERHAAVE distilled 500 times gave 14.110.

(II) It volatilizes in heat and changes into vapor.

Note 3. Quicksilver expands 15/1000 from 0 to 100 degrees of heat; boils at 350 degrees of heat [357 C]; contracts from boiling to solidifying, irregularly at first, but much below 0 very strongly, and altogether about 1/10 of the original volume.

The solidification of quicksilver , according to Mr. BROWN'S account, has about the same amount of expansion and sound as lead, but cuts more easily. The tensile strength has not been determined and can hardly be tested.

(III) It can be changed into a red calx with the aid of heat.

Note 4. This calx is improperly called mercurius praecipitatus per se. The operation requires many months of continuous heating and is best carried out in flasks with flat bottoms, into which the quicksilver is poured and the necks are then drawn out by being held over a lamp flame. They are placed in a sand-bath, and the heat should be as high as possible, without vaporizing the quicksilver, whereupon the surface of the metal gradually turns into a red powder. This can, as Mr. Dr. WEIGEL reported, recover its lost phlogiston in acid of salt and dissolved alk. volatile in the wet method; this can also occur with heat without addition, whereby it appears that quicksilver is an intermediate between precious and non-precious metals. [Here Bergman was again flirting with the discovery of oxygen. For mercurius praecipitatus per se was nothing else but mercuric oxide, which both Scheele and Priestley heated to produce oxygen.]

(IV) It dissolves in aqua fortis, but not in acid of salt, if it has not previously been dissolved in another liquid; acidum vitrioli, both before and after rectification, attacks it.

Acids cannot completely calcinate quicksilver, without [first] obtaining Merc. subl. corr. and other [compounds] combined in strong heat with part returning to the fluid state [running mercury]; therefore not all can do this, since some must absorb the acid, which alkali, calx, etc., produce, without imparting some fatness or the least sign of it.

Note 5. Concerning this solution, see §§ 80, 95, 108, 124, and 146.

§ 271. Quicksilver dissolves all complete metals and forms amalgams with them, with the exception of iron, which however can be combined with it if a solution of iron vitriol be added to it. Various semi-metals can also be amalgamated.

In general the operation is carried out as follows: 4 to 6 parts quicksilver in a glass mortar are ground with one [part] of the dispersed metal. With many metals this may not be enough, and more quicksilver is taken to form a buttery consistency. Metals which melt below glowing heat, such as lead, tin, bismuth and zinc, combine immediately in this way with quicksilver.

Lead, tin and zinc are amalgamated most easily with less quicksilver, and are volatilized with it over a flame; next [most easily] gold, then silver, and thereafter copper, which latter is left behind when the quicksilver evaporates in heat. All non-precious metals can eventually be rubbed out [i.e., amalgamated with quicksilver]. Metallic calces cannot be amalgamated. Some other [properties] of this kind appear in various descriptions of individual metals.

Note. OSIANDER used flame, amalgamation and grinding, used alternately, to bring gold itself to a yellow irreducible powder.

To Purify Quicksilver

§ 272. Quicksilver can be mixed with so many foreign substances that it seldom is obtained entirely pure.

(a) If it is only some earthy material mixed in, it can be separated sometimes by pressing through thin chamois, which the dirt cannot penetrate.

(b) Grease and salts can be washed away in a glass mortar with vinegar and common salt.

(c) Gold, silver and copper can also be separated from quicksilver through chamois, but not lead, tin or zinc, which penetrate [the chamois] with it [quicksilver] more closely combined, and this is often used for adulteration. In this case the quicksilver must be distilled with iron filings, alkali or unslaked lime, and then is forced through the leather; but if we want to be very careful, it should be sublimed with sulfur, when the metals mentioned do not follow with it, and afterwards the quicksilver is driven out of the cinnobar.

Note. Whether lead, tin and zinc can be driven over similarly on distillation of quicksilver, and if it is prevented by addition of iron, alkali or lime, I have not had time to investigate up to now. That bismuth is not separated from quicksilver by chamois is known.

To Distill Quicksilver From Cinnabar

§ 273 (a). This cannot be accomplished without adding something which attracts sulfur more strongly than quicksilver; examples of this are many, such as copper, iron, burned lime, alkali fixum, etc., each of which can be used; but if, in regard to the quantity of quicksilver, we choose which of the additives is to be preferred because it has the strongest action, this is alkali fixum, followed by iron filings.

(b) Alkali fixum and cinnabar alone are ground together very thoroughly in a glass mortar until the mixture gets an even color; this [mixture] is put into a retort which is properly heated in a blast furnace. The receiver need not be cemented on, since although they are volatile, the fumes are collected in the neck of the retort; some water should be there, so that the heat of the quicksilver may not be harmful.

(c) The heat is increased little by little until the retort glows, after which the set-up is allowed to cool. The quicksilver condenses like dew in the retort neck, but col- lects into drops which fall down into the receiver.

(d) The residue is a liver of sulfur which almost always contains some residual quicksilver which is detected by dissolving [the residue] in water, filtering, and precipitation with acidum nitri, since the precipitate gets black, whereas sulfur alone otherwise turns gray. This occurs principally with plenty of alkali, so that the sulfur is well saturated, since quicksilver rises [vaporizes] without heating to glowing. Now and then the cinnabar contains some acid and this rises similarly to cinnabar if burned lime is used, whence one part of the acid is saturated and thereby becomes insufficient to absorb all of the sulfur.

Note. For decomposition of cinnabar, iron filings are both the most serviceable and the least expensive, when no quicksilver is left in the residue.

Quicksilver ore is assayed for pure quicksilver or cinnabar by simple distillation, but we must know the correct amount of quicksilver held in the iron filings.

To Prepare Cinnabar

§ 274. Cinnabar is a compound between sulfur and quicksilver. For its production we first prepare aethiops mineralis.

(a) This can be made in 3 different ways: (1) when 3 parts quicksilver are ground together with 1 part sulfur in a glass or stone mortar until it all becomes black and no more quicksilver is visible. (2) To well melted sulfur in a retort is added 3 times as much quicksilver, [the mixture] stirred well with a pointed iron rod, the heat increased somewhat and the stirring continued until it finally begins to thicken and blacken, and finally is ready when it becomes dry and crumbly. If, despite all care, the mass catches fire, it should be immediately poured out on a flat rock, and must be burned again, in case it is not already black. Both these preparative methods are difficult, and in the latter much sulfur is lost. (3) Still another method is given, but little known, yet most convenient of all, which is based on the fact that sulfur and quicksilver over mild heat do not mix until the sulfur melts, but then the combination occurs, with a strong light and a sharp report, in an instant, after which it turns into a black mass. We need to put together 4 parts quicksilver in an open retort which is placed in a blast furnace and heated up. A little sulfur is added but not all of it, which might lose the retort; the moment of combination is noted well and then more sulfur is added accordingly; combination now occurs with strong heating, but without report. The sulfur being more volatile is sublimed to a small degree, before combination occurs, but this is before the retort becomes so warm that it [sulfur] melts and runs down.

(b) If either of the first two [methods] is used sublimation occurs in an open retort without an alembic; if the third [method] is used, aethiops is added only at the end, and when it is ready the heat is increased. Both quicksilver and sulfur are volatile with less than glowing heat, but their combination is much more incombustible. Therefore the heat must be increased until the bottom glows, after which the furnace is allowed to cool.

(c) The impure aethiops has gleaming needles, and is in some cases somewhat dirty and black from the aethiops, which puffs up to the brim. If sublimed again, it all becomes more beautiful and delicate, but requires stronger heat.

(d) In the bottom of the retort after the sublimation is found a brownish residue, which contains some kind of earth, but seems to be nothing other than partly some impurity existing in the quicksilver or sulfur, and partly phlogiston burned to an ash, because all the sulfur never combines with the quicksilver, especially if the heat is strong, without some of it being lost and acid of sulfur fuming away. As often as we sublime cinnabar we will always find such residue, which can only be attributed to the abundance of impurities; it cannot be an earth either, since this would all be separated in the first sublimation.

Some residual quicksilver is usually found in the neck of the retort or deposited on the bottom, which only occurs when sulfur is deficient.

Note 1. The author is not always referring to the subtle behavior of phlogiston, which now is usually noted under that name, among some coarser materials, which are burned to ash, and are obtained from volatilization of water, etc. (§ 201).

(e) Holland is the only place where cinnabar is sublimed in quantity, and its manufacture has become such a secret that we hardly even know where it is done. The pro- cess must be carried out at very low cost, since cinnabar and quicksilver are inexpensive here, although the former should at least be more expensive [here] on account of wages [for workmen]. The third method for making aethiops, as the least costly, would most likely be that which is used in Holland.

In native cinnabar we have the better and cheaper kind; the finest costs over 20 plate per pound, but the cheaper kind is not as good as the artificial, which costs 2 plate per pound. [1 plate = 20 pence.]

Cinnabar can be used with both oil and water, but gets a better appearance with the former. It must first be ground dry, then with water until it no longer sparkles; then it is dried and ground again until no gritty sound is heard, and finally with the water or oil.

Note 2. The higher color of natural cinnabar comes from its lesser content of sulfur than in the artificial, since it often contains 7-8 parts quicksilver to 1 [part] of sulfur.

(f) Cinnabar can also be prepared from Merc. subl. corrosivus and antimonium crudum mixed and sublimed, when the quicksilver, which has a stronger attraction to sulfur than antimony, finally is sublimed to cinnabar, which is still more expensive than that prepared in the previously described manner. It is usually used for epilepsy, pain of childbirth, etc., but is no better than other kinds of cinnobar.

Note 3. In addition to the extensive use of quicksilver occasioned by Medicine, and its compound with sulfur used in the art of painting, it is also useful because of its particular liquidity in Barometers and Thermometers; for its fluidity occasionally in injections; for its volatility, added to its ease of combining with gold and silver, for certain gilding and silvering processes; for its strong attraction to gold and silver, to separate these precious metals from stone and earth; for its firm combination with surfaces of tin and glass, for making mirrors; etc.

Sweden does not have the least indication of quicksilver ore, and thus must satisfy all its needs for this metal from abroad.

Concerning Lead

§ 275. Lead is : (I) in weight compared to quicksilver as 11,345 is to 14,019; but compared to water as the same number is to 1,000.

Note 1. The specific gravity of pure lead is somewhat greater, namely 11,386. [11.337 at 20 C.]

(II) It is of all metals the softest, and has no clang.

Note 2. LEMERY once found a clanging lump of lead; DE REAUMUR in 1726 has more closely investigated this property and found that bowl-shaped or elipsoid segments of lead do have a clang. If it is fused and molded into a ladle it behaves similarly, but gets soundless if it is jolted after cooling, or stirred with a steel wire; hammering takes away the slight clang. Lead sounds also in a bell-form, but lower.

Its [tensile] strength is such that a filament of 1/10 of a Rhenish inch diameter breaks at 29-1/4 pounds, which is less than any other complete metal.

(III) Of all but tin it is the most easily melted and it flows before glowing [heat is reached].

Note 3. Lead needs less heat to melt than zinc, but more than bismuth, about 300 degrees [327 C].

(IV) It is calcinated very easily to a gray-green powder and is soonest brought to a slag; in fire we also obtain a redness, which appears as litharge and red lead.

Note 4. Lead gains about 10 percent in weight in calcination.

The blue-white color of this metal darkens quickly in air, but resists longer than both iron and copper before it is calcinated.

(V) It is dissolved by all weak acids with a sweet taste, but not sharper than them. It is calcinated white by acids.

If lead white is boiled with some pressed oil it dissolves with effervescence until its saturation is attained, when it becomes as hard as wax, but darker, and it is called Balsamus saturni, which is the best external plaster for open wounds and contusions.

Note 5. Lead, and the more its calx, is dissolved with the help of heat in fatty oils, and makes a natural base for plaster. For the rest of its solutions see §§ 80, 83, 96, 112, 125, and 135.

To Granulate Lead

§ 276 (a). A piece of lead is melted in a ladle with mild heat, and a little tallow is added to prevent burning.

(b) When it is all well melted and the tallow burned a- way it is poured into a wooden trough, which is coated with chalk to prevent its burning, it is shaken back and forth just when it begins to solidify, and is fanned up and down so that it forms into small granules.

(c) When it is all hardened it is poured on a strainer, and that which does not go through is remelted and granulated in the same manner.

To Prepare Redlead

§ 277 (a). Lead is placed on a dish in a heated test oven, and when it melts it must be immediately and continuously stirred. After some hours a light yellow lead calx is obtained, which is further calcinated, but at a very mild heat, otherwise the calx is clotted and the color destroyed; it darkens gradually, but not at a proper red heat in the muffle, but with a flame.

Note. Redlead is prepared in England at two places, from 15 hundredweight of lead all at one time in a single oven, which I have a sketch of, and which by the flame of stone coal produces all grades of calcination.

English redlead therefore always contains a small amount of acid of vitriol. The color is at first yellow, from which the calx gets the name of lead-gold, yellow mastichot, minium flavum, etc.

(b) If white lead is burned, redlead is obtained in one hour, but not fully reddened, for the just mentioned reason.

(c) Redlead is used on sores, and mixed with wax as a very good plaster for gout.

§ 278. Lead mixes with all metals except zinc; cobalt and nickel have not yet been studied.

Note 1. Lead does not combine with cobalt; and slowly with nickel and iron, but still does not let itself do this [completely]. Zinc can be fused together with lead; equal amounts of both give a somewhat ductile mixture, which is whiter than lead; 10 parts zinc volatilizes 1 part of lead.

Lead is amalgamated very easily, and according to the account of Mr. SAGE has already been made to crystallize in a mixture in which 1/2 times as much quicksilver is added.

Lead is added in the working of tin, concerning which we will have more in its place [i.e., under tin}.

Note 2. Since the ores of lead and all the following metals have been tested in a crucible, I believe now is the proper place for the author's [Scheffer's] general notes on this subject. They concern the concentrating of ores by washing, roasting, and finally smelting down.

To Wash Ores

§ 279 (a). The ore is ground and sifted very fine, and then some of it is placed in a washing or sifting trough, a little water poured on it, the mixture stirred by hand, more water poured on and stirred in until it is all drenched and no lumps remain.

(b) Then the trough is held with the left hand and beaten on the large end with the other hand, but only gently, so that it is only shaken back and forth a little, and the water does not wash out, since it always carries some ore with it. By such stirring the ore is collected at the back of the trough, and the mineral [dross] at the front end.

(c) When now a pile of dross is found collected, by the shaking of the trough, somewhat to one side, it [dross] can be separated from the ore better; [the trough] is tilted more and more forward, so that the water runs away and takes the dross with it, and this should be continued until it does this. Some residue sticks in the trough and it is washed away, but carefully, so that no ore goes with it. The beating on the side should be done carefully so that the trough does not tilt backwards, since then the ore is stirred up, and some is lost. When the ore is well separated from the dross, the trough is tapped gently on the edge, to empty out the ore.

(d) In the first product, the ore is never pure, but all the dross must be separated by repeating the process, after which the washed [ore] is rinsed into a glass and dried.

The same process is continued with all [the ore] which is to be washed, which is then called schlig [German: schlick - slime], which is then washed one more time, to make it purer still. Although all this be carefully carried out, some ore is always lost, which goes off with the dross.

To Roast Ore

§ 280. Besides litharge, hearth-stone, certain iron ores, etc., which contain no sulfur or arsenic, such easily melting ores as those that stick to the container and make analysis uncertain, such as lead glance, and those which melt to pure stone [skärsten], must all be roasted. The last named contain more sulfur well mixed in than is necessary, but this excess alone does not separate with cer- tainty, so it is best removed by roasting.

The reason the sulfur should be removed is that it will otherwise form liver of sulfur with the alkaline flux on blasting, which [liver of sulfur] is dissolved by the metal, and in stronger heat is vitrified with the glassy material.

Roasting occurs as follows:

(a) The ore is ground up and sifted to a flour, is put into a roasting pan, and placed in a moderately heated test oven, at the front at first, and stirred with a stirring hook so that it is heated evenly and does not form clumps. When the ore is hot it begins to burn, like coal dust, and smells of sulfur while it is stirred evenly.

<u>Note.</u> The more easily fused [ores] should not be so finely pulverized, since they lump easily and often unavoidably. The author [Scheffer] notes this himself further on under antimonium crudum.

(b) When the ore is burned out its weight decreases in the pan like sand, and it is moved into stronger heat. Here we continue as above (see (a)).

(c) When this has also burned out, it is moved into the strongest heat, and meanwhile the oven has been heated to the fullest. The muffle door is closed, while the vent hole is opened, and the pan left thus for 1/2 to an entire hour, after which it is quickly removed, stirred and smelled; if no sulfuric odor is noticed it is fully roasted, but otherwise it is returned and the test repeated.

(d) The weight of the ore at the beginning compared with that at the end gives the loss on roasting.

(e) If notwithstanding all precautions the ore still has lumps, it must be reground, but as some is necessarily lost thereby, this should be taken into account when the loss on roasting is determined. Call ´a´ this weight before grinding, ´b´ the same after, ´c´ the weight after complete roasting, and ´x´ the correct weight which should have been found if no grinding occurred; then must:

$$b:a :: c:x$$

To Blow Off [Afblåsa] A Test Crucible

§ 281 (a). This takes place on a forge, which should be well walled up, and an overturned crucible laid on its base as a mold, 3 crossed fingers [space] between, so that the coal can fall in between, and high enough so that the bellows blows evenly on all of it.

Note 1. Mr. J.G. GAHN has devised a clay-daubed slab of pig-iron to close the forge. High crucibles are set along the upper edge so that they allow a suitable opening; a hole with a lid is installed for putting in the coal. This device is excellent; we can obtain a stronger heat and also a great convenience for observing the test, which in an open strongly heated forge presents great difficulty.

(b) The ore and flux should be finely sifted and then weighed. The latter [flux] is first put in the crucible or finger-stall, then reweighed and mixed well. The ore is put in on top, but no granules should come to the edge, which is also avoided when the ore is now finally mixed up a little with the flux, but sticks to it and is evenly spread out; anything which helps the mixing during fusion brings the strongest clods together.

(c) If the flux is alkaline and it swells up vigorously during fusion, a little common salt is laid on top, whereupon the regulus deposits and is collected, which is prevented if the alkali sputters; this also prevents the edges being corroded by the flux. But if no alkali is present, no salt is needed.

Note 2. Common salt also makes a fused layer on top, which prevents entrance of air and the accompanying burning.

(d) Then an overturned crucible is cemented fast, which should be large enough so that it measures just 1/2 inch in front, and the clay can lie evenly. French clay is used here, and finely powdered crucible meal only, which is used with just so much water that the mixture becomes stiff enough that it can hardly be worked with the fingers, because otherwise it breaks. Then the cementing is carried out, and rubbed smooth with a little water, and sprinkled with red sand, which glazes the surface when fused. When the cementing and other processing are completed well enough that the crucible is not tilted, by which the common salt can be mixed with the ore and scattered around with crackling. The crucible cover can be either whole or broken off to the width of an inch, whereby two kinds of burn-offs occur, with closed or open crucible.

(e) Then the lower crucible is dipped in water so that it does not break from excess heat, and placed on its base in the ash with one flat side toward the mold, when the blowing works strongest on the material. The test with the crucible cover closed is begun in cold ash, coal is filled around it, and on top are placed some small glowing coals, with the wall furthest from the mold; when all the coals are aflame, as is best seen through the mold, the blowing is begun. The test with the open crucible cover is not as easily started in cold ash, if we expect it will take longer than to crackle the salt, because the flux is usually burned before the blowing can begin. Cold coal can still be placed nearest around it and a loose crucible turned over it until the common salt crackles.

(f) When the blowing is begun most of the ash is filled with coal, and also the space between the crucible and the mold, the the blowing plays immediately on the crucible and cools it. When the material begins to bubble and melt, two weights are laid on the bellows to increase the heat strongly. During the blowing the odor of acidum salis from the salt is very noticeable.

Note 3. Vigorous and sufficient heat for self fusion are important for this test.

(g) After completion the test remains a certain time in the ash, calculated from the beginning of the blowing, but it is taken out and opened as soon as the material seems to remain still and clear, otherwise combustion occurs which must be watched for particularly with rich ores.

Note 4. The time for closed blowing of the test must be observed not only after the strongest part of the blowing, but also after the time when the ash and the coal are burned.

(h) A test [sample] is taken out and placed on an iron plate, on which it is beaten with a hammer several times hard, but then more gently until the regulus is collected; if all blows are hard the smelted metal is scattered.

Note 5. To avoid both too hasty cooling and upsetting, the test [sample] after removal can be put into a prepared burning ring.

(i) When the crucible loses its redness it is cooled in water, except with lead-, tin-, and skärsten-tests, which cool by themselves, because the regulus in the first two cases [lead and tin] frees itself and is cast out; skärsten falls apart again.

(k) Finally the crucible is broken, the regulus taken out and weighed. The slag is ground to look for loose kernels [of metal], in case all is not collected in one place.

Note 6. The test weight is usually so set up by us that 1/4 lod is taken for 100 marks or 1 quintal, which is also the usual amount for a test. Finally if the regulus is found to weigh 1/8 lod, the ore is said to contain 50 percent, 1/16 corresponds to 25 [percent], 1/32 to 12-1/2, 1/64 to 6-1/4, and 1/128 to 3-1/8 percent. Therefore we can use the ordinary lod weight, in place of only hundred weight, which are still necessary in certain cases for accurate assaying. In our test weight, 1 mark = 1/400 lod.

To Test Lead Glass For Lead

§ 282. Glete (lithargirium) is lead vitrified on a blast forge, and it has properties similar to redlead. Its powder has a sooty taste in the mouth and is very unwhole- some, so people who shovel this and pack it in barrels, can- not endure this work more than 2 to 3 years, without getting a hitherto incurable disease, called Hyttenkats [cottage cough] in Germany.

To 1 part lead glass are taken 2 [parts] blackflux, or tartar flux (§ 37 (b)), preferably the latter, because the former is too weak and is blown off more safely in an open, rather than a closed crucible. In a closed [crucible] it is blown off in 2 minutes, whereas 5 minutes are required with a weight on the bellows [for blowing off in an open crucible].

To Test Leadglance For Lead By Slagging Down

§ 283. To 2 parts [leadglance] is taken 1 part borax, which is put on the bottom in an open crucible and the ore on top, but yet not brought to the brim. It is then set in ash and a long piece of iron placed with one end in the ore, after which the blowing off is begun. If it is not well fused, a weight is put on the bellows; if too much heat is used, the lead will stick to the stone. When the crucible gets red hot and the material is fused clear, the crucible is taken out, and also the piece of iron. The regulus is found when the crucible is broken apart.

Because leadglance cannot be roasted (§ 280), iron is used here to slag it down, which attracts the sulfur more strongly than the lead, forming a stone, which fuses upon it and keeps up the burning off. Borax slags the mineral if some is present, also the iron which is calcinated out of the skärsten, from which the slag becomes pure and separ- ates itself well from the regulus.

To Test Leadglance Mixed With Pyrite For Lead

§ 284 (a). This kind [of ore] can be roasted, but this should still be carried out cautiously, since it is very apt to concrete itself (§ 280).

Then 1 part of roasted ore is weighed out, 1 [part] borax and 2 [parts] tartar flux, which is blown off in an open crucible. To add glass powder is unnecessary because enough is vitrified from the iron in the pyrite; the lead should not contribute here because when present it causes error.

(b) If we wish to know the silver content, 1 part ore is weighed out, 1/2 [part] borax, 1/2 crystal glass, 2 redlead, and 2 tartar flux, which is blown off in an open crucible. The glass makes the slag purer, and with its help a part of the lead is also vitrified, which is not made certain here when only the conservation of silver is desired. So that no lead may be lost, redlead is added, but lead kernels are less useful, because they do not extract the silver as quickly from the roasted ore as redlead or lead glass, which extract it just as it is being reduced. The regulus should then be slagged on a pan, since it some- times happens that some of the iron reduced from the pyrite enters the lead; if no iron is found to be present, no further slagging is required than is suitable for the size of the cupel, in which finally it is driven off, the kernel is weighed, and the [added] redlead content is subtracted.

To Mineralize Lead

§ 285. A wet crucible is placed in hot ash, and when it has been heated it is blown off without weight [on the bellows] until it is red hot; when the sulfur and lead glass have been finely ground and well mixed they are folded in paper and dropped in. Immediately after this a weight is laid on the bellows, and pieces of sulfur are dropped into the crucible little by little, so that the lead may be well saturated. If the lead glass is impure, so that it does not fuse, a little borax is added. When all is melted and clear the crucible is removed, another overturned over it to prevent combustion of the sulfur, by which the artificial lead- glance crumbles, which also occurs if it is cooled off in water.

Note. Lead, although a metal of poor appearance and malleability, is found to have many uses. In other countries it is used for roofing large buildings. Water is carried away in lead gutters, and kept in cisterns made of the same material, which would sometimes seem to be detrimental to health. Paint manufacturers make a linseed oil product with this metal, which also takes different colors, such as white lead, slate white [skifverhvitt], masicot, and redlead. For its weight and ease of melting it is very useful for bullets and shot. In the mining business and in assaying it is used in winning, boiling [ansjudningar], making tests, and slagging. It is used for alum- and vitriol-containers; for enamelling clay vessels, and for certain kinds of glass; and in Medicine for plasters, sugar of lead, etc.

Several hundred pounds of this metal are imported annually.

On Copper

§ 286. (I) Specific gravity [of copper] is to [that of] silver as 8843 is to 10535.

Note 1. The proper weight is usually given as 8,726-8,843 [8.92]; the Japanese [kind] is somewhat heavier, about as 9,000, and indeed our coinage goes to 9,324. The difference must be attributed in large part to differences in smoothness, purity, etc. The hammered [kind] is heavier than the fused.

(II) Of all copper, the Tyrolian is both the most malleable and retains its clear redness the longest, without blackening or turning green, is therefore cemented with chopped gall [gallmeja] for artificial gold braid, since in both malleability and persistent clarity it surpasses all Pinschback.

Note 2. Its hardness is greater than our silver, but less than platinum. By heating to glowing and hammering, its hardness is increased, but it can still never be brought to [that hardness of] steel up to now. Its elasticity is greater than our silver.

Its toughness is less than our silver, indeed than our yellow-brass, but about 6 times greater than our tin. A copper wire 1/10 inch in diameter requires 299-1/4 pounds to break it, but one of brass 360.

Its malleability is greater than our iron, but less than our silver; the purer the copper the more malleable.

(III) It requires a high heat to melt it, but it never can be cast solid by itself, since with mild heat it becomes so lumpy inside that it does not submit to forging; but with high heat and the combustible substance it becomes bubbly outside. It can be calcinated in red heat, darker than [the heat] for quicksilver and lead.

Note 3. Copper melts before iron, but after gold; it requires white heat, about 788 degrees [1083 C]. The surface of melted copper displays a blue-green flame; if it cools slowly, different glistening figures appear on the surface. The scale falls off on hammering, and is nothing other than more or less calcinated copper.

(IV) It is dissolved by all mineral and vegetable acids, and also by alkalis. With acid the solution becomes green or blue, and then after calcination in fire [it becomes] black, but some comes away gray.

Note 4. Concerning copper solutions, see §§ 80, 84, 97, 113, 126, 136, 140, and 157.

To Calcinate Copper

§ 287. This can be carried out in many ways:

(a) If thin copper scales and sulfur are stratified in a crucible, and this is put into heat strong enough that the sulfur is ignited; but coal should be avoided. Little by little more sulfur is added until it all forms a lump which is ground finely and fully roasted on a pan.

(b) Thin copper sheaves are calcinated by holding them in strong heat, but not strong enough for melting, for a long time. The best way is with window-glass, if it is laid in the mouth of a [relatively] cool oven, from which the flame plays evenly; after 2 to 3 weeks it becomes brittle and can be ground.

(c) If we are in a hurry to get copper ash, Spanish Green is burned to a powder in a crucible, which is then roasted in a pan, boiled in water, and further extracted with it [water] until syrup of violets is not colored green [by it]. Burning and roasting drives off the vegetable acid, but after this there is always left some alkali fixum, which cannot be separated in fire, so extraction is used.

§ 288. Copper mixes [alloys] with all metals, although nickle has not yet been investigated. Quicksilver is amalgamated with it readily, if a little aqua fortis is added; the surface gets black, which is mostly calcinated, otherwise it is prevented and combination is made very slowly.

Note. With copper many useful mixtures [alloys] are made. In minted and worked silver it is added to increase hardness. Ten parts copper and 1 part tin is called Styckgods [Piece goods] (metallum tormentarium), particularly if some yellow-brass [mässing] is to be made; it is hard and brittle. Copper 16 [parts], tin 4, yellow-brass or zinc 1, gives klockegods [bell metal] (met. campanarum, aes caldarium). These proportions are not always used. The old klockenspeise [German: glockenspeise; bell metal] contains about 100 parts copper to 12-1/2 tin. For metallic mirrors we usually take copper 3 to tin 1. Also 1 part copper is used with 3 tin and a little arsenic for burning mirrors; it is dense, hard, smooth and steel-colored (met. chalybeum) [Latin: chalybs; steel]. Mixtures of copper and tin become harder, more bell-like, and more resistant to air than copper. Copper 4 parts with tin 1 has no remaining trace of copper color. Sixteen parts copper-tin alloy [kopparbleck] are interlayered with 1 [part] tin and fused in strong heat to obtain finally the gold-colored mixture reported by Mr. POTT. Nickel melts together with copper readily, about which something more follows, and also about its combination with zinc.

Copper is readily amalgamated if it is finely divided and very pure, as when just precipitated with iron, corroded by vinegar or aqua fortis, etc.

To Test Gray Copperglass

§ 289. This consists of just copper and sulfur, roasted carefully, of which is taken 1 part, then 3 of tartar flux, which is put first into the crucible, 1/2 [part] crystal glass added, and finally the ore. It is blown off open, and taken out when it is well fused, since if it remains ever so little longer [than necessary], the more copper is burned up the richer the ore is.

To determine the amount of unroasted ore correctly, a determination should be made of the loss on roasting. Let us assume that 1 centner [100 pounds] loses 20 marks [pounds], so if 1/2 centner is to be tested then to half of the residue, or 40 [pounds], is added 3 times 40 of flux and 1/2 [times 40] of glass; but since this brings in errors, it is better to weigh out 1/2 centner of roasted ore, and the same of flux and glass added, but then at the end to make a correction. If 1/2 centner gave 37-1/2, so must a whole [centner] give 75, and the following calculation gives the correct content, namely:

100:75::80:60.

Note. The percentage should be calculated from [the weight of] unroasted ore in the way the author [Scheffer] has shown.

The ore, which contains only pure blue or green copper calx and pure copperglass, free from sulfur and arsenic, need not be roasted, but can be melted down immediately with combustible material.

To Test Yellow Native Copper Ore

§ 290. This must first be washed (§ 279) and then roasted (§ 280). The darker yellow copper ore becomes on roasting the richer it is, and the redder it is the more iron it contains. To the roasted ore is added 3 times as much black flux and 1/2 pulverized crystal glass.

Since the ore is yellow, it contains a large amount of iron, which loses phlogiston on calcination and should be slagged in fusion, for which metallic calces are suitable, both with and without glassy materials [being added]. The more phlogiston a metal contains the more strongly it attacks the flux, and the more difficult it is to reduce its calx. In the present instance no more reductant should be added than the copper requires to restore it to its metallic state; meanwhile the iron, which cannot get enough of the black flux, goes to slag, although some follows along into the regulus when the iron is abundant.

If the test goes properly, (1) the slag should be black and lustrous, but if it appears green some copper has been burned; (2) the regulus should be black on the outside, redness indicating calcinated copper.

Note. If the ore is not well and fully roasted, regulus is often found within a husk of hard stone [trotsten].

Since copper calx is dissolved by alkali volatile, copper can be extracted from roasted ore in this manner, whereupon so little iron accompanies it that the regulus which is contained in the solute is not attracted by the magnet, according to Mining Counselor BRANDT'S research. Aqua fortis, he has found, takes so much iron calx out of a calcinated ore that the impure copper kernels precipitated from this [extract] are attracted by the magnet.

STAHL mentions that tartar and common salt or alum alone, boiled in water with roasted copper ore, attack the copper and dissolve it.

To Fuse Native Yellow Copper Ore To Skärsten

§ 291. This is the safest way to separate ores, since then they are not lost in washing. For this operation we take 1 part unwashed and unroasted ore with 2 parts borax, which is put into an open crucible and covered with a lid, which when the crucible is red hot is taken off and blowing started. When the contents are fused and clear they are removed. When boiling is observed, and it fuses and clarifies itself, the crucible should be taken out immediately, otherwise the skärsten is burned.

In this way [different kinds of] ores are separated, since skärsten consists of only sulfur and metal. Calx goes in well mixed, as at Tunabergs and Sala skärsten works, but it is accidental.

To Test Skärsten For Copper

§ 292. This can well happen through full roasting and ordinary melting down, but the regulus becomes somewhat mixed with iron. At least the impure kind is roasted very little, so that only the sulfur gets burned off at once by itself in the mouth of the muffle furnace; it is then fused again with a similar amount of borax to the skärsten, which becomes very coppery, but weighs less than it did at first, in proportion to the sulfur burned off.

This is fully roasted, fused with 3 parts black flux and 1/2 [part] glass, in an open crucible, when the regulus becomes mostly iron-free.

The reason for the foregoing [procedure] is that copper attracts sulfur more strongly than iron, so the latter consequently must let its [sulfur] loose first on roasting, and then in the fusion of the skärsten it goes to slag. If we could just deprive the iron of all its sulfur the result would be good, but it is impossible to do this. In reference to this uncertainty in cold-roasting, the following procedure helps get the desired result more closely.

To Concentrate Copper in Skärsten

§ 293. Unwashed ore is fused to skärsten, this is fully roasted and fused again to skärsten with 1/3 raw ore compared to the weight of the original raw ore, and an amount of borax similar to the weight of roasted skärsten and raw ore together. This is fully roasted and fused in the usual manner, when a regulus of lower iron content is obtained.

Note. To collect the regulus from a low-grade ore much better, we use the addition of fine silver, or at least silver of known standard. I have also used added nickel, which is slagged slowly, in the method which is described in more detail in the iron test.

To Refine Copper

§ 294. The impure copper after smelting is called Raw-copper, and always contains other foreign impurities such as sulfur, iron, or lead, which cannot be separated except by burning off: this operation bears the name of refining, which is performed on a small scale in the following way.

(a) Because a stronger heat is required than can be afforded by a test oven, a muffle furnace must be used, or in place of this, a large crucible is set down in such a way in ashes that its back side leans on the usual crucible foot, and the opening faces outwards; the stones which are taken out on the one side are fitted around and on top of it. Consequently the blowing [from the bellows] comes to play upon the closed end, and the test is put in. The ash is filled with coal, and some red hot coals are laid on it, so that all is heated up little by little. A pan is put in the farthest part of the muffle, to be warmed with it.

(b) When all the coals become heated the blowing is begun, and when the pan is well heated, the weighed raw copper is laid on it, partly in small amounts, moved into the strongest heat, and blown on very strongly so as to melt it quickly.

(c) When the copper melts, 1/10 [part] of lead is put in with it, which immediately combines with it; the mass should immediately become clear, various colors and blooms moving about and playing on it. If this does not occur immediately, another 1/10 of lead is added along with a little borax, so that it may clarify so much quicker.

(d) With the blowing continuing until the copper remains entirely clear and unmoving without change, it is then taken out and cooled in water, and the copper is [thus] refined.

(e) If the dose of lead is sufficient and the copper is not taken out too soon, it always becomes refined and appears red, both outside and inside, dry in the fracture and very brittle. During strong heat the copper loses some phlogiston, but gets malleable again if it is laid on top of a crucible filled with small coals and melted down into it with blowing.

If the copper is not completely refined, it is put on another pan without lead, and blown upon until there is no more motion.

(f) In refining there is always some loss, because impurities and the lead burn off their own weight, together amounting to at least 1/9, so that 1/10 of the whole loss, the lead being undetermined, could be compared to the weight of refined copper in order to determine the correct content.

(g) If copper has half or more impurities in it, which sometimes can be inferred from its appearance, then we first place on the pan a like amount of refined [copper], then the raw [copper], and lastly lead in sufficient amount.

(h) Attraction to sulfur is so strong that old copper, which has been both forged and fatigued, gives off a sharp silver blackened odor of sulfur, when some score of pounds are melted down together. The refining is certainly aided by lead, although Fahlun copper always is malleable, and the ore contains little lead. At Ädelfors, and also at Gladhammar, investigation is being made into how to refine non-malleable copper from sulfur through the addition of 1/40 to 1/20 [part] of aged cast iron on the coppersmith's forge to make it good [malleable?], since the sulfur attaches itself to the iron and is burned off, with the phlogiston of the latter metal, in bright sparks over the forge during cooling. The non-malleability of sulfur-containing copper arises from the fact that the sulfur dilates itself during cooling and makes the metal loosely packed [otät].

To Prepare Gray Copperglass Artificially

§ 295. In a crucible with the cover off, thin beaten copper is laid, and when it glows, sulfur is cast upon it a little at a time, it is blown off [with a bellows], and is taken out when the material is clear melted.

To Prepare Yellow Copper Ore Artificially

§ 296. Like amounts of thin slagged iron and thin scales of beaten copper are laid in a crucible, the former on the bottom, after which another [crucible] is cemented on bottom-side up. When the metals glow, sulfur is added a little at a time until the material remains quite clear.

Note. Copper is a metal much in demand, which is used sometimes for coins, for roofs of large buildings, for cooking vessels, for Spanish Green, for bellmetal, for ordnance material, brass, pinschback [a kind of brass], and which is used for pictures, statues, clocks, pots, and many other uses. As for its use in the preparation of food, without doubt this is dangerous and has caused many deaths, because copper is attacked not only by vegetable acids, but also by the acidity which exists in fats, if the contents of the pot are allowed to cool in it. Coating with tin will prevent contact with copper, but the covering [with tinplate] is seldom complete, and soon wears off; whether this covering itself is harmless will be discussed under tin. Arch. ELLER has found that 8 lod of common salt boiled with 5 pounds of water in a copper vessel dissolves 20 grains [from the vessel]. In France they have started to coat successfully the inner side [of copper vessels] with a thin coat of 16-standard silver, and thus this otherwise serviceable metal [copper] is freed from danger of poisoning.

Sweden, which once could produce 15,000-21,000 pounds [of copper], now barely reaches 6,000 in a year.

On Iron.

§ 297. This metal is: (I) compared to gold in weight as 7,852 is to 19,636, and to water as 7,852 is to 1,000.

<u>Note 1.</u> Cast iron has a density of about 7,100, but when forged it reaches 7,795 and above; steel is about 7,704; malleable [iron] 7,738; and that for springs 7,800.

(II) Bar-iron from sulfur-containing ore can be forged cold, but not warm without breaking, and it is called red-short [iron]. Cold-short [iron] behaves just the opposite.

After different kinds of forging the iron gets different names. It is called <u>pinched-iron</u> [Knippjärn] when it is beaten into hatchets, spades, hammers, shovels, nails, etc. <u>Band-iron</u> consists of long thin bands, which are used to wrap barrels and casks, but also sometimes for other iron implements. <u>Loop-iron</u> [ögeljärn] is made into small bars, a finger thick, 1-1/2 quarter broad, bent into the form of a figure 8, which are sold abroad. <u>Ordinary plate</u> is broad plate bar-iron which gets the name of <u>pattern-iron</u> [Schamplun järn] when it is of a certain thickness and breadth. <u>Square</u> [Fyrkant] is squared on all sides like bar-iron. <u>Bolt-iron</u> is made in bars with 8 sides. <u>Osmund's-iron</u> is produced from bog-iron-ore [myrmalm] which is smelted in small ovens to fresh, i.e. viscous, but not liquid masses, which are taken out with hooks put into the oven, because there is no opening underneath; they are beaten into small bars with iron clubs, since the factories seldom have over 2 water hammers, which are not enough; the bars are then bent by splitting across, almost round, and are sold in this form. This iron is neither cold-short nor red-short, but is very good and therefore much prized, either in the bars described or made into scythes and other small merchandise, under the name of Osmund's-iron-ware. Granje's-iron, which is entirely cold-short, is forged very tough and good by Osmund's smithys. Because of negligence by the smiths, and welding, Osmund's-iron has lost much of its desirability in recent times.

Note 2. Red-short iron can be hammered cold and heated to white heat, but not red hot; cold-short [iron], on the contrary, is only malleable when red hot; good iron of dry-stone [torrstens järn] is malleable both red hot and also sometimes when it is cold. That red-short [iron] results from acid of sulfur is pretty certain, but to what cold-short [iron] can be attributed is not yet determined. A malleable bar-iron can simply turn short by melting down. This property of the metal, to aspire to a certain structure for its particles which makes it swell into a momentary structure, is the most useful of all for castings, because the smallest corner of the mold is filled; its brittleness should also arise from this expansion after full melting and slow cooling.

Its hardness is greater than some other metals of the same kind, greater with the cold-short than the red-short, and greatest with steel, which consequently takes the best gloss, and freshly heated approaches that [hardness] of flint; this is increased by rapid cooling, but can be tempered again by heating to glowing. Elasticity is considerable and greatest, of those named above, with steel.

Iron is next to gold in toughness; a wire of 1/10 tum [inch] diameter requires 450 pounds to break it. A pound [of iron] can be drawn out to one mile in length, as Director and Knight RINMAN has shown. Iron is much more malleable than tin, but is not as extensible as copper. Cast-iron is rarely malleable, if not specially processed in a small smelter.

(III) Next to Platina, [iron] is the most difficult to melt; yet cast [iron] itself can be made fluid in fire, but not malleable, without addition of phlogiston.

Note 3. Iron is, next to platina, the most difficult metal to melt, and requires about 880 degrees of the thermometer [1535 C]. With a burning mirror it melts with strong boiling, during which a smoke arises, which is really a flame underneath; finally it forms a black slag left behind. Steel melts more easily. If well fused iron hardens slowly it gets different shapes on the surface.

Forged iron is dark gray at fractures, glistening and fibrous; red-short likewise, but less glistening and many-colored on the outside; but cold-short [iron] is white, glistening and coarse.

(IV) It [iron] is burned up most easily of all metals and can be calcinated in the heat of a flame without melting. The reason is that it contains more phlogiston than some other metals. In fire it becomes entirely black, which looks like forge slag. When fully calcinated iron gains 36 percent in weight.

Note 4. If a polished steel sheaf is laid on hot coals, it acquires in order [of appear- ance] a whitish, yellow, burned yellow, purple, and finally a blue surface, which last color also occurs if the heat gets to react further. Similar colors arise from electrical sparks and can probably be attributed to certain degrees of phlogiston, since iron shows more clearly than other metals its different states due to different amounts of the combustible substance. Iron glows easily, even through simple friction; heated to white heat, it becomes scaly and sparkles in fire, especially red-short. The red scale, which falls off during hammering, forms a red powder called hammerslag.

Iron calx is slagged with slow gentler heat than is needed for reduction, just the opposite to that used with other metals.

In a closed vessel iron does not change, and KUNCKEL also determined that when it lies 4 weeks in a glass oven, it becomes more dense and more glistening, but as soon as it is removed to the mouth of the oven it swells up and is calcinated in 8 days.

(V) {Iron} is dissolved by all acids, indeed it is attacked by water and is changed into rust or crocus.

Iron calcinated in acids becomes red, which resembles brick.

<u>Note 5.</u> Concerning solutions of iron, see §§ 85, 98, 114, 127, 149, 158, and 163.

Acids provide an easy way to distinguish iron and steel, along with their different hardness and composition. Steel dissolves in aqua fortis much less than iron, when all conditions are otherwise alike; the former [steel] takes on from the attached iron calx a dark gray surface, the darker the harder; but the latter [iron] becomes white, the more [white] the softer. Consequently we can get information from simple appearances;

corners and streaks of another color indicate uneven composition. If the dark film is scraped away or scoured away, the steel appears gradually brighter than the whiter iron, all of which Mr. RINMAN has disclosed with careful research. This is the basis of the etching and damasking on guns and swords.

Gray so-called nödsatt [trouble-built] cast-iron has more phlogiston than white hårdsatt [hard-built], but is dissolved with more difficulty in aqua fortis and leaves a black sediment, often of a volume similar to the [original] iron, which [sediment] is a substance like black-lead and an earth of iron saturated with phlogiston. The phlogiston is decreased by the smelting to white iron, or cementation with absorbents, and the iron is then dissolved without residue.

Iron, which lies for a while under water without exposure to air, falls apart to a black powder, which is attracted by the magnet, and is soluble in all acids, and is called aethiops martialis Lemerii. In moist air it is changed to rust; steel resists the longest, particularly when polished, and cold-short [iron] next longest.

(VI) [Iron] has the power of contraction and astringency. If iron filings are ground together with sulfur alone and calcinated in a crucible, we get crocus martis aperiens Villisci, which opens, rinses and cleans the blood similarly to the crocus from iron vitriol, which has the same nature. If water is boiled with crocus Villisci we get the same taste as with iron vitriol or mineral [surbrunns] water .

Note 6. The finer crocus gets the name aperientes, and the coarser adstringentes; they are prepared in different ways, through changing to rust in fire, in air by means of moisture or vinegar, by precipitation from vinegar, etc.

(VII) [Iron] is attracted by the magnet.

Note 7. Iron does not only have the extraordinary property of attracting the magnet, which [the magnet] is one of its ores, but also can win for itself the magnetic power; soft [iron] takes on this power readily, but loses it quickly too; cold-short [iron] must be hammered a lot and heated to glowing in between; steel becomes magnetic slowly, but then preserves this state best. A rust or iron calx does not follow the magnet; it must previously have been reduced to a certain degree, but it still does not require fusion, but merely heating to glowing with enough phlogiston.

Note 8. Steel is a very noteworthy improvement on iron, by which 1 pound can be brought to a value of 80,000 copper dollars [c.d.], worked with fine tools; indeed made into artistic objects and instruments to 400,000 c.d., according to Mr. EKSTRÖM'S calculation. Steel has a geater specific weight, is harder, more elastic, and more difficultly soluble in acids than iron having the same fine grained texture. In general there are 2 kinds of it [steel], namely fused-steel and burned-steel, according to Mr. RINMAN'S classification, who has made himself so well known for his profound knowledge of iron. The former [fused-steel] is obtained by fusion and hard working, and not directly from the ore, like cold-steel [vargstål], or from cast-iron like Raw-steel [Råstål]; raw-steel rolled together gets the name of refined-steel [garfstål], to which general class belong the German, Hungarian, and Milanese. The latter is obtained by cementation of bar-iron; here belong cast-steel [gjutstål], which is the densest, finest and most serviceable with regard to [holding] edges [as on knives], which after burning is fused in crucibles with retention of its malleability. For the rest, many different kinds arise from different processes, different materials, and varying fuels, such as pit-coal, charcoal, wood, etc.

The nature of steel seems to depend principally on the fixed amount of phlogiston [contained]. Cold-short iron yields wretched steel.

§ 298. Iron lets itself blend with all metals except zinc. Research has not yet been done with bismuth. Quicksilver does not combine with iron unless a solution of iron vitriol is poured on it, when an amalgam is obtained, but without which the iron can be ground and nothing dissolved [by the mercury].

Note. Iron is combined with other metals by fusion, even with bismuth, but with zinc [ir- on] hardly does this because it [zinc] burns up so easily. With tin is obtained a philosophical device [lana philosophica] for a burning mirror, from which HOMBERG believed that zinc was a blend of iron and tin; if the iron makes up only 1/15 of the mass, the magnet is still attracted. Iron is not used in compositions other than usually with tin and arsenic, when a silvery blend will be obtained from them, which nevertheless tarnishes in time. Arsenic and antimony in certain proportions deprive iron of its power of attracting the magnet.

FIFTH SECTION, CHAPTER II, Ignoble Metals

To Test Iron Ore

§ 299 (a). Since not all iron ores contain sulfur or arsenic, and some are a dense iron calx, so they do not all need to be roasted. To determine if an ore contains sulfur, a piece is laid on a low fire. When it is somewhat heated it is taken out and smelled; if it does not smell of sulfur it is heated some more, and if it still does not now smell it is heated fully to glowing, after which, if it does not smell of sulfur it needs no roasting, in case there is no zinc present. Sulfur-rich iron ore does not give up any sulfur if it is immediately put in strong heat without being warmed slowly, whereas a weak [ore] does not release its small amount [of sulfur] without being heated strongly. All mica-containing ore should be roasted after previously being heated with coal, from which it becomes somewhat softer.

If the ore contains zinc it is often known [to have] no sulfur, before the finely ground [material] comes into the strongest heat. Such [an ore] should be lightly roasted at first, but then very strongly at the end, since otherwise the mixture does not lose all its sulfur. When an ore calcinated in this way is mixed with half [its weight] of coal dust, is put into a crucible to which another [crucible] with a hole in the bottom is cemented on, is set in ashes and blown upon [with a bellows] until it becomes hot for an hour or so, after which it [the blowing] is stopped, and the coal dust burns off with a violet flame at the opening, whereupon the zinc begins to show itself with light blue to green drops. When the flame dies down it is blown upon, but when it again begins to be sustained it is continued until no more flames come out. No weights are put on the bellows, partly because they are not needed, and partly because the fairly small iron kernels might run together in the crucible at the end. Finally when the zinc has gone a- way, the crucible is removed and the ore is laid on a pan, and the coal dust burned away in the test oven.

(b) When the iron test is blown off with closed crucible cover, the following addition is made. To 1 part ore, black flux 2 [parts], tartar flux 2, broken glass 1, borax 1, pure soot 1/2, and coal dust 1/2. If the ore is mixed with quartz an additional 1/2 [part] of unslaked lime is taken, which vitrifies it; but if no rock material is present, the lime attacks the ore, making the slag viscous, and consequently is undesirable.

For poor kinds [of ore] the soot and coal dust are reduced in quantity; not so much is needed for the reduction and the excess remains unfused. Iron is difficult to reduce and therefore much phlogiston is needed; if there are other metals in the ore they go into the regulus, since the iron is slagged most easily.

Blowing off is carried out for 5 minutes with no weight [on the bellows], 10 [minutes] with a single weight, and 15 to

20 with double weights, all according to whether the ore appears difficult to fuse or not.

Note 1. Mining counselor VON SWAB recommended the following amounts for the test with closed crucible cover, namely to 1 Centner of roasted ore, black flux 2 [Centner], tartar 1, sal ammoniac 1, glasgalla [bile-glass, green-glass?] 1, borax 1/2, glass 1/2, pure soot 1/4, and coal dust 1/4; yet this appears to be much the same composition as I have found for entirely different ores, such as bog-iron ore, granite from Vestanfors, brunsten [manganese ore] from Nerke, Eisenman, greenstone, black drawn sand from Medelhafvet, Noreberg's slatey iron-colored bloodstone, steelstone, sulfur-bearing iron ore, etc., and these not only produce a regulus, but also usually more than with some others. The time required for blowing off a closed test is most difficult to regulate. I have found for a single test, which in all circumstances was made with most careful measurements, that 5 minutes for short blowing off produces about 15-20 percent separation into the regulus, but much longer barely half the mentioned loss. Different fluxes with the same ore give entirely different percents for the most part.

(c) For a test with uncovered crucible we take for 1 part ore, tartar 4, tartarflux 2, borax 1, crystal glass 1, finely sifted dry white lime 1, and white arsenic 1/2.

Note 2. Arsenic is properly believed to form a flux quickly, so that the regulus has the opportunity to collect well, but this can likewise be accomplished just as well with a little lead glass, and without fear that the composition will be disturbed.

(d) When an iron test goes successfully the slag should be brown and glistening in the fractures, like colophonium, but the least black indicates burned iron.

Note 3. The ready combustibility of iron has the result that the slag seldom avoids be- coming black. The test kernel should be well con- tained and collected.

(e) Ores which contain 50 percent [iron] are classified as rich, some contain 90, almost 100, without being attracted by the magnet, and are consequently not the best. Lake-ores give abundant material and cold-short iron, mostly the latter, which still does not always come from the ore, but from greenstone, which is used for flux and blowing off.

Note 4. Those which contain less than 30 percent do not pay to be worked.

Mr. CRONSTEDT has produced iron from heavy stones [tungstenar] when 1 Centner of well-roasted [ore] is mixed with coal dust 1, soot 1/4, sal ammoniac 1, egg shell 1, glass 1, and borax 1/2, and blown off in a good flame [to produce] 5 [Centner] of residue. Afterwards he has found that borax and alkali fixum slowly dissolve these materials, but sal microcosmicus solidifies them, which therefore should be used in the flux and borax excluded.

To Test a Low-grade Ore With The Help Of Nickel

§ 300. This semi-metal [nickel] is vitrified very slowly, since when it is fused with 5 parts lime and the usual fluxes of the iron test, 1-2 percent is lost. Therefore if nickel is added to a test of low-grade ores, not only is fusion facilitated but the regulus is also collected together and is ready for burning off.

Preparing for the test, we take 3 parts nickel with the ore, the weight of which is based on the regulus, but [at least] 2 percent, or as much as investigation has indicated, is added for burning off.

In place of the nickel, white arsenic can be taken in the same proportion, but then also only lime must be added, to attack and bring it into reguline form with the iron. If the impure smooth regulus is attracted by the magnet, there is 1/16 or less arsenic in it, but if it is not attracted, there is more than 1/16 present.

Note. I have made a similar study, which was quite successful. One centner of such ore, which otherwise left 48-1/2 percent at the most, was mixed with a similar amount of nickel 137-1/2 marks, which when 68-3/4, as residue, when 1 centner of the same nickel with similar flux and similar length of time being held in the flame, is attracted, it shows the iron content to be 68-3/4.

To Prepare Artificial Pyrite

§ 301. Into a crucible, heated to glowing in ashes, are put thin scales of iron, and on this when glowing, sulfur is cast little by little, until the iron is saturated. As soon as the iron melts like the whole mass, but without becoming fluid, the crucible is removed, some sulfur put in, so that there is no deficiency [of sulfur], and the whole allowed to cool by itself.

Note 1. If white glowing iron is spread with sulfur over a pan of water, it melts immediately and changes to pyrite, which can be collected in the pan. If the iron [so] sulfurated is calcinated and reduced, about 30 percent is lost.

Note 2. Iron is among the most indispensible of metals and also seems to be the most plentiful in Nature, distributed over the entire Earth. It is used according to its different qualities for all sorts of tools and instruments, salt pans, pots, roof plates, nails, anchors, and almost innumerable other necessities. Sweden has produced in later times about 400,000 pounds per year of bar-iron, but does not reach that now; Norway makes 80,000; England, 225,000; and Canada about 7,000; after which Russia, Germany, and the rest of Europe produce a considerable amount, which all bears witness to the demand for this metal. By using large ovens much wood is spared, but the iron is less suitable for fine forging than that from small [ovens].

On Tin

§ 302. This metal is:

(I) One of the very lightest, being to gold as 7.321 is to 19.636, and to water as the same is to 1.000.

Note 1. Tin's proper weight is 7,180 to 7,400 [modern value, 7.31]; the latter is the purer. This metal is the Plumbum album of the ancients, and has been known from the oldest times. It is harder than lead but softer than gold; it hardens somewhat during hammering, as do all whole metals, and can be made malleable again through heating not quite to the melting point but nearly, which also can be said of lead.

(II) It is malleable but makes other metals more brittle, except for iron which in combination with tin becomes more malleable than by itself.

Note 2. Its malleability can be judged somewhat from the so-called tin-foil, which is used for mirror-coating; its [malleability] is less than that of copper , but greater than lead.

Pure tin has hardly any elasticity, but its addition makes other metals more bell-like than before. A wire of 1/10 tum diameter breaks at 49-1/2 pounds, and consequently its toughness is greater than that of lead, but much less than copper. When tin is bitten or broken it creaks.

JUNKER says that tin detonates with saltpeter and when then reduced does not creak.

(III) It melts without previous heating to glowing and produces fumes very dangerous to the lungs, yet which seem to be nothing other than phlogiston volatilized during calnation, like that which comes off from hot coals, but whose fumes are dangerous to the heart even if no arsenic is present,.

Note 3. It [tin] melts at 213 degrees and does not then have heat enough to ignite paper, hair, or gunpowder. With a burning mirror it immediately becomes liquid, producing filamentous fumes like white lime, which form small fibres or crystals.

If the amount of copper nitrate in the form of a dry powder held on a knife is laid on a piece of tin foil, moistened quickly and folded up together somewhat compactly with all possible dispatch, after an hour there breaks out elastic vapors, and the tin is ignited and sends sparks around. This experiment was devised in London and was communicated to me by Mr. GAHN.

(IV) It is calcinated readily to gray-gold, but the calx does not allow itself to be melted per se, unless by a burning mirror.

Note 4. When tin stays fused, it is calcinated on the surface to a grayish powder. GEOFFROI has noted that, if it is heated to glowing, and the calx surface layer has also broken open, small white flames come out. The more phlogiston the calx loses the whiter, the more difficult to melt, and the less reducible it becomes; it is used to polish glass, to make enamel, etc. This calx can be made into a red-gold, transparent glass in strong heat with no additions.

In free air the color of tin darkens somewhat, but it is not further attacked.

(V) It is dissolved by all acids, except acidum nitri, which corrodes it to a white-gold calx.

Note 5. See §§ 87, 99, 115, 128, 150, znd 159.

§ 303. Tin blends with all metals in fire; but with nickel and cobalt it has not been investigated.

Note. Tin is seldom free from arsenic. It melts together readily with both nickel and cobalt, and also with one of the latter it makes a malleable mixture, which is quite worth noting. 3 to 4 parts of quicksilver are combined with it readily by simple rubbing, and up to 10 [parts] with the help of warming; in crystals, 3 [parts] quicksilver enter into 1 part tin. Quicksilver is driven out of lead more easily than tin. This metal combines with copper, as has previously been related (§ 288), and in addition to this it has been used as a cover plating to prevent rust. Tin plating is carried out in 2 ways in general, either with or without addition of lead; the latter [without lead] with pure tin is accomplished by means of sal ammoniac and without scraping, and only this method should be permitted, since lead is also as easily attacked and injurious to the health as copper; but our coppersmiths get to know this method very well. Sheet iron is usually tin-plated for certain uses; the plates are scoured, corroded bright either in the vat with acidifed bran, or sal ammoniac, after which they are dipped into molten tin.

All tin which is made into vessels is usually cast with lead. In Sweden 3 mixtures are allowed, namely 4-stamped which contains only 3 percent lead; 3-stamped which contains 17, both with 1 percent leeway; and 2-stamped in which lead makes up half, which latter [alloy] is only used for small flasks or the like for peasants. In Germany 9 to 14 percent lead is added.

To be able to judge with certainty and handily whether the composition of the alloy is entirely right, the specific gravity is still the only means we use, which assumes unchangeability in volume, and in addition that different heats of melting can remain uncertain.

The calculation, given by Mr. SCHEFFER in 1755, and used in control-accounting, has the following basis. Call a the weight of a pure metal, b the weight of a similar volume of a heavier metal, c the weight of an alloy of the 2 foregoing metals having the same volume, x that which goes in of the heavier metal into that last mentioned [alloy], and y that of the lighter [metal], so then:

$$x = (c-a)b/(b-a)$$

and:

$$y = (b-c)a/(b-a)$$

which in the present case only seeks the percent of the ingredients; so a and b are set at 100 parts, from which:

$$x = \{(c-a)/(b-a)\}\{100b/c\}$$

and:

$$y = \{(b-c)/(b-a)\}\{100a/c\}$$

Further, as the real weights of lead and tin are to each other as 23 and 15, so 23 can be set for b, 15 for a, and 8 for b-a, then finally:

$$x = \{(c-a)/c\}\{287-1/2\}$$

and:

$$y = \{(b-c)/c\}(187-1/2)$$

Consequently when enough weight of pure lead and tin shot is taken, cast in the same form, we can find from their weight in the alloy a similar volume percent. Mr. SCHEFFER has also calculated tables, which give the weight of shot in certain form immediately, without further calculation, and give the percent of excess or deficiency.

To Test The Zinc Group For Tin

§ 304. To 1 part ore we take tartar 2 [parts], tartar flux 1, and colophonium 1/2. Blowing occurs with open crucible cover, and after this the ore usually contains 60-70 percent.

Note 1. The [amount of] flux given here by the author [Scheffer] is better than some others which I have investigated up to now. For a successful operation, speed and proper heat are required, and to remove the test at the proper time. This method is the one which proved best for me: that the centner of roasted ore with its proper amount of flux is divided into 3 portions, each of which is folded up in paper. When the crucible reaches white heat the first [portion] is put in and covered with the lid; when no more flames come out the second [portion] is put in, and in the same manner the third, and not over 4 minutes is needed for this. All the while the bellows is going evenly, and also then must boiling up stop, which with a good heat does not continue over 5 to 7 minutes.

Note 2. Tin is a metal which is used for all sorts of flasks and vessels, for tin plating, mirror foiling, and for light metal compositions; the calx is used for enamels; tin solution for purpura mineralis; for enhancing the color of cochenille, etc. Tin 2 parts with quicksilver, sulfur and sal ammoniac each 1 [part], gives aurum musivum after 16 hours heating; this contains 10 parts tin to 7 sulfur.

In Sweden we have not found any worthwhile occurrence of tin up to now, and this metal is rare elsewhere also, but in England, Saxony, Bohemia, East Indies, and sundry other places are found notable deposits of it.

THIRD CHAPTER

Concerning Semi-Metals.

§ 305. The malleable [metals] are called whole [eg- enteligen] metals, and those which are found to be brittle under the hammer [are called] semi-metals.

Note. The semi-metals generally occur less frequently than whole metals, and cannot be used for so many applications because of their lack of malleability. They lose their phlogiston in fire and so belong to the non-precious [ignoble class], and regarding all of which it is agreed that friction or warming causes each to give off its own odor, and often each similarly reveals its own individual taste. At one time we believed that volatility in fire was a common distinguishing mark of semi-metals, but this was entirely in error, since among the whole metals are found different ones which fume, and among the semi-metals some that are non-volatile.

On Bismuth

§ 306. This semi-metal:

(I) Is somewhat like lead, but the color leans toward gold, and its specific gravity is much less.

Note 1. Its specific gravity is 9,625 to 9,926 [9.80 at 20 C.]

(II) Is very brittle and glistening in the fracture, but not as hard as the regulus of antimony.

Note 2. The fractures occur in sheaves, in confusion with each other. Bismuth is soft but has some clang.

(III) It melts readily and fumes in fire.

Note 3. It melts sooner than lead, but later than tin, and before glowing [heat], at about 250 degrees [271 C]. It is sublimed in a closed vessel, but requires stronger heat than zinc.

(IV) It is calcinated quickly and then melts to a slag, not unlike lead-glass; it glazes the inside of the crucible if bismuth remains molten for half an hour. It behaves like lead on a sand bath, and leaves a little silver kernel for the most part, since most often it contains silver, but in such small proportions that the cost of the operation is not recovered.

Note 4. It fumes in fire more strongly than lead; the fumes are inflammable. The calx yields a redlead, but paler than the ordinary [redlead].

Neither water nor air attack the surface of bismuth, although after a long time the color changes a little towards purple.

(V) It is dissolved by acidum nitri and aqua regia.

Note 5. See §§ 88, 100, 116, 129, and 151.

§ 307. It blends with all metals in fusion, except zinc and cobalt; iron has not been investigated.

Bismuth is used as an ingredient in ordnance [styckegods] or bell-metal [klokegods], but these uses are not well known; in Sweden they do not occur. For ordnance it is quite unsuitable, but it would seem to be serviceable in bell-metal, but it would be expensive.

Note. Quicksilver 2 parts combine readily with 1 [part] bismuth; this much also enters into crystals. NEUMAN reports that bismuth amalgam sticks to iron, and after fuming off the quicksilver, makes an artificial silver plating. Bismuth 2 [parts], lead 1, and tin 1, combined give a noteworthy mixture which melts in boiling water. Bismuth 1, lead 2, and tin 2, give an easily melting solder; bismuth, lead and tin taken up with a little quicksilver give a material suitable for injections on certain occasions; bismuth 2, tin 1, lead 1, and quicksilver 10 give reflecting globules [spegelklot]; bismuth , tin and antimony serve pretty well to stamp coins in; bismuth, tin and quicksilver give argentum musivum. Zinc can fuse with bismuth but with difficulty; tin and lead get from bismuth both more shine and more hardness; it [bismuth] does not combine with arsenic.

To Test Bismuth Ore

§ 308. To 1 part roasted ore is taken 2 [parts] black flux, or better tartarflux. Blowing off is carried out better with an open, than with a closed crucible cover, just as with leadglass (§ 282).

§ 309. In most cases bismuth is smelted along with cobalt. A wood fire is built over a hole, and on it is laid unroasted cobalt ore, which often contains bismuth, which comes out with heat. Also usually the ore is put into clay pipes, which are heated, and when the bismuth runs out it is collected in a gutter.

Then the blueglass is melted and we get the so-called Speisen, often mixed with bismuth on the bottom of the glass pots, from which it can then be separated.

When bismuth ore is dissolved in aqua fortis or aqua regia, black cracks form which are phlogiston, if some sulfur was present, but often the ore contains nothing else but arsenic.

Note. At Schneberg in Saxony this metal is collected in inclined pipes in domed ovens. According to HELLOT, quite a few have mixed bismuth and cobalt, although these could not be combined by melting alone; but in speisen nickel often serves to bring them into a lump. The ore is most often obtained [relatively] pure.

Bismuth very rarely is found in Sweden. Out of the mixtures which were mentioned previously, it is usually used for paint; LEMERY, the father, begetting a multitude, was himself and his family engaged in the preparation of Blanc d´Espagne [Spanish white] (§ 100), which in those times was a secret.

On Nickel

§ 319. Nickel is:

(I) A distinct semi-metal which is obtained from an ore called kupfernickel [copper-nickel].

Note 1. Kupfernickel was [believed to be] composed of a mixture of iron, copper, arsenic, and cobalt, until Master Miner CRONSTEDT, in a report of his in 1751, published research maintaining that it was an ore of a distinct semimetal which he called nickel, and he reported that it had the following nature: namely to be calcinated green, often with rising vegetation [mineral growth]; to produce a hyancinth color in glass; to be dissolved in acids with green color and in alkali volatile with blue. The specific gravity differs somewhat, approximately over 8,500, and sometimes under 7,000 [8.90 at 20 C]. It is purifed with difficulty from all sulfur, cobalt, and arsenic, but so far no method has been found to separate it from all iron; it has always been very tough, malleable, and very difficult to melt. When it is not attracted by the magnet, it signifies nothing else than that arsenic is present and reacting to prevent it. It is fairly heat-resistent, but not in strong and continuing heat. The regulus is attracted always by the magnet, if arsenic is not present in a certain amount; indeed it is found to be magnetic itself. Its color is white, sometimes bordering on red; the fracture is usually like steel.

(II) It must be slagged.

Note 2. On this basis it can be used to test low-grade iron ore (§ 300). This semi- metal both releases phlogiston slowly and takes it up quickly.

For its solutions see §§ 89, 101, 117, 130, 142, and 151.

§ 311. Nickel does not mix with silver, but does mix well with iron, bismuth, cobalt, and arsenic; with others no investigation has been made.

Note. Nickel combines with all metals except silver, quicksilver, and zinc.

To Test Kupfernickel

§ 312. The Swedish [ore] is golden, fine-glistening, and mostly stained with white veins which contain iron and cobalt, from which the regulus of the former [iron] becomes somewhat tough and porous in the fractures. German kupfernickel is usually compact and yellow, not unlike pyrite, containing not as much iron, but also some silver, which clings to the regulus in small kernels when it is melted out.

The greener the calx becomes in roasting the richer it is in nickel, but the redder it becomes the more iron, and it is almost never free from iron. At the start of the roasting sulfur comes off, and then arsenic, from which latter substance a little effervescence has occurred as soon as it reaches the heat.

Roasted ore is blown off with twice as much tartar flux in an open crucible; it usually yields 50 percent.

Note. This regulus, which is so obtained, is far from being pure, and can still be freed from arsenic by calcinattion with coal dust and by slagging freed from the largest part of the cobalt, but still not free from iron. Nickel is found in pure quantities, and is used as it is in most cases, but also gets some use in the so-called white copper, from which various small articles are manufactured in China. As a coloring for oil or water, nickel calx is pale and ordinary, but can be aided by addition of blue. From glass and argentum salitum can the silver be precipitated by means of nickel, but not very pure.

On Arsenic

§ 313. The regulus of arsenic:

(I) Does not melt by itself, but only in combination with a second metal.

Note 1. This semi-metal melts at about the same degree of heat as antimony, catches fire easily, burns with a small weak light blue flame, and repugnent odor of garlic, rises in a closed metallic vessel, but is calcinated in an open [vessel]. The semi-metal is more volatile than its calx.

The regulus has a specific gravity of 8,308, is very brittle, and more or less leaf-like in the fracture; it is similar in color to lead, but tarnishes quickly in air, first yellowish- and then black-colored.

Its solutions have already been mentioned in §§ 90, 102, 118, 131, and 152.

To avoid ambiguity it should be noted that with arsenic its regulus is always indicated; the calx is indicated by an additional term such as white, or otherwise. The same is true of cobalt and antimony.

(II) Its calx, white arsenic, is a common poison; it fumes away over fire with the odor of garlic, without melting; it is soluble in acids, particularly aqua regia, spiri- tus [vini], oils, and all liquids; combines with alkali fixum in fire to a water-soluble mass, and is precipitated with a reagent.

Note 2. Mr. SCHEELE has discovered that it is nothing else but an acid itself, combined with so much phlogiston that it retains a dry nature. It is not as volatile as sulfur, and appears not to yield the odor of garlic unless it can take up more phlogiston. The dissolved calx has a proper weight of 3,750, and when reduced to a solid mass or glass by heat, 5,000. [Arsenic (metallic) has a density of 5.727 at 14 C].

White arsenic is also outwardly effervescent and [the metal] probably still more acid by itself, because phlogiston usually neutralizes [dulcifierar]. But, regardless of the combustible [phlogiston], the calx is dissolved in all liquid substances, and thus reveals its saline nature.

White arsenic's volatility is attacked by fire, by lime, clay, and also flint; alkali fixum attacks it also.

§ 314. White arsenic is combined in fire with all metals except quicksilver; but zinc has not been investigated.

This is carried out in the following manner: to [1 part of] the metal is taken one [part] white arsenic, and double [this amount] of alk. fixum. The alkali and arsenic are pulverized, mixed, and placed in a crucible, and on it the metal. A cover is glued to the open crucible, and this is blown off in ashes until its matter is well fused, then it is allowed to cool. All such combinations become white and fragile, but yellow in air, and lose both color and gloss.

We get an arsenicated copper from Germany variously worked, such as spoons, etc., We also get a mixture of brass and tin which resembles the material mentioned. All metals mixed with arsenic are dangerous and leave something dissolved in warm liquids or foods.

Note. Arsenic also lets itself melt together with zinc, but not with bismuth.

To Test Ore For White Arsenic

§ 315. This is found in iron-, copper-, silver-, lead-, and cobalt-ores, but principally in a peculiar whitish pyrite, made up of iron, a little sulfur, and mostly arsenic. Whether auripigment, which contains only arsenic and sulfur, and which is dug from mines and sold, should be considered an arsenic ore cannot be settled exactly, because arsenic is never distinguishable from auripigment in regard to cost and the inconvenience which are associated with it.

The ore is ground up to the size of nuts, placed in a retort, which is put in a sand bath, to be heated up so that white arsenic [fumes] rise up and condense in the upper part; but finally, when the fire becomes stronger, the sulfur comes over, by which Rauschgelb [reddish yellow, German name for orpiment or auripigment] is formed, and at this point the heating should be stopped immediately, and when the retort cools off the arsenic calx is collected and weighed.

To Reduce White Arsenic

§ 316 (a). Pure regulus is obtained if sublimation occurs in the mixture with alk. fixum and phlogiston, especially indigo or dried blood.

Note. Mining Counselor BRANDT suggests making a dough with oil, which under increasing heat is brought to combustion, and by its charcoal gives phlogiston for reduction. The regulus, which first starts to rise up, readily becomes somewhat sooty and is purified by a new sublimation with the addition of a little potash.

We can also win the regulus by fusing, if 1 part white arsenic is mixed with 2 [parts] German soap and 2 [parts] potash, fused quickly, and poured out. By this method most of the arsenic vaporizes and only part of it is combined with alkali.

(b) Regulus arsenici martialis is obtained if 1 part iron filings is taken with arsenicum album and black flux, 2 parts each, which is fused in a crucible and cast in a mold.

To Prepare Artificial Auripigment and Rauschgelb

§ 317. Sulfur and arsenic sublimed together are readily combined in whatever proportion they are mixed. If we take 1 part sulfur to 3 of arsenicum album and the subli- mation is carried out in an open glass retort until the bottom glows, we find a proper Rauschgelb has deposited itself, the lighter the higher up. The greater the proportion of arsenic, the stronger the heat required, and the more times it is sublimed anew the redder the sublimate obtained.

Native Rauschgelb is called Sandaraca mineralis and re- algar, which latter word is still used in general for all native mixtures of sulfur and arsenic, whether they be yellow or red.

We could believe that the volatile arsenic would evaporate in an open vessel, but the sulfur slows it down, and glowing heat is not needed for sublimation.

At the bottom of the retort are found the impurities from the [original] ingredients, and some small part of the unsublimed; if the residue is pulverized we get brick-color.

Note. One part sulfur with 5 of white arsenic gives Rauschgelb, Risigal or red arsenic, which it is also called; but if the amount of arsenicum album is doubled the sublimate becomes yellow or auripigment.

White arsenic, antimonium crudum, and sulfur melted together well enclosed give a red-brown glistening mass, Lapis de tribus or Lapis Pyrmison so-called, about which the chemists have much to say.

The specific gravity of white arsenic is 3,223, but of auripigment 3,313.

§ 318. From auripigment is made atramentum sympatheticum [sympathetic ink, invisible ink], when it is boiled in water with a like amount of unslaked lime, with continuous stirring and subsequent filtration.

If we then write with a clear solution of crytallized sugar of lead, we get a writing which disappears on drying, but the vapor from prepared symp. ink forms a dark brown permanent color. The vapor is so penetrating that if 2 to 3 sheets of paper are placed between the writing and a blotting paper steeped in the symp. ink, it [the writing] still becomes clear and visible.

The solution of auripigment should be warm to be useful, or otherwise it does not work; it can be used for a long time.

Note. This ink also gets the name Liquor vini probatorius, because it detects lead in wine. Why wine turns sour is something about which nothing is yet known, nor what can take away the acid without altering the taste and color; except for lead, litharge, or other saturnina; but this in turn makes the wine extremely dangerous [poisonous]. Liquor vini probatorius is nothing else but a solution of hepar calcis [calcium sulfide]; as soon as this is dropped into wine it always causes precipitation, but if lead is present the deposit appears dark or black. Saline liver of sulfur gives the same reaction.

Auripigment also is present in the familiar Psilothrum or Depilatorium, which is used so much in Turkey, along with unslaked lime, strong lye of burned beanstalks [bönstjelkar], sulfur, saltpeter, Iris root [radix Ireos], and oil of spike. Good soap-making [såpsjudare] lye serves the same purpose.

§ 319. If auripigment is dissolved with merc. subl. corrosivus we obtain butyrum arsenici, just as easily melting as the ordinary; the sulfur makes cinnobar with quick- silver.

Note. We can also get butyrum arsenici from white arsenic, common salt, and calcinated vitriol, distilled together. Arsenicum album cannot decompose mercurius subl. corr., but the regulus can, which gives its phlogiston to the calx of quicksilver and is united then with the acid of salt.

FIFTH SECTION, CHAPTER III, Semi-metals

On Cobalt

§ 320. Cobalt is a distinct semi-metal.

<u>Note.</u> Mining Counselor BRANDT in 1735 first demonstrated this, and it is strange that LEHMANN, writing in 1761 on the history of cobalt, goes back as far as PLINIUS, indeed into mythology itself, though not for its name.

Its specific gravity is about 7,700 [8.9 at 20 C]; the color is gray-blue and faint, but becomes somewhat darker in air; bell-like; brittle, and fine-grained in the break, almost as dense as steel; the hardness is about that of Antimonium; it melts after glowing [heat] at about the same heat as silver; it fumes but does not burn; it is calcinated black and seems in the beginning to adhere together somewhat, with the same glowing as coal dust.

Its solutions are described in §§ 91, 103, 119, 132, 137, 141, and 153.

§ 321. Cobalt blends in melting with copper, iron, nickel, and arsenic, but not with silver or bismuth; with platina, lead, tin, antimony, zinc, and quicksilver, investigations have not been made. Cobalt and white arsenic fused together give a blue flame.

<u>Note.</u> Platina, tin, and antimony unite with cobalt, but not quicksilver, not lead, nor hardly zinc. Cobalt freed from arsenic makes with iron a tough and malleable mixture.

To Test Cobalt Ore For Azure [Schmalts-blått]

§ 322. To 1 part roasted cobalt ore are taken 3 of refined potash and 5 of washed flint. The potash, being lightest, is laid in first, then the flint, and the ore on top; then [the mixture is] stirred well in the crucible and a lid cemented on. Blowing off occurs for 5 min. without weights [on the bellows] and 25 with weights. The crucible is taken out and dipped in water immediately, whereby the glass breaks and is separated more easily from the brim; it should be entirely dark blue.

To Wash Azure

§ 323 (a). The glass is finely ground in a glass mortar with a little water, then [the mortar is] filled with water, the ground mixture stirred with the pestle, and as soon as the coarse [particles] settle out, it is poured out into a glass, wherein the fine powder settles itself and has a pale color, which is called Eschel, and is sold under the name of Blue Starch [blå Stärkelse].

Fresh water is poured into the mortar and [the coarse particles] are ground for one hour; then [the liquid] poured off into another glass when the coarse [particles] settle out, and so is obtained a somewhat darker Eschel. We continue in a similar manner until it all goes over. That which comes after both kinds of

Eschel is entirely dark blue and is called couleur; the darker and finer, the better, more desirable and more expensive it is. The longer the grinding, the lighter the color, so that the trick is to get hard stones which quickly make the glass into the finest powder, whereby the color is better preserved. The Dutch understand this best; they buy blue-glass at cobalt forges in Germany or elsewhere, that which comes out the most beautiful, grind it fine, and sell it to the Chinese, who paint their porcelaine with it under the name of Indigo.

In commerce the goodness of the color is denoted by F; that is, the more [F's] it is marked with the finer it is. In testing we denote how much sand the color is made with, at the same [amount] of color itself. But sand is measured as parts of flint. The potash is not calculated, since we always take half as much of it as cobalt ore and flint together. Consequently FFFC of 3 sands [means] best color, in which are taken 3 parts of flint [C stands for sand?]; FFC of 3 sands is somewhat weaker, and FC of 3 sands is the poorest. Similarly with Eschel, FFE of 3 sands and FE of 3 sands.

If the ore is considered weak, we only take 2 parts flint [sand], but on the other hand if it is reckoned the strongest [we take] 4 or more, which is determined by the test. In commerce the barrels [of ore] are marked only with the intensity of coloring FFFC, etc.

When iron goes into the glass it becomes foul and dark, which can be predicted for the most part if the ore is melted to pure stone [skärsten], when the greater part of the iron goes to slag in place of sulfur (§ 325 (b)). A little iron does no harm, since its color is driven off in flame.

To Test Cobalt-Ore For Regulus

§ 324. The ore is roasted cautiously, in consideration of the amount of sulfur and arsenic which are found in it. Then 1 part of ore is blown off with 2 of tartarflux in an open crucible. Its richness is 50 to 60 percent.

To Purify The Regulus From Iron

§ 325 These metals must be separated from each other, and although the cobalt regulus also is attracted by the magnet somewhat, it allows itself to be separated.

(a) If it is refined like copper with borax on a cupel so long that it is not attracted by the magnet, we then lose too much in the slag with the iron.

Note 1. One part iron with 3 of cobalt is attracted by the magnet.

(b) If it is fused with sulfur to skärsten, and of this only 1/2 is roasted completely, and the other [half] not at all, and both [halves] are then fused together, then the cobalt, which has a stronger attraction for sulfur, takes the remaining [sulfur] to itself and the released iron goes to slag with the added borax. The new skärsten is roasted fully and fused to the regulus, which, if it is still attracted [by the magnet], is fused again to skärsten. If roasting could be so regulated that sulfur becomes equal to cobalt, the purpose could be achieved quickly.

It could well be possible otherwise that cobalt itself is attracted by the magnet, and consequently that which is not attracted contains arsenic, which also is difficult to drive off, being fire resistant, as long as it has something to attach itself to. Thus iron should be able to be freed from arsenic if it is first driven into the slag and then reduced.

Some have used cobalt for blending with iron, copper, and arsenic, which should give blue glass, but depends on the proportions used, or we find that this occurs because cobalt is a distinct semi-metal. Moreover no whole metal gives color to glass which persists in fire the way cobalt does.

Note 2. Cobalt is a much sought after semi-metal, principally for the property which its calx possesses, that of coloring glass, not only with high color, but also very peristent in fire. The Saxons carry on a very profitable trade in it.

In Sweden it is found at Los, Thunabewrg, and a few other places.

On Zinc

§ 326. Zinc is produced at Gosslar [Germany], in Cornwall, and a little in Sweden in Tuna parish; it comes also from China under the name of Tutanego.

Both blende and redslag contain zinc.

Zinc is:

(I) Brittle.

Note 1. Zinc cannot be finely powdered with a hammer without a small part being beaten into small scales; consequently it is considered midway between a whole metal and a semi-metal. That which is distilled with coal dust can be beaten quite thin.

Specific gravity varies: that from Gosslar 7,065-7,215; that from the East Indies 7,212-7,240 [7.140]. Its color is white, tending to blue, almost like lead. It creaks during fracturing, like tin.

(II) It burns in fire with a beautiful blue-green flame, and is calcinated at the same time. The calx has the name flowers of zinc, is very heat resistant and so difficult to melt that it slags difficultly with additions, but tutia, which is also nothing else but zinc calx, stays unchanged when added to glass and gives it an opaque burned gold color. Flowers [of zinc] are good for weak eyes, drying and healing of sores, and to absorb acids; similar to calamine lotion [gallmei].

Zinc ore, which contains sulfur, does not burn in fire unles it is roasted away.

If we wish to know if a mineral contains zinc this is easily determined, if it is roasted, packed into a crucible with coal dust, an open crucible cover cemented on, and [the crucible] set in fire, since in a short time blue flames show themselves, if zinc is present. Rammelberg ore is brown and very heavy, and is therefore called Brown lead ore.

Note 2.. Zinc melts at 370 degrees of heat [419 C], consequently after lead, but before antimony. As soon as the heat advances beyond glowing, the metal catches fire in free air; it is calcinated and expels a white wool at the same time; but in a closed vessel it rises up in metallic fumes, with less heat than bismuth or regulus of antimony.

Concerning zinc solutions, see § 93, 105, 122, 134, 142, 155, and 161.

§ 327. Zinc blends with all other metals except iron, lead, and bismuth; with cobalt, nickel, and arsenic it has not been investigated.

With copper it combines in various proportions which have different names, namely: Brass, Pinschbach, Tombak, Prince Robert's metal, etc., all being more or less brittle, malleable, yellow or red.

The most copper which can be taken up is 89 percent, when it is saturated and can no longer be kept in the heat which burns zinc. This composition is brittle, white, tending toward reddish at a fresh fracture, but tarnishing in a short time in air to a golden color, which however can be partly removed merely by hand rubbing with a cloth. Different objects are cast from it, since they cannot be made by hammering.

When zinc only makes up 5 to 6 percent compared to copper this combination resembles gold work, with copper designated as 20 to 21 karat; this is the most beautiful of all [brass] compositions, and approximately as malleable as copper.

The copper which is used for such compositions should be free from sulfur, otherwise it does not retain its polish, but blackens when it comes in contact with the hands, or gets wet. The Gosslar zinc, somewhat purer, is also better for this than the Chinese.

The mixture is prepared in that manner in which newly refined and brittle copper is melted down with coal in a crucible (§ 294 (c)). When it runs down to the bottom the coal is blown off with a hand bellows, the zinc is slagged, common salt which has been melted in another crucible is poured over it, after which it is blown off until the material melts, when the crucible is removed [from the fire]. Some burning away of zinc should always be expected.

Zinc is amalgamated easily, which was mentioned before under quicksilver.

Note. Zinc combined with copper gives a more or less yellow mixture, which it does not produce with any other metal. The yellowness does not seem to come from thinning out of the redness of copper, since then silver should produce the same effect as zinc, but nothing else causes this yellowness, except tin. GEOFFROI gives only copper and zinc with 1/16 iron for that mixture which approaches the color of gold most closely, but it becomes brittle; Mr. LEVIS [LEWIS?] has found that copper 10 [parts], zinc 8, and iron 1, gives a denser, harder composition more like gold. This is called Pinschbach or Similor. No mixture, although like in appearance to gold, has been found up to now which does not appear paler in comparison [to gold]. Tombak is malleable and gold-red, but Prince Robert's [metal] or Bronco metal is brittle, and is believed to contain about 16 percent zinc, when similar [alloys] up to 57 percent do not lose all malleability. The so-called white Tombak contains arsenic. Or de Manheim [gold of Manheim] is a gilded Similor.

If zinc is to be crystallized from quicksilver, it requires 2-1/2 parts of the latter, to one [part] of the former. Zinc is united quickly in mass with iron, but can, as Mr. MAIQUIN discovered in 1742, cover it over and serve the same purpose as tinning.

To Prepare Flowers Of Zinc

§ 328. Zinc is put into a crucible, over which another whole one is arched without cementing, and this is put into [hot] ashes. As soon as the zinc melts, a white floss arises, which hangs together like spun cloth. After an hour the crucible cover is removed and the flowers of zinc are collected both from above and below [the crucible]; whatever has not yet vaporized is treated in the same manner as long as any of it remains.

The flowers [of zinc] can finally be burned to be certain that all metal has been separated from it.

Note. Flowers [of zinc] are also called Pompholix, nihilum album [white nothing]; and by

RESPOUR [was called] Jason's aurea catena [Jason's golden fleece], protected by flame-spouting oxen, to conceal its other alchemical name. Zinc gains 1/10 in weight during calcination. The calx is fire resistant, yellows under the blow pipe, but becomes white again during cooling; it can be melted to a yellow-brown glass and reduced by a combustible substance.

Yellow-brass [mässing] added to glass produces avanturin-flux.

To Prepare Yellow-brass [Mässing]

§ 329 (a). For this the zinc ore must be well roasted, and it is necessary to completely drive off the acid of sulfur in a very strong flame. Gallmei [callo- mine] is the name given to naturally or artificially calcinated zinc. A mixture vitrified in air, on the other hand, is more easily roasted than the entirely fresh [ore].

Tutia, or Cadmia fornacum, is collected in quantity at Gosslar on the inner side and upper part of the front of the smelting ovens. It is a collection of flowers of zinc, which, after pulling down [off the ovens], is cast on the slag piles, and through weathering becomes all the more useful for forging of brass the longer it lies there, since some sulfur seems to remain with it when fresh.

(b) The copper to be used is beaten thin and cut into small pieces, the better to be acted upon.

(c) To 1 part copper is weighed out 1-1/2 [parts] of fine coal dust and 1-1/2 gallmei. The latter is mixed, moistened so that the ore, by its weight, shall not sink so easily and separate. The mixture is packed into a crucible, copper laid on top, and then a thin layer of coal dust, more copper, and so on until all the copper has been added. Then the crucible is nearly filled with coal dust, and another [crucible] with a round hole in the bottom as wide as a goose quill is cemented on.

(d) After the crucible is moistened, it is put into a well heated oven, then some cold coal is laid close around it and [the whole] heated up.

Note 1. We can also conveniently carry out this operation in [hot] ashes.

(e) When the crucible gets warm enough the coal dust burns first, and then bluish and greenish flames show themselves in the opening; when this begins to abate an iron spit is put through the opening in the crucible, which falls to the bottom of its own weight, thus giving indication that the material is melted, and also stirs it well, so that kernels may collect together. If there is any obstacle for the spit, the test is left in the heat until it [the spit] dis- appears.

(f) As soon as the material melts thoroughly, the crucible is taken out so that no burning may occur, it is placed on a flat plate, is tapped gently and left to cool.

(g) If the gallmei contains less than 30 percent [zinc?] it is useless, and is replaced in the mixture with some much richer.

(h) The brass [mässing] is more flexible than copper, but does not bear forging warm, unless it is well cooled after heating to glowing, before it is forged.

Note 2. Mässing has a specific gravity of about 8,000, and the forged kind 8,349, and contains 20-30 percent zinc; it is an improvement on copper, so pliable that it can be drawn out in trade like a hair.

We usually test zinc ore as described above; the more the copper is increased, the richer the ore, but it is always burned heavily; if therefore we wish to know the content more closely, the roasted ore should be mixed with coal dust and distilled in a stone retort, to which is attached a receiver containing water.

Note 3. Zinc is the most useful semi-metal, particularly for combinations with copper; it is used in fireworks for its beautiful flame, and also for improving tin. Flowers [of zinc] are used in medicine. That blende, gallmei, and white vitriol contain zinc was discovered by Mining Counselor BRANDT in 1735. Previously Mining Counselor VON SWAB in the 1730's began to distill zinc in Tuna Parish, but this establishment soon stopped operating, and no other has taken its place, although we have good resources in ore, which is sulfur-bearing for the most part.

On Antimony

§ 330. Antimonium was often called by the ancients Saturnus or plumbum philosophicum. The name Antimonium was given to it by the accident having occurred, that BASILIUS VALENTINUSs began to use this material in medicine and thereby helped many of his cloister-brothers into the grave, so that this medicament came to be called antimoine, or contra monacum [against monks].

It is:

(I) A semi-metal.

Note 1. Its specific gravity is 6,734- 6,852 [6.684 at 25 C]. The color is glistening white, quickly becoming paler in air, but is not noticeably calcinated.

In fractures it appears as thick scales or sheaves cemented together, placed irregularly with each other. Its hardness is greater than that of lead.

The regulus of antimony is called simplex when it is prepared with the addition of salts, and compositus when other materials are used. The so-called regulus antimonii medicinalis is an impure, half-glazed, brownish, and hepatic [i.e., sulfurous] mass.

(II) It does not melt unless the crucible glows, but the semi-metal itself does not appear to be glistening. It is volatile, and fumes away entirely in flame.

Note 2. It melts later than lead, but before silver, at about 430 degrees [630.5 C]. It forms a thin liquid and is therefore useful for type. From freezing to boiling point, antimony expands less than some other metals, and 3 times less than zinc, but during congealing it contracts more than zinc.

When the regulus is well fused it gets a pointed form [of crystal] when congealed, the radiating rays seeming to extend from an axis, growing on top of each other in parallel rays like a star; thus it gets the name of Regulus antimonii stellatus.

If no more heat is used than just to melting, the regulus begins to fume vigorously, and if these fumes are condensed, silver white glistening needle-like crystals are formed, called flores argentini, which are volatile and slightly soluble in water, as Mr. BAUME´ reports.

(III) The regulus obtained loses 3 to 4 grains both above and below; otherwise it has almost the same effect as arsenic. Ant. crudum is not quite so hazardous. ETMULLER and his followers prescribe roasted antimonium as a suporific, but an earthy tasteless substance cannot produce sweat, since it does not dissolve well in the body, so that it becomes an emetic, and in the right doses is a complete poison.

§ 331. Antimonium is readily melted together with all metals, but nickel and cobalt have not been investigated yet.

To amalgamate it, boiling water must be poured on the quicksilver, and then melted antimonium added, when they can then be combined through grinding. When melted metals are mixed with quicksilver at glowing heat, they extend themselves and form slag violently, as hazardous as grenades and bombs bursting. Hot water reduces the glowing heat immediately in the present case, so that combination occurs quietly; but with metals which need a higher degree for melting this procedure is not suitable.

The regulus fused with calx is amalgamated immediately through grinding; similarly the pure [regulus].

Note. Antimonium also melts readily together with nickel and cobalt. In combination with iron alone it prevents the mixture from attracting the magnet. Copper and antimonium alone produce in some Apothecary's shops the useful Regulus antimonii cupreus; if this is melted together with Reg. ant. martialis we obtain a reticular surface (rete vulcani); R. ant. martialis and tin alone, halved or doubled, give R. ant. jovialis or electrum minus; R. ant. martialis with tin and copper alone give Regulus metallorum or electrum majus, as it was called by the ancient chemists.

To Smelt The Regulus [of Antimony] Without Roasting

§ 332. Antimony ore is always sulfur-bound, and is seldom or never free from lead. Antimonium crudum is a pure stone [skärsten] of the same nature but obtained artificially and free from mineral.

(a) To the newly obtained unroasted regulus is taken a similar amount of antim. crudum, saltpeter, and red tartar, which are ground together in a glass mortar and well mixed. At the same time a crucible is heated to glowing in [hot] ashes, and small portions of the mixture are dropped in little by little, which detonate sharply and rise up in the crucible when too much is added at a time.

When all is burned up in this way, it is blown off until the material is melted clear, and is then poured out into a mold.

In this way the regulus is obtained purer than by almost any other method, but also with considerable loss, since about 9 parts of ant. crudum go to slag; that which comes from the corners of the mold is hard (it does not become so when cooled in the crucible) and is called Regulus antimonii simplex.

Note. If a proper amount of regulus is cooled in the crucible it also becomes hard. Hardly more than 27 percent regulus is obtained in this method, described in the text [Scheffer's]. Materials are usually taken in the proportions of 8, 3, and 6.

(b) The alchemists have placed much importance on scrapings [scoriarne] after the smelting. They are dissolved immediately in hot water. If this [solution] is filtered and distilled vinegar added, a red-brown precipitate deposits, which is extracted with water to free it from all salts, and is called Sulphur auratum antimonii. The scrapings consist of antim. crudum and alkali from nitrum and tartarus; since now an acid cannot precipitate alkali, it is clear that sulphur auratum is nothing but antim. crudum with so much more sulfur that we get regulus in proportion. 3 to 4 grains cause vomiting.

Note 2. We usually also separate sulphur auratum praecipitatonis primae, secundae, etc., since everything does not come down all at once, but when the first [precipitate] is removed, fresh acid is added, whereupon more deposits, etc.

To Separate [Antimony] Regulus By Precipitation With Iron

§ 333 (a). For this we weigh out 2 parts iron in the form of filings or turnings, put them in a crucible, put this in ashes and heat it up gently. When the iron glows, add 4 parts of coarsely pulverized antim. crudum folded in paper, blow on it, and lay weights on it [the bellows] as soon as the crucible permits it. When the material melts, 1 part of white flux is added, which causes such boiling that the mass will overflow, so we are often obliged to pour some out before its time, which otherwise should not be done sooner than when it becomes clearly melted and sparks rise up. Pouring out then occurs in the usual manner.

The impure semi-metal is called Regulus antimonii martialis, is almost free of sulfur, which the iron takes up, but is much contaminated by iron. Addition of alkali here should produce a liver of sulfur, and this dissolves the metal, but it should be noted that although hepar [liver] is produced, only the iron is attacked, which by the power of stronger attraction takes the sulfur to itself and antimonium is conserved, which also would seem to occur with other similar mixtures; indeed in the present case the regulus does not separate itself from the other mass without addition of alkali, which still should have the proper proportions, otherwise everything sticks together and little or no regulus is produced.

Note 1. If 5 parts iron are taken to 16 antimonium cr. and the melt is brought to full flux, and finally one part saltpeter is cast in in pieces, a few at a time, the regulus, according to Mr. BAUMÉ, is obtained immediately free from all iron and sulfur.

(b) To purify the regulus from all iron and sulfur contamination, it is pulverized finely, mixed with 1/10 saltpeter, poured into a glowing crucible, fused with bellows to clarity and poured out. The saltpeter burns up the phlogiston in the sulfur, and then acidum vitrioli combines with one part of the free alkali, and when the combustible part of the iron has been consumed, it [iron] goes with the excess alkali to slag. We must take so little nitrum that no regulus may be attacked, which nevertheless can hardly be completely avoided. If the regulus is entirely pure it should then be radiating from a central point; but it is a certain test to investigate a small portion of it with a blowpipe, since iron betrays itself readily in the slag which forms a residue on the coal [carbon block used with the blowpipe], and finally the antimonium fumes. If all the iron is gone we can be certain that all sulfur has likewise been removed, since the regulus has less attraction for it.

If iron-bearing slag shows itself it is remelted with 1/10, 1/16, or 1/20 saltpeter, until the end result is accomplished.

The scrapings contain some antimonium, which melted down with half as much tartar, can be reduced.

Note 2. Antimonium has less attraction for sulfur than iron, copper, lead, silver, and tin, and the latter consequently can be used to precipitate the regulus out of antimonium crudum. The specific gravity of Regulus antimonii martialis is 7,500.

To Smelt Out The Regulus From Roasted Ore

§ 334 (a). Antimonium crudum is beaten into small pieces, laid on a planchet, placed in the mouth of a muffle when the oven has low heat, and the pieces turned with a stirring rod until no more fumes come off. The pieces should now be broken into smaller bits and placed again in the same place; when they are smoked out they are pulverized very fine and roasted again in the muffle mouth, as long as any fumes are observed, after which they are completely roasted in the usual manner. Without such caution, ant. crudum melts, sticks to the planchet, and does not release the sulfur so easily.

(b) To 1 part roasted ore are taken 2 of tartar flux, and this [mixture] is fused in an open crucible. The regulus becomes completely brittle and crusted.

Ant. crudum yields 45 to 50 percent regulus, more or less mixed with lead.

Note. I have in the above described manner lost 23 percent during good roasting and obtained 60 to 78-1/2 percent regulus.

To Purify Regulus Antimonii From Lead

§ 335. The regulus is remelted a few times with ant. crudum until a small amount evaporates entirely with the blow-pipe, without leaving any residue, so we can be certain that the lead has been removed.

The sulfur takes up no more regulus than to saturation, and it has this amount in ant. crudum, does not react with the addition, but the lead is taken out by the ant. crudum, which can take both more and less of it. Some sulfur contamination still sticks to the lead-free regulus, which can be destroyed with a little nitrum (§ 333 (b)).

To Prepare Vitrum Antimonii

§ 336. Roasted ant. crudum is fused with a closed crucible cover, using a bellows for 5 minutes without weights [on the bellows], and 20 minutes with weights, after which the crucible is removed and contains a dark-burned, half-transparent glass, like cheap colophonium. Covering is necessary during blowing, partly to reduce loss in fuming away, but principally to contain the coal dust, since if some of this is present a regulus is obtained on the bottom.

The property of forming glass without additives shows an abundance of earth, since anything which contains some earth goes to glass, although more or less slowly according to the amount [of earth] present, and the extent to which it resists fusion.

Note. To prepare vitrum antimonii is a simple operation, but one which is very often unsuccessful, if we are not familiar with it. Roasting should be continued for a long time, but not longer than necessary to produce some glistening sparkles in the calx. The other condition which should be observed is that it should not be fused more than once, since otherwise when all of it becomes completely fused, when it first gets enough heat it gets burned, whereupon the glass gets a poor color; indeed the most beautiful [color] often turns completely opaque from overfusion.

Vitrum antimonii ceratum consists of 8 parts vitrum antimonii which happens to be well worked with 1 [part] melted wax.

§ 337 (a). If ant. crudum and nitrum alone are mixed in a powder form, put into an iron pan, ignited with a glow- ing iron spike or a small burning coal, a hefty detonation occurs with a high flame. At the bottom is found then a white scale, consisting of alkali nitri with acid of sulfur; but underneath lies a liver-colored mass, which contains antimonium and sulfur; this is so named from the color which ant. crudum takes on when the sulfur is reduced by smelting, i.e., Hepar antimonii.

Note 1. We can spare nitrum, since if antimonium crudum is roasted somewhat and then fused we obtain a similar hepar.

(b) The hepar is ground and leached well with water, the regulus remaining with some sulfur, which the nitrum was insufficient to burn away, and this the alkali cannot dissolve and consequently is saturated with acid of sulfur. The powder is called Crocus metallorum, but still contains sulfur, and no metals except antimonium. It also gets the name Terra auri Rulandi, Terra sancta, or rubra Rulandi. It takes away both above and below, when 3 to 6 grains are taken. [The expression 'takes away both above and below' must refer to the action of antimony compounds as both an emetic and a laxative.]

§ 338 (a). If 1 part ant. crudum is burned off with 3 parts nitrum in an iron pan, we get a white mass which is called Ant. diaphoreticum nitratum, and is a sweat produc- er [sudorific]. The dose is 1 drachma. BAS. VALENTINUS is its originator.

Note 1. This preparation otherwise gets the names of Diaphoreticum non lotum, Fondant de Rotrou, etc.

(b) If the foregoing preparation is leached out well it is called Ant. diaphoreticum vulgare, officinarum, or dulce. It is found in all appothecary's shops, and from the name it should be sudorific, but is an earthy, calx-like material, which does more damage than good in the body. All sulfur is driven off, which is evidenced by the white color; it is not dissolved by acidum vegetabile, but if the vegetable acid is mixed with acidum salis and distilled, antimonium is sublimed as a white skin in the neck of the retort.

(c) When antimonium diaph. nitratum is neutralized and the wash water evaporated to dryness, nitrum antimoniatum is obtained, which serves as a diuretic.

(d) If distilled vinegar is poured into the just mentioned wash water in (c), a white powder precipitates, which when neutralized is called ant. sulph. fixatum or materia perlata. It is believed to be good for fevers and the plague, but is nothing else than a fine ant. diaph. dulce.

Note 2. Materia perlata is soluble in alkali.

There are very many preparations of antimonium, and among them is Kermes minerale, the secret of which was bought by the Regent of France, the Duc d'Orleans; it consists of antimonium crudum boiled 2 hours with a like amount of acrid lye, after which it is filtered and allowed to cool for a day, so as to settle out a quantity of antimonium and sulfur, which does not dissolve in the cold, and produces the aforesaid kermes, which otherwise used to be called Pulvis Carthusianorum. The residue is usually reboiled 2 more times yet.

In addition to the use of antimonium in the Medical Arts, it can be used in typecasting, for some metallic alloys, to purify gold, for stars in fireworks, etc.

Sweden is not completely devoid of antimonium ore, but the resources are very small.

Note 3. Without now going through the 14 metals, we have lately gotten knowledge of a fifteenth [metal]. That something metallic is found in brunsten, which makes up its essential part, I have reported in KVAH [Kunglige Vetenskaps Akademien Handlingar: Transactions of the Royal Academy of Science (of Sweden)] proof of its density, its power of coloring glass, and its precipitation with bloodlye. Mr. J.G. GAHN, not knowing of my reports, has at the same time actually brought out of it by reduction a semi-metal, which in tensile strength approaches nearest to platina, and on the other hand does not resemble any of the earlier known [metals], whereupon he soon learned to publish his research. I have also afterwards obtained the regulus of brunsten by reduction, but could not purify it from iron; it is called Magnesium, to avoid excess verbiage, and to distinguish it from the earth magnesia. [Bergman is referring here to manganese. His subsequent publications are confusing to us today when he continues to refer to manganese as magnesium.]

This semi-metal is known both for coloring glass and for taking away its color, according to differences in amount and processing. It is a quaint phenomenon, that a blowpipe in a few minutes can produce a borax-glass [pearl], which dissolves a little brunsten, and at different times is hyancinth-colored and colorless, and finally by application of no light [produces] a flame, brownish and less plain on the outside and blue and well defined on the inside; the former, or outer [flame], which is immediately surrounded by the air, quickly loses most of its combustible element [phlogiston], because the air on one side attracts it strongly, and the heat on the other [side] increases its volatility, and consequently this can easily divert phlogiston from the glass pearl, whereupon the color of the brunsten becomes visible; but in opposition to this, the lighter part is much richer in the same fine fatty material [phlogiston], and must therefore, if it does not increase, at least not decrease what [phlogiston] is added to the pearl from the coal, whereupon consequently colorlessness can result if not too much brunsten be added (§ 176, Note 2). This occurs not only with brunsten, but also with several [other] metallic substances.

Like nickel, cobalt and magnesium [manganese] cannot be freed completely from iron by any method, and consequently are always attracted by the magnet, if arsenic does not prevent it by its presence; so we could easily reach the conclusion that they would seem to be only different modifications of iron. This conclusion gains support when we consider that with iron it is more evident than with some other metals, that we can discern a great diversity due to different amounts of phlogiston. Iron and steel have manifold varieties. So it results that up to now the metals mentioned all become more difficult to melt the more they are purified, and likewise with their pliability and malleability. Finally iron also can be made red, blue, green, gold, brown , and black, of which colors the three metals each produce 2 or more in the wet or dry method. Nickel yields a green calx, green solution in acid, blue in alkali volatile, and hyacinth-colored glass; cobalt [gives] red solutions, black calx, and blue or violet glass; magnesium [manganese] [gives] a red solution with acid, and green or blue with alkali, black after long heating, a green calx, and a more creamy than red-gold color to glass. That iron gives green, yellow or brown-red solutions with differing amounts of phlogiston is known; with alkali phlogisticatum it [iron] yields Berlin Blue; if saltpeter is fused many times in strong heat together with iron, blue, blue-green, and purple-colored flowers spread throughout the crucible; in glassy materials it [iron] yields, according to conditions and surroundings, also green, yellow, brown and reddish colors. Soapstone, lapis nephriticus, jasper, Terre verde, etc., get their greenness from iron; azure-stone gets its blue from iron; etc. When there is added to it at least nickel and cobalt, purified as much as the art permits, the tensile strength and specific gravity are increased, and also the iron retains phlogiston so much that it can hardly be calcinated in flame in the usual way, from which we deduce information which is very credible. But until we can prepare nickel, cobalt, or magnesium [manganese], purified as much as we please from iron, these bodies [metallic elements] must exhibit differences, and yield fresh information about stable properties; but this should not discourage closer investigation with all diligence for its own sake.

SIXTH SECTION: CONCERNING DYEING.

Introduction

§ 339. Among all colors known up to now, there are only three primary colors, which exist by themselves and cannot be produced by combinations of others; these are red, yellow, and blue. All others can be made up from these 3.

The primary colors should be treated first, then black, and lastly the fashionable colors.

Note. The art of dyeing includes a science and practice which wholly and entirely depend on chemical principles. To dye is actually nothing other than to saturate the surface of a body with particles which reflect the kind of rays of the color we desire. It reduces then principally to finding a substance which contains that kind of useful particles which attach and fasten themselves to the recipient material.

In regard to their nature, the dye-plants sought are of many kinds. They are called extractive when they dissolve by themselves alone in water; there are slimy (mucilaginous) kinds, salty kinds, and also fatty kinds, which combine with salts of a soapy nature. Of such are vau [wold, dyer's weed], ängskär [saw-wort], genista, krapp [madder], gurkmeja [turmeric], bresilja [brazil], blauholz [logwood], chermes, cochenille, and many which because of their ready solubility must be fastened in combination with salts, or at least be changed into a less soluble form, for which alum is most often used. Mr. MACQUER uses the name refino-extractive or refino-terrestrial for those which, quite insoluble by themselves such as earths and resins, which however in combination with extractive soapy substances can become quite soluble in water. Here we group sumac, fantalum [alum color], albark [alder bark], hassel-rötter [hazel-red], nötfnas [nut shells], purpur snäckan [cochineal purple], etc., all of which, except for the last named, give dark so-called root-colors, but remember that they do not take up either water or soap, and require that either goods or dye-stuff receive some particular preparation. When a mixture of 2 substances is put into a liquid which only can dissolve one of the ingredients, one of two things happens: (1) the solvent does prevail, taking all the soluble [part] to itself, because one part is surrounded and protected by the insoluble part; (2) it forms an intimate compound, which is made up of the soluble and insoluble parts, so that some portion of the latter [insoluble] is taken up and retained by the liquid, as long as it is connected with the former [soluble], but if it gets separated it immediately precipitates. Pure resinous or oily dye-powder cannot be extracted by water alone, [even] using extractive soap. Examples are indigo and the red in safflower, which therefore cannot be utilized without the help of alkali.

Besides the beauty of a dye, its durability is an essential property; those which can be washed out by rain and washing, or are bleached out quickly by sunshine, are obviously of little value, but can be used for wool which seldom gets exposed to washing; so have dyes properly gotten the name false [oäckta], which are destroyed by means of daylight or sunshine; and those are called genuine [äckta], which undergo little or no change.

With dyeing we also find another notable difference in appearance due to the nature of the goods themselves. Animal materials such as wool, fur, etc., take up dye most easily; vegetable materials such as linen, cotton-goods, etc., most slowly; and silk, which is actually from both kingdoms, from an animal but one which feeds on mulberry leaves, seems therefore to have an intermediate nature, since the dye goes on better than on linen, but poorer than on wool. We can also make linen more adaptable to being dyed by processing it beforehand with an animal substance.

§ 340. To show how other colors and shadings of them can be produced from the 3 primary colors, we choose pieces of cloth of similar strong blue, which still have all diff- erent shades of color, from lightest to darkest. These are boiled with wold blossoms and madder, in several different mixed proportions, so we get brown, gray, green, black, etc., of all possible variations.

Note. Here is disclosed an important difficulty concerning the description of research in dyeing. Namely, because the shades of dyes are infinite [in number], and only a few have separate names, so we arrive at the dilemma of how to distinguish most of them properly by a name, with which to identify [that dye] used by the investigator. This difficulty occurs in all descriptions in Natural-History and the manifold substances in natural science. Dr. SCHAEFFER proposes to name the variations of the 7 prismatic colors after natural bodies which display them, such as wax-yellow [vaxgult], pale wax-yellow, bright wax-yellow, goldfinch yellow [steglitsgult], straw yellow, etc. But for distinguishing some which arise from the 7 different principal colors and different variations , he proposes 3 ways, namely: (1) to enumerate scholarly, artistic, scientific, etc., such as already occur with Pompadour, Orange-yellow, Isabelle dye, etc.; or (2) a collective name after the mixture itself, such as reddish yellow, blue-green, etc., (with the prevailing name last), and this seems to be the most natural method from which similar compositions are known, but becomes cumbersome when 4, 5, or more materials are blended; or (3) to incorporate and enumerate the standard dye [färgmönster], which [method] without doubt is the easiest, since a complete standard [name] would be obtained from the dye, and all variations could be based upon it and consequently become the standard name.

Now what the Author´s [Scheffer´s] research and thinking was, with respect to the principle dyes, I do not know, whether it only had reference to the art of dyeing in its practice, or whether he also believed that of the 7 prismatic colors, which are seen in the rainbow, only three should be considered for simplification. Mr. MAYER proposed this last theory in 1758 and came up with the idea of representing these [3 primary colors] in the form of a pyramid. Messers LAMBERT, EXLEBEN, etc., have since worked on improving the pyramid of colors. MAYER adopted cinnobar, royal yellow, and mountain blue [bergblau] as the basic colors; Mr. LAMBERT chose carmine, gamboge, and Berliner Blue, as of equal importance; Mr. EXLEBEN carmine, royal yellow, and Berliner Blue to have equal space, from which also the purest basic colors seem to arise.

SIXTH SECTION, CHAPTER I, Red Dye

FIRST CHAPTER: On Red Dye.

To Dye Madder-Red on Wool

§ 341 (a). The [woolen] goods is weighed and boiled one hour in water with wheat bran, to extract substantially all the oil, then it is rinsed in cold water until all the adhering bran has been washed away.

(b) After this we weigh out, compared to the weight of the goods, 1/4 [its weight of] Roman alum and 1/8 tartar, which are dissolved in enough water with boiling, and then the goods is boiled in this [solution] for 2 hours, taken out and hung up overnight to dry a little, and the day after it is rinsed in cold water. The goods is not put in until the water boils, which is necessary for all boiling except that with the bran. Hence we do not dye immediately but must put it [the dye] on the wet cloth to keep it moist.

Note 1. The dye can be put on just as well without this soaking but it becomes very soluble. The proportion of alum to tartar used is not always the same; some people use up to 5 times more of the former [alum], and only 1/16 the weight of the goods of the latter [tartar]. The tartar is used only so that the dye becomes dark cinnamon brown, but good and permanent. Alum, with saltpeter in place of tartar, gives an artificial cinnamon dye.

(c) For dyeing we take 1/4 good madder, put it in cold water, and the goods itself in with it, which is continually stirred about during boiling; it is taken out as soon as the lather begins to subside or loses its red color, since otherwise it becomes brown; it is rinsed in cold water and hung up to dry, after which the color is found to be dark red, but a little shiney.

Note 2. HELLOT uses 1/2 madder, puts it in when the water is not hotter than the hand can stand, stirs it well, then adds the goods, keeps it in about one hour at 90 degrees of heat, and finally boils it 4 to 5 minutes. With longer boiling an abundant yellow is produced with the powder, which establishes itself and makes the color poorer.

Note 3. Madder [krapp] is called Red of Rubia tinctorum; it is called Garence in France, and Färberröthe or Grapp in Germany. The root is long, narrow, harvested after the second year in September, dried, ground, packed into casks and stored, usually for 2 to 3 years, while it changes within, because it is believed to become better and richer in color; 7/8 of its weight disappears in drying. In open air it loses its strength. The pure pith is ground and is usually called Grapp in commerce, and Färberröthe, if bark and narrow root comes with it. That which comes from Zeeland is considered the best. In Turkey we also obtain excellent madder, expecially around Adrianopel and Smyrna. Mr. DAMBOURNAY has found that 4 parts fresh madder root have the same strength in dyeing as 1 [part] of the ground [root], by which half can be saved.

SIXTH SECTION, CHAPTER I, Red Dye

Madder is boiled gently in water to obtain a ruby-red decoction with a musty odor, and a little bitter and astringent taste. By dilution with water it becomes at length entirely yellow; a little acid revives the color, but more turns it yellow; alkali makes it darker, but more beautiful; alum makes it bright red; salmiak somewhat yellow-red; blue vitriol cinnamon-colored; and the green [vitriol] a hearty yellow-red; tin solution a bright golden yellow, according to Mr. PÖRNER'S research. All these soluble additives, except alkali fixum, produce precipitation, and if the color is observed, the plateau occurs. Saltpeter and common salt produce a good precipitation, but otherwise no notable change.

When an animal eats madder, within a few days its bones become red, chiefly the hardest [bone], but not the same in different animals; the dye gives the usual test, but entirely disappears if the animal gets other food for a time.

§ 342. Since it is found through research that acid prevents madder dye from producing the proper effect, we can get the idea to absorb alum acid with chalk.

(a) If to [the weight of] goods we take 1/4 [this weight] of Roman alum, which is boiled with water then allowed to settle, and finally the clear mixture is used in the usual manner for soaking, and finally [the goods] is dyed (§ 341 (c)), the dye becomes quite beautiful, but very uneven, for which the appetite for fatness of tartar can be useful, which improves the dye and helps it go on more evenly.

(b) If therefore the foregoing operation in (a) is carried out with no changes other than [to add] 8 [parts] tartar, the dye becomes more beautiful and fuller.

(c) It becomes still more lively if crystalline tartar is used, because the earth of tartar, which attaches to the goods, darkens it the same.

Note. Only purified tartar with the madder, dissolved in the broth, with no other preparation than full maceration in water, gives a brown-red fast dye; doubling makes the dye still redder.

§ 343. What soaking in tin solution accomplishes can be gathered from the following test.

Tin dissolved in aqua regia and diluted with water is boiled for 1-1/2 hours with goods previously boiled with bran. With this mixture is dissolved 1/4 alum over a fire, then removed [from fire]; the goods is put into it,and immediately taken out again from the tin solution and allowed to cool; the alum water immediately becomes darkened by acid of vitriol, which combines with the excess tin, that which is not attached to the goods.

When it is all cooled, the goods is hung up overnight, rinsed, and dyed with 1/4 madder, and so gets a yellow-red color, which is caused by the tin.

Note. The composition of the tin solution was not found to be reported [by Scheffer].

If the goods is not completely soaked, but only boiled with 1 part tin solution and 2 [parts] madder, a cherry-red is obtained, which becomes darker in air.

§ 344. The effect of green-vitriol on madder-dye is learned from the following study.

(a) Green vitriol, 1/4 [part], is boiled with the goods for 2 hours, after which it [goods] is rinsed and boiled out with 1/4 madder, which like the goods is put in cold water and brought to boiling for 1 hour. The dye becomes coffee-brown at the end but lighter.

Note. If the goods is not soaked, but boiled with 1 part vitriol and 2 [parts] madder, the dye becomes brown, turning to red, and even redder if used alone; both darken some in air.

(b) If the process goes on as it did previously (§ 342, (a)), only with the addition of chalk, the dye gets darker but does not produce the same effect.

With addition of 1/4 tartar in the soaking the dye goes on better, but becomes lighter and less atractive.

(c) If the goods is first treated with alum (§ 341 (b)), and when it is boiled with madder 1/4 green vitriol is added, and the whole kept boiling 1 hour, the dye gets somewhat darker, but more shimmering in red.

§ 345. Blue vitriol used for soaking like the green [vitriol] (§ 344), produces a yellow tending towards light brown with madder, and it is lighter in the beginning.

Note. If the goods is not soaked but only doused well in hot water, (which is necessary in the dyeing, since otherwise it will not go on evenly), and is then boiled with 1 part blue vitriol and 2 [parts] madder, the dye becomes light brown tending towards yellow, and somewhat darker and leaning towards green when [madder] is used alone; both darken somewhat in air.

§ 346. White vitriol used for soaking in the same way produces brown, dark and tending a little towards red.

Note. The usual zinc-vitriol [white vitriol] is contaminated with iron, which also influences the process somewhat.

§ 347. If aqua fortis 1/4 [part] is saturated with white lead, and then diluted with water, and boiled with the goods for 2 hours, [the goods] taken out and immediately put into a solution of one quarter alum, which has recently been finished ahead of time in heat, [the goods] left to cool in it, rinsed and boiled again with 1/4 madder, a red dye is obtained when the acid of saltpeter is not fully saturated and consequently can attack the tin pan somewhat in which the trial is usually carried out. If a little sugar of lead is added and boiled 1/2 hour, it becomes red-brown.

Note. Since madder is the cheapest of all permanent red dyes, it is also used as an additive with others more costly. Consequently it is mixed 1/2 [part] to [1 part] pine berries [grana kermes], to make so-called half-breed scarlet [ecarlatte mi-graine], but not in the first but in the second boiling; similarly with 1/2 [part] cochenille in the last boiling to make the ordinary half scarlet [demi-ecarlatte] and half-crimson . Mr. HELLOT has also found: that a 1/2 ounce piece of cloth boiled with 10 grains of Roman alum and 6 of purified tartar for 1/2 hour, taken out, squeezed out, cooled and again put in the same liquor, then 24 grains of madder are first added to it, and when the dye is extracted, 20 drops of bismuth solution, made with only water and spiritus nitri, makes after 1/2 hour almost as beautiful a purple as with cochenille; if it is boiled a quarter [hour] more the purple becomes brighter. If the goods remains wet several days after soaking, is dyed then with madder in the usual manner in new liquid without salts, until it becomes highly cinnamon-colored, and afterwards bismuth solution is poured in, only a dark brown is obtained, which shows how much the results depend on small changes. Mr. HELLOT´S research merits repeating since, if he considers bismuth and cobalt as one and the same, then we are uncertain whether or not the latter semi-metal [cobalt] was at work, especially since nothing is mentioned about the color of the solutions.

Madder-red is usually made brighter with fernbock or sorrel [oseille], which are soon lost in air.

To Dye Silk With Madder

§ 348 (a). It [silk, 1 part] is first soaked with 1/4 [part] alum, which had been treated with chalk, as before (§ 342 (a)). The silk is put into the clear [solution] after it is well cooled, and allowed to stand in it for 18 hours, after which it is rinsed and hung up to dry.

(b) The boiling out takes place with madder alone, and gives the goods a fairly handsome red color, but dark.

(c) If the silk has not been previously boiled with soap, this should occur before the soaking is undertaken. 17 lod of silk cloth is boiled in 1 kanna of water with 4-1/4 lod shaved Venetian soap for 2 hours; as soon as the water evaporates it is diluted anew and the silk turned frequently. Finally it is rinsed and dried. If it is not found to be soft after this it is boiled again with 2 lod of soap.

Note. All raw silk is stiff and takes the red more or less as an overlay, which is neither gummy nor resinous, but oily and insoluble in spiritus vini. The stiff [silk] is used for light-colored, but all that is to be dyed and otherwise processed must be boiled. For this we ordinarily use soap of the best kind, since poorer kinds require larger amounts and the silk becomes weakened by this, slow to dry and yellowing with age. That [goods] which is to be dyed is boiled once with 20 percent soap [solution] for 3 hours; for light [red] we take 30 [percent soap] and for bright red 50. That which is meant to be white is first treated with 30 percent soap dissolved in pure boiling water, which is evaporated somewhat in the cold [state], and the skeins are suspended in it and rotated on a stick until they becomes white; then the silk is wrung out, left in this rough state and boiled again for 1-1/2 hours in a similar solution as before. The Chinese use a fruit in combination with mineral alkali to make an excellent liquor, from which silk gets a higher luster than the European [silk]. The Academy in Lyon put forth a prize question in 1761, asking for a better method for taking away the rawness of silk. Mr. RIGUAT won the prize by proposing a dilute solution of soda, but with it a great deal more care is necessary than that needed with the handworked method, and in addition the goods does not become quite as white.

§ 349 (a). Silk is treated with gall in the following way: to each lod of gallapple powder is taken 2 quarts of water, the powder is put in when the water boils, and [the solution] immediately removed from the fire. After a little while the folded goods is put in to be treated, and the vessel covered for 14 hours, after which it [goods] is taken out, rinsed, and dried.

(b) The silk treated with gall is put in cold alum water tempered with chalk (§ 350 (a)), and left there 18 hours, rinsed, and dried.

(c) The silk is finally boiled with a like amount of madder, until the soap begins to boil up completely and the foam becomes pale, but produces only a light color; if just a little soap is poured in from a black pan the silk becomes dark brown.

To Dye Cotton Goods With Madder

§ 350 (a). If it [cotton goods] is boiled with 1/8 Venetian soap for 1 hour, rinsed, and put into cold alum water, in which is dissolved 2 lod of alum per quart over fire, it is found that after rinsing, drying, and ordinarily boiling with madder alone for 1-1/2 hours, that it is not only fully [dyed], but also retains most of the dye during 5 minutes boiling in weak soapy water.

(b) Printing a pattern goes better on cotton. The goods is treated with gall first, just as was described for silk (§ 349 (a)). Meanwhile 2-2/3 lod of alum is dissolved in water over fire, and after cooling is precipitated with chalk. The galled dried goods is dipped into [the solution], wrung out, and hung up to dry overnight. Boiling out occurs finally after rinsing with madder alone, by which we obtain not only a fairly handsome red color, but [one] which also resists 10 minutes boiling in soapy water with no change.

(c) If the procedure in (b) is repeated with this change, that 4-2/3 lod of alum is used, and 5-1/3 gum per quart, and the goods dipped into this thick mixture after precipitation with chalk, and finally boiled out as before, the dye becomes more handsome after 5 minutes boiling in soapy water, but if left in the dye broth 1 hour, the soap acts completely, and both stay in after 1/2 hour.

(d) It succeeds also without the gum, if 3 lod of Roman alum is taken for each quart (Swedish [alum] makes the dye poorer and less resistant to tvål [hard soap]), and we proceed with the rest as in (b). Usually the crystals settle upon the galled goods which are dipped in, but they are rinsed out, as often as necessary for no more of these [crystals] to be found. This dye becomes very handsome and resists soap in boiling in soapy water.

Thread also gets a handsome dyeing in this manner, but does not resist 5 minutes boiling in soap.

(e) If the goods is not treated with gall, but otherwise treated as in (d), the cotton thread gets a beautiful rose red dye, but one which does not resist boiling with soap as well. Consequently, treating with gall makes the dye more permanent, but darker, though the darkening can be bleached out.

(f) If the same proportions of Roman alum, chalk, gall-apple, and madder are used as in (d), but in this manner, that the thread is first boiled for two hours in gall water, then 1 hour with alum, and finally with madder, the dye becomes as full as possible.

(g) If the goods treated with gall is put in alum water for 24 hours, alum water made with 4-5/8 lod per quart, precipitated with 1/2 lod of powdered chalk (which is the proper proportion to prevent crystallization when cooled) and then clarified, and then [the goods] is rinsed and boiled with madder alone, it gets an ugly color which does not withstand boiling with soap; the same thing occurs when Turkish madder is used.

§ 351. White arsenic has some use in dyeing cotton with madder, since if 2-1/2 lod of alum is dissolved per quart of water, and then white arsenic added with boiling until saturated and the solu- tion thereafter taken from the fire; if goods treated with gall is put in and left there for about 24 hours, rinsed, and boiled out, either a more handsome or a fuller dye is obtained.

§ 352. Calx does not give any particular improvement. For testing this, 2 lod alum is dissolved per quart (which is also retained [in solution] after cooling), and the cotton goods treated with gall is steeped in it overnight, then the solution is cooled; after which [the goods] is hung out until almost dry, then boiled for 1 hour with about 1/3 [part] sieved calx to [1 part] water, calculated to saturation. Meanwhile the alum-water is poured into the vessel to be used, and 2-2/3 lod alum [added] per quart, after which the pan is filled up with pure water and 4-2/3 lod [more] alum is added per quart of this [solution]. When the solution boils it is taken from the fire and the goods dipped in it as the solution wells up; then it is put in cold alum-water overnight, which contains 2 lod [alum] per quart, it is rinsed, and boiled with madder. That [dye] which is taken up when the foam begins to get pale, is not particularly attractive, nor is that which is left in longer, but both resist boiling with hard soap [tvål].

§ 353. If soda-salt is boiled with sufficient water and powdered sulfur, until no more of it dissolves, and when it is removed from the fire it is allowed to settle, and the clear [solution] collected, we get a lye with which vain attempts have been made to fix madder dye. The goods, boiled in it for 4 hours until the soda-salt begins to let loose the sulfur and this fastens to it [goods], it is hung up to dry, rinsed well and dried again. Then it is left in gall-water 24 hours, hung up to dry, left thereafter for a like amount of time in alum-water, made from 4-5/8 lod [alum] per quart and precipitated with 1/2 lod of chalk. The goods is again hung up to dry, rinsed well and dried again. Then if this is boiled out with madder alone we get a full dye, which is still more undesirable the longer it is boiled, but resists fairly well 3 minutes boiling with soapy water.

And Turkish madder, called Kiockboya, gives no better dye color.

Note. When cotton goods, with no other preparation than a thorough soaking in water, is boiled with madder in combination with different salts, we get all kinds of varieties of pale red, among which are some which resist boiling with both lye and soap, but are lost in the sun. The same result is obtained if a firm combination is used with potash, common salt, salmiak, alum, or gallapple, but the colors often get brighter, and even somewhat like Turkish red in particular cases, such as when the layered goods is left [to soak] 24 hours in potash lye and gall water, and finally in alum water, and it should dry between each treatment.

How to dye with Turkish red is still generally considered to be a secret, but it is in no wise difficult to understand the chemistry and gain some practice of the dyers' art. It has already been noted that the dye goes on better on wool than on linen. Consequently there must be something in the animal part which aids in this and which is missing in the vegetable [part].

This can be proven in a moment, since when cotton goods or linen is properly blended and treated with certain animal substances, it becomes so to speak animalized, after which consequently the dye penetrates into it and becomes permanent, when previously it could not be sufficiently attached in any manner. The above applies also to the above-mentioned goods with Turkish red.

To Dye Wool Scarlet With Cochenille

§ 354 (a) The goods is boiled well with wheat bran.

(b) One-fourth [part] tin is dissolved in 1 [part] aqua regia, and 3 lod of solution is weighed out per pound of goods, which is put into a similar kettle with only white starch, previously softened in water, and also a like amount of crystalli tartari. The starch makes the dye smooth.

Note 1. The tin solution is usually called the "composition". For it, aqua regia is required, but it is better made in so-called spiritus nitri, diluted with water alone and provided with 1/16 [part] pure salmiak to 1 [part] of the acid, in which 1/8 [part] English [salt?, $MgSO_4$] and granulated tin is dissolved bit by bit. HELLOT used to add 1/32 [part] saltpeter, so that the dye would go on more evenly. One eighth [part] of this "composition" is taken to 1 [part] of the goods, and it should be poured in a moment before the cochenille, and all this before the goods. Water is made up to the mark at about 4 kanna. According to Mr. SCHEFFER'S text there is used for each 1 pound of priming 6/10 lod tin and

2-4/10 [lod] acid, but from HELLOT, 1/8 lod tin and over 7/8 undiluted acid.

(c) After this the pan is placed on the fire with sufficient water, and when it nearly boils 1/4 lod of cochenille is added per pound of goods, the solution is stirred well and the goods is put in and boiled for 1 hour, after which it is taken out and rinsed immediately.

Note 2. At the removal [of the goods] no more dye should be in the bath solution, which requires about 1-1/2 hours. The cochenille should be finely pulverized and sieved; also the tartar.

(d) Boiling out takes 1/2 hour with 3 lod of starch, 1/2 lod of tin solution, 1 lod crystalli tartarti, and 1-3/4 lod cochenille, all calculated per pound of goods, and well stirred about. Finally the goods is rinsed in cold pure water.

Note 3. Mr. HELLOT takes for the first boiling, per pound of goods, 4 lod cremor tartari, 1-1/2 drachm cochenille, and 4 lod of "composition" [see Note 1 above]; to this next is added 1 lod starch, 6-1/2 drachm cochenille, and 4 lod "composition", and all boiled 1-1/2 hours. Above all the operation should be carried out in a kettle of fine tin, but if you have nothing but copper, it should have been previously cleaned carefully and the goods suspended in a net so as not to touch the sides. The brightest scarlet dye was formerly called Hollander, but since it was also made in the greatest perfection in France, it got the common name Ecarlatte de Gobelins or ecarlatte couleur de feu. Since the strength of the acid was not measured, there results great differences in the dye; too much "composition" makes it yellow, which for the most part can be helped by boiling in water alone, or at least by the addition of Roman alum. If the dye turns blue the goods is taken out and a little more "composition" is mixed in; we also use boiling with turmeric, which is a counterfeit dye, but the deception is easily detected when the cloth is cut, because the inside is not white but yellow.

Saved water must undoubtedly be avoided, since it reddens or takes it into blue; if approximately 1/10 acidified water with bran is added and boiled, the impurity can be separated in the scum; slimy red egg-white serves the same purpose.

Cochenille by itself produces a redness deepening to blue, but partly the acid in the "composition", and partly the white tin calx with which the red cochenille particles are blended, cause a change to yellow and an agreeable brightening in appearance.

Note 4. Cochenille is a small insect [coccus] which in America lives on a cactus. Two kinds are classified; the wild [kind] is the lesser, makes a more permanent but lower grade dye, indeed 3 to 4 times weaker than the fine [kind], which has a resistance to severe exposure to rain, and is usually called mestek after a region in Mexico which is overrun with it. The females of the Coccus stick fast and become immovable when they are very small, and are then made pregnant by the drones, which swarm about continuously, and after the last metamorphosis get wings. The females swell up after this from a considerable amount of young, and die at last in bearing their young directly. These pregnant females are harvested three times a year; of those which are collected the last time a part are stored in nests built inside the leaf of the cactus, which consequently keeps itself fresh, although separated from its root. When the rainy time is over they spread out in small nests of moss on the growing cactus, but do not remain long feeding there inside, after which the young spread out and the dead mothers yield the first crop of cochenille; after 3 to 4 months the second [crop] occurs, and after a similar period the third. Those which are meant for dye are killed with hot water or drying. I do not know if the males in their maturity are fit [for dye]. In these creatures is a wealth of dye, which is not lost over time in dry storage, since we have found 130-year-old [samples] as good as fresh. 700,000 to 900,000 pounds of fine cochenille comes annually into Europe.

A saturated solution of cochenille is ruby-red, somewhat bitter, but noticeably astringent; if it is diluted with water the dye tends more and more to blue; a considerable amount [of dilution] is required to make the dye unnoticeable. Almost all known salts precipitate a filtered cochenille decoction, and the dye behaves as follows after filtering. Acids draw it into yellow, a small amount causing a bright firey color, but a larger amount destroying it; acid of vitriol contaminated with iron makes it rosy until all color disappears; alkalis bring it more into blue, but after a few days into yellow-red; solutions of saltpeter, common salt, salmiak, and alum make it lighter; a little tin solution produces the brightest scarlet red, and a precipitate settles out, and if enough is poured in, almost all color is lost; similarly with green- and blue-vitriol, but the precipitate with the former becomes red-brown, and with the latter violet.

SIXTH SECTION, CHAPTER I, Red Dye

§ 355. If wool is boiled for 1-1/2 hours in dilute tin solution, and thereafter put in boiling hot alum water made with 1/4 [part] to [1 part] goods, and is left there until it is all cooled, rinsed and boiled with 2 lod cochenille per pound [of solution?] for 1 hour, it gets a beautiful crimson, caused naturally by the alum.

Note. The usual method for getting a crimson [color] with cochenille is to boil with 5 lod alum and 3 white tartar for 2 hours; boiling out occurs for 1 hour with 1 lod cochenille per pound, which is put in when the water is somewhat more than lukewarm, but the goods themselves at boiling. To get a true crimson color as well as handsome, HELLOT has, after the usual boiling, laid the goods in a fresh solution with a little salmiak, whereupon the dye tends immediately to blue with volatile alkali, and maintains all its luster. Fixed alkalis do as well as the above [volatile alkali] but the goods gets a pale appearance. Usually it is made rosy with canary-weed [orseille], which is an arificial dye.

§ 356. If the goods is first boiled with 1/4 [part] blue vitriol for 2 hours, and then boiled with cochenille, 2 lod per pound, for 1 hour, in the beginning it becomes somewhat reddish, but when a little white vitriol is added we get a gray-green color.

Note. If the goods is boiled without soaking with like amounts of blue vitriol and cochenille, it gets dark olive-green.

§ 357 (a). If the goods is treated in the same manner with green vitriol and cochenille, a gray-black color is obtained.

Note 1. If the goods is boiled without soaking with only green vitriol and cochenille it becomes dark gray.

(b) If 1/4 [part] green vitriol and 1/8 [part] tartar are dissolved together over heat, and then the goods is boiled with the clear [solution] for 2 hours, and finally cochenille, 2 lod per pound, for 1 hour, it gets dark olive-brown.

Note 2. HELLOT has found that the color of cochenille with tartarus vitriolatus becomes gray-agate-like; with Sal Glauberi the same, but impermanent; with gold solution, chesnut brown and streaky; with silver solution, cinnamon brown and streaky; with copper dissolved in acid of saltpeter, dirty crimson; with bismuth solution, a handsome gray; and with arsenic, crimson-brown.

(c) If the goods is boiled in the same way with white vitriol and then with cochenille, it gets reddish in the beginning, but when some [more] white vitriol is put into the brew, a beautiful gray color is obtained which tends towards brown with longer boiling.

§ 358 (a). If 1/4 [part] aqua fortis to [1 part] goods is saturated with lead, diluted and put to boiling for 2 hours, the goods rinsed and boiled with cochenille, 2 lod per pound, for 1 hour, we get a beautiful ruby color.

But if some bits of sugar of lead are put into the solution and boiling continued for 1/2 hour, a beautiful violet color is obtained.

(b) If the goods is dyed in the manner described in the beginning of (a), [then] boiled 1 hour with 1/4 [part] gray vitriol, it becomes almost black, or darkened.

(c) If the goods is dyed in the same manner as in the beginning of (a), [then] first boiled 1/2 hour with some sugar of lead, and then for the same length of time with green vitriol, although the former salt [sugar of lead] brings it towards blue to a like degree, as in (a), yet the color gets a bit darker than in (b) from the latter [green vitriol].

(d) If the goods is boiled 2 hours with 1/4 [part] lead solution, is put without rinsing into boiling hot alum water, made with 1/4 [part alum], left to cool, hung up overnight, rinsed, boiled out with cochenille 2 lod per pound for 1 hour, and finally [boiled] with the same brew darkened with 1/4 [part] iron vitriol for 2 hours, the color becomes as in (c).

To Dye Ponceau-Red On Silk With Cochenille

§ 359 (a). Four parts aqua fortis is mixed with 1 [part] common salt and in it is dissolved 1 [part] tin. Then the solution is diluted with 2 times as much water, and the silk is left in it for 24 hours, taken out and rinsed in pure water until it stops getting paler. The silk is then almost pale yellow, with none of the silk s own natural color, or the boiling off which occurs with tvål.

<u>Note 1.</u> The silk seldom yellows in tin solution if the aqua fortis contains no iron, of which there is more when we take up dyeing silk yellow. Tin solution is used up through soaking in it a little more [of silk] than its volume, can seldom be reused for new soaking, but should still not be made up in larger quantities than can be used immediately, since some tin precipitates with time. The silk should be treated with tvål immediately after soaking and be boiled out the same day, otherwise the acid has time to make the process go badly.

(b) The soaked silk is boiled out with 5/16 [part] cochenille for 15 minutes, without diluting, and with very little water, because then the acid can do its work better, to draw it to yellow, whereas to make a crimson color we should use more water. The brew which even yet still contains much dye, can be used for priming or boiling out woolen goods.

<u>Note 2.</u> The residual dye solution can also be used to dye silk lightly, and for a true crimson color, if 2 to 3 lod of cochenille are added per pound, and the goods previously soaked in alum.

(c) If the silk does not get yellow in tin solution, it gets then a bright rosy red color.

<u>Note 3.</u> Mr. MACQUER has discovered in 1768 how to give a scarlet color to silk, in a method almost exactly that which Mr. SCHEFFER knew about in 1751. The former [Macquer] carried out the following procedure: for 6 pounds of goods he made up a solution of 4 [pounds] aqua fortis and 2 [pounds] spiritus salis, in which was first put 1 ounce granulated Malacca tin, and then [more

tin] gradually until 3 pounds of it becomes soluble, so that no heat forms in the mass; the last 5 to 6 ounces can then be put in all at one time. The solution should be dark brown and clear, otherwise more tin is added. The silk is put into this, after being primed with orleau, and left in it for 1/2 hour, wrung out, and [rinsed] in water until it is no longer turbid. It is immediately boiled out, and to it is added 4 ounces of cochenille per pound, and 1/4 ounce of white vitriol; the solution is boiled, diluted with cold [water], until the hand can be held in it, the silk put in, heated, and taken out after 1 minute boiling and rinsed. This dye withstands washing, gets a little rosy when exposed to the sun 5 to 6 days, which also happens with wool, while silk bleaches out almost completely to a saffron red during this time, and besides is similarly expensive and more troublesome; with cochenille it is increased up to 1/4 the weight of the silk, which is not done with saffron.

(d) If silk, which has lain in tin solution, is wrung out and then left overnight in cold alum water, made with 2 lod per kanna, and finally wrung out, rinsed and boiled out, it gets only a weak Ponceau [color]

<u>Note 4.</u> Tin solution dropped into pure water loses so much acid that the metal precip- itates and is tinted [tingeras]; this scarlet colored lake, well stirred up into suspension, settles on wool, but not on silk, which later either by preparation with animal juices, or soaps made with animal fats, can be made to ad- here.

(e) If the silk is then left overnight in alum water, as in (d), is well rinsed until all loose tin calx is washed away, and then laid in alum water 6 hours, and finally boiled out in the above described manner, we obtain a weak crimson color.

(f) If 1 part tin is dissolved in 4 [parts] aqua regia, diluted with 8 [parts] water, and the silk soaked in it for 2 hours, rinsed, dried, and boiled out as before, the color becomes pale crimson.

(g) If it is similarly soaked, as in (f), but the wrung out silk put in cold water, acidified with a very little of acid of vitriol, left overnight, rinsed, dried, and boiled out, it becomes almost like the foregoing, but tends a little more into blue.

(h) If after similar soaking the silk is immediately put into cold alum water, made with 1/8 [part alum], left overnight, rinsed, dried, and boiled out, as before, and then boiled again in somewhat stronger cochenille solution with a little acid of vitriol in it, we obtain a beautiful ponceau.

Note 5. Soaking in cold tin solution in the manner which was described in Note 3, im- proves and dyes it a red wood color, which is otherwise unattractive. A strong brazilnut decoc- tion gives yellow silk a scarlet color, much the same as with cochenille, but much prettier and more durable, than after mere soaking with alum, and resists testing with vinegar, like crimson and ponceau. A beautiful violet is obtained in this manner with logwood.

(i) Boiling in dilute tin solution deprives the silk of its luster, which is deduced from the following experiment.

If this procedure is carried out with 1/8 tin solution (but in case there is some tval remaining after boiling out it is better to use only 1/4) for 1-1/2 hours, rinsed and boiled out with 1/16 cochenille for 1/2 hour, the color becomes at the end a bit paler than in the beginning, but is restored a bit if a little acid of vitriol is poured into the solution; however it remains weak and without luster.

(k) If the silk after boiling in tin solution, as in (i), is wrung out, laid in cold alum water, made with 1/8 [part alum], is left there overnight, dried, and boiled out with 3/16 cochenille for 15 minutes, it becomes rosy red, but also without luster.

(l) If silk, which lies 24 hours in such a tin solution as in (a), is rinsed, dried, laid again for one day in similar [solution], but in which is dissolved as much tartar as possible in the cold, is rinsed and boiled out with 5/16 cochenille for 1/2 hour, it becomes ponceau, but without luster.

(m) If the silk is left 24 hours in tin solution, saturated cold with tartar, wrung out and laid overnight in cold alum water, made with 2 lod per kanna, is rinsed and boiled out with 5/16 cochenille for 15 minutes, we get a crimson color, but with no special luster.

§ 360. If silk is soaked overnight in cold alum water, made with 1/8 [part alum], is wrung out and boiled out with 1/10 cochenille and a little aqua fortis for 20 minutes, we get a beautiful purple color.

Note. Soaking with alum and tartar similarly gives much less advantage than with either alone.

To Dye Cotton and Linen With Cochenille

§ 361. If the goods is first soaked for 24 hours in cold tin solution, wrung out, rinsed, and boiled for fifteen minutes with 5/16 cochenille, cotton goods gets a beautiful light red color; linen becomes somewhat darker and resists sunlight well, but not washing.

Note. To dye linen properly is not difficult, since this material takes up and retains the dye particles slowly, but it is also exposed to washing many times, indeed it is of such low price itself that expensive dye is not used with it, nor is extensive or difficult treatment. So consequently if a method can be discovered for giving linen various colors at little cost, which resists sunlight and washing, a useful problem will have been solved.

To Dye Wool With Grana Kermes

§ 362 (a). This succeeds best if soaking is carried out just as it was with cochenille for scarlet [dye] (§ 354), when a beautiful red color is obtained, but tend- ing a bit towards yellow.

Note 1. This tendency towards yellow is increased by greater amounts of Kermes. See Note 3. The redness is modified, as with cochenille, to yellow by acid, and to blue by alkali.

(b) If the goods is boiled out with 1/4 alum for 2 hours, and then with Kermes 2 lod per pound for 1 hour, the color becomes pale crimson.

(c) If tartar is used in place of alum, and the process is carried out the same otherwise, the color becomes paler but tending less to blue.

Note 2. If tartar is used without alum, and as much "composition" as with cochenille, we get a bright cinnamon color with 1/4 kermes in a single boiling; the goods is then put into alum water, again getting something of redness. If tartar is taken, [then] Roman alum, "compositi-on", and kermes to make the same decoction, we get a little of various alterations according to the proportions used; alum should always be more plentiful than tartar.

(d) If 1/4 [part] tin is dissolved in 1 part aqua regia, and 6 lod of this [solution used] per pound of goods, and a like amount of purified tartar, and [the solution] taken to boiling for 2 hours, the goods being put in when it begins to boil, and if it is boiled for 1/2 hour more with 1/8 [part] kermes, which is put in when the solution begins to bubble up, and immediately thereafter the goods [is taken out], the dye goes on well, and becomes fairly attractive, but tending to yellow.

If the boiling is carried out with about 1/4 [part] kermes in 3 quarts [of solution], the color becomes fuller and more attractive.

(e) If after boiling, as in (d), the goods is then boiled 1/4 with addition of cochenille, the color becomes about as before.

(f) If the boiling is carried out, as in (d), but the boiling out is done with 1-1/2 [part] kermes, 1-1/2 [part] cochenille, and purified tartar in the same amount, for 1/2 hour, the color becomes quite beautiful.

(g) If the boiling is carried out as in (d), but only for 1 hour, and then for another hour with 1/10 Roman alum, and finally boiled out with 18 percent kermes, 1-1/2 cochenille, and the same amount of purified tartar for 1/2 hour, the color becomes almost as in (f).

Note 3. Grana Kermes is a kind of insect [coccus] which lives on the Quercus coccifera [insectiferous oak] in Southern Europe. The females of this species attach themselves fast, become immovable, and resemble peas, when they for the last time shed their skins, lay a considerable amount of eggs, and die. The first generation in the spring settle on two small branches going out from each other, but the second [generation] for the most part [live] on the leaves and are also smaller. Collection is done, before the young hatch out, by women, and one person can gather about 2 pounds per day. Those who purchase grana kermes sprinkle them with vinegar and dry them in the sun, whereupon they die and produce the red dye. In the Levant it is much used, but in Europe [it is used] for little else than for dyeing wool for certain tapestries, except in Venice, where a darker scarlet is fancied, which is called ecarlatte de Venice, ecarlate de graine, and also oxblood dye; it does not have as much fire as the usual [dye?], but holds up better in the sun, and resists stains and soiling better. Yarn is boiled first 1/2 hour with bran; then it is soaked in water, with addition of 1/5 acid, and likewise to [1 part] goods [is added] 1/5 [part] Roman alum, and 1/10 [part] red tartar, in which [solution] it is boiled for 2 hours, the solution pressed out, [the goods] put in a bag and left in a cool room for 5 to 6 days; for boiling out it is put in water, and when it warms up, 1/4 kermes for producing a full dye, and indeed if it is old, also at the first swelling up of the goods, and boiled for 1 hour. Clothing is so compact that it [dye] cannot penetrate, as it can in yarn; therefore with this dyeing [of clothing], the amount of salt and dye should be decreased by about 1/4; and it does not need as long a time for boiling. Lighter variations are obtained with smaller additions of Kermes, and always the palest first.

Ecarlatte migraine is used for a blood-red dye, and is very fast. In this manner we obtain Venetian scarlet, but with only 1/2 kermes and the rest madder.

After soaking with tartarus vitriolatus, kermes gives an agate-gray, beautiful and fast dye; Glauber's salt causes a less attractive and non-fast gray; green vitriol and tartar gray-brown; blue vitriol and tartar olive-like; the same with cuprum nitratum; bismuth solution added during boiling with kermes gives violet, etc., as Mr. HELLOT observes.

We can also produce a red dye with gummi lacca, held in baculis, which is found in insects fastened to branches of Croton lacciferum, and which contains a red dye soluble in water. To determine accurately the proportion used by Mr. HELLOT, we similarly extract the dye with mucilage from symphyti root in water, then precipitate it with Roman alum, and then collect and dry the same. In this way we get slightly over 1/5 the weight of the lacca; thus it is not particularly economical to use it in place of cochenille, since the color is not as bright, but because it is faster it can be used to advantage for priming.

Coccus polonicus, which is collected right after midsummer in certain parts of Poland from the root of Sceleranthus perennis, is much used in Turkey for dyeing wool, silk, horse-hair, nails [finger nails?], etc., but is relatively unknown in European dyehouses. It produces, like the foregoing, only crimson by itself.

§ 363. If the goods is boiled 2 hours with 1/4 blue vitriol and then boiled out with 1/4 kermes, we get a full yellowish olive color.

§ 364. If the process is carried out similarly with green vitriol we get a brown color, which is more handsome if it tends more into yellow, which probably can be obtained by a small addition of blue vitriol.

§ 265 [365]. If the silk is first soaked in tin solution, as in § 359 (i), and then left 18 hours in cold alum water made with 1/8 [alum], is dried, and finally boiled with 1/4 kermes for 15 minutes, it becomes quite pale red, but in addition without luster.

Note. Sufficient research has not been carried out up to now with kermes on silk and linen, so nothing can be reported yet with certainty about it.

SIXTH SECTION, CHAPTER I, Red Dye

To Dye Silk With Safflower

§ 366 (a). The safflower must first be washed in this manner, that it be enclosed in a bag, and then trampled in water until no more yellow color comes out than can be forced, and those dried petals which were yellow before become red; about 1/2 its weight is thus lost.

Note 1. Safflower is the petal of Carthamus tinctorius, which contains a yellow dye extracted with water, which must be separated so that a red dye can be obtained quite pure by means of alkali. This dyestuff is also called Safranum, probably for its resemblance to genuine saffron, which however does not contain any similar red dye.

(b) The washed safflower is put either wet or dry into a clear solution of alkali fixum with as much water as is needed for complete maceration of the dyestuff. The dye usually takes 2 lod of alkali for each pouund of unwashed safflower; too much not only uses more acid for useless purpose, but also darkens the color, so that it is better to take less in the beginning, since if it is not all extracted we can afterwards use more added [alkali?].

Note 2. Usually the washed safflower is interlayered with 6 percent sieved ash or soda, and trampled; thereafter it is leached out with water until the safflower becomes colorless, when if the recommended amount of alkali is not sufficient, more must be added.

(c) After 3 to 4 hours the foregoing [solution] prepared in (b) is used for dyeing. The safflower is pressed out through a collander and washed out with water as long as there is color. The first lye [leachate?] running out appears quite brown.

(d) Then the lye should be saturated with acid, whereby it becomes quite red. To this should be added old citrus juice, which has been fermented and which has liberated an abundance of sediment, whereby it becomes clear and very sour; so it is customary to add unripe fruit [ankartals] from France and Alicante. Fresh is no good. We use also distilled vinegar, but the dye is not as bright. But rowanberry juice can serve just as well as citrus juice, if the berries are well crushed with a wooden crusher, pressed out, and left to ferment, after which it clarifies and is very sour, is diluted in bottles, and becomes better with age; this operation requires several months and should necessarily occur in the summer.

As soon as the foam becomes light red and a little purple, it is a sign that enough acid has been added; too much takes the color into crimson.

(e) Into the red solution so prepared we lay the silk, and turn it frequently, until it is dyed out, when it is put in new [solution], and so on, until all the rosy color shall become bright.

If the silk was previously boiled out with tvål, it needs no rinsing, but only wringing out and putting in [new solution].

Note 3. The silk should be boiled until white, and for a bright firey color it is primed with orleana, or the brighter yellow dye of safflower.

(f) This beautiful dye is changed both by sunlight and washing, but when rowanberry juice is used, the silk can hang in the sun for a day with no apparent change, when at the same time that dyed with 3 times as much citrus juice can be quite leached out.

To Dye Cotton and Linen With Safflower

§ 367. This is carried out in the same manner as with silk.

Note 1. Herr Prof. J. BECKMANN in Göttingen has recently published very useful research on this subject. Linen, which is doused well in an alkaline solution of the dye, and then rinsed in water acidified with citron juice, acid of vitriol, alum- or tin-solution, gives different handsome [color] changes, which become fuller and heavier through repetition [of the dyeing?]. This is also useful for priming with the yellow dye of safflower, which draws into red through boiling with purified tartar and alum. It is primed with madder and alum, in which the goods is left for a long time while the solution is kept well heated, then the alkaline solution is saturated with acid of vitriol to produce a beautiful violet dye tending towards red. If the alkaline solution is saturated more or less with acid of vitriol, and the solution along with the raw or primed goods is gradually brought up to boiling, we obtain in proportion to the acid, water, etc., and after the time of boiling, manifold varieties of red, violet, dark brown, and yes, even black. We can also dye both yellow and red on wool with safflower.

Note 2. In addition to the above described method, which gives a red color, we can also likewise obtain such [color] by Fernbock (chips from an as yet unknown wood from South America); Brazil; Braziletta; and Sappan (the wood of caesal pine); Canary-weed (Lichen roccella, L. parellus), of which there are two kinds, which are both prepared with urine and burned lime, the first and best growing abundantly on the Canary Islands, the second in Auvergne; goat-hair dyed with madder and then dissolved in alkali fixum; and more, which also give artificial dye, except for Canary-weed, which by means of tin solution can produce not only a beautiful redness but also a fairly permanent one.

SECOND CHAPTER: On Yellow Dye.

To Dye Wool With Wold [Dyer´s Weed]

§ 368 (a). The goods is first boiLed with bran.

(b) Then it is soaked with alum 1/4 [part] and tartar 1/12 [part] in water, during 2 hours boiling.

(c) The [dye] solution is prepared from wold blossoms alone, through boiling in water 1/2 hour, after which it is obtained clear, the goods laid in it and vigorously stirred about, until the desired color is reached. With boiling we must proceed cautiously here, otherwise the dye becomes darkened. It is best therefore to boil the water, put in the wold, take the vessel from the fire, cover it, let it sit overnight, boil it again the next morning, separate the wold, and put in the goods. All boiling which follows with wold to the desired yellow, ought to be undertaken so that the solution is boiling when the goods is put in, but then gets to boil no longer.

(d) The dyer uses added lime, which is called Schoua, when most of the strength is taken from the solution, to make the solution quite yellow again, but this is not only unnecessary, since the wold gives up its dye still from itself, but also detrimental, since the goods is thus stained by all acids, in fact it hardly resists wine and sour beer. Those who wish to see what enhancement alkali can otherwise accomplish, need only dip the goods, boiled a little with wold, in cold lye to see the dye immediately brighten.

(e) After the boiling out in (c), the dye solution is still quite yellow, from which we can deduce that the dyers´ weed alone, as it is usually used, is too much, which is further confirmed when the soaked goods is boiled out in the same manner with only 1/2 [the amount of wold], and we get a beautiful yellow dye.

Note. The dyers´ weed [wold] is usually called Vau or Vå by us, Wiede in Germany, Gaude in France, Woad in England, and is like Mignonette [Reseda]. It gives a good and fast dye of different varieties after differing soakings, or addition of alum, common salt, salmiak, gypsum, etc. It works also on cotton goods, often in such a way that neither sun nor boiling with hard soap [tvål] can take it out.

§ 369 (a). If the goods is soaked in 1/4 tin solution and 1/4 crystal of tartar for 2 hours, rinsed and boiled out with wold alone for 15 minutes, we get a beautiful color.

(b) If after similar soaking, boiling out occurs, with 1/3 wold for 1-1/2 hours, we obtain a handsome color, but weaker. Although both of these foregoing colors are bright, they do not penetrate the cloth and we find on cutting that it is white inside.

(c) If the goods is soaked in 3 lod tin solution and a similar amount of crystal of tartar per pound for 2 hours, and we then boil it out with 1/8 wold for

1 hour, the color becomes fairly handsome, but somewhat pale, and does not penetrate, from which consequently there is hardly anything to be learned.

(d) If the goods is soaked with 1-1/2 lod tin solution and a similar amount of crystalli tartari for 2 hours, and then is boiled out with 1/16 wold for 1 hour, the color in ugly, and neither does it penetrate very much.

To Dye Silk With Wold

§ 370 (a). If the silk is soaked in cold tin solution made with 4 parts aqua fortis, 1 [part] common salt, and 1 [part] tin over heat and saturated with crystalli tartari, left 24 hours, rinsed and boiled out with wold blossoms alone for 1/2 hour, we get a handsome straw color.

Note. Usually silk is dyed with wold, after boiling in 20 percent tvål and soaking with alum, in a filtered decoction in which 1/2 [part?] grass [wold] is boiled 15 minutes, and cooled until the hand can be held in it. After this the same [amount of] wold is boiled a second time and poured into the used decoction, whereby only half [as much wold] is used. Finally 1/20 [part] potash is mixed in with the remainder of the second boiling, and when it clarifies the goods is stirred in it until the desired height [of color] is obtained. For citron-yellow boiling should occur with 35 to 40 percent tvål, since the whiter it is the more beautiful it is. A little orleana with the just mentioned lye draws the color into yellow-gold.

Grains d'Avignon, or the berry of Rhamnus minor, can be used in the same way as wold, but gives a more soluble dye.

Orleana, or dough of Bixae orleanae seed, is called rocou by the French, and is a resinous dye, which must be dissolved with 1/4 [part] potash; too little takes it into a brick color. With it we get an aurora [orpiment color?] without soaking, but if this is to serve as a basis for Mauredoré it is soaked then, because boiling out finally occurs with yellow-wood [gelbholtz] and campeche wood; if it gets too red it can be changed more into yellow with a little green vitriol. Alkali changes this otherwise red dye into yellow, and with acid, which saturates it more or less, we recover the desired degree of change into red, such as pomeranian-yellow, etc.

Gold color and chamois is gotten from the decoction which previously served as aurora. This dye is not permanent, but becomes brick-colored after some time.

(b) If the foregoing process is repeated only with this change, that crystalli tartari is excluded, the color becomes less attractive, from which is shown the usefulness of tartar in the present case.

(c) If the goods is soaked as before in tin solution and then is put into cold alum water overnight, made with 2 lod per kanna, is rinsed, and boiled out with somewhat more than enough wold for 1/2 hour, the color becomes not especially attractive, from which we observe the unfitness of alum for this process.

(d) If the goods is boiled for 1-1/2 hours with 1/8 tin solution, is rinsed, hung up to dry, and finally boiled out with wold, the color gets quite pale and also without lus- ter.

This dye, when it is handsome, has the advantage over other yellow [dyes] which it replaces, in that it resists acids well, which is no small advantage.

To Dye Wool With Saw-wort [Ängskär]

§ 371 (a). Goods boiled with bran is soaked while boiling for 1 hour in alum water, which contains 2 lod per kanna, and 1/2 crystalli tartari, after which it is hung up overnight and well rinsed before boiling out.

(b) The saw-wort is put into cold water and then kept boiling for 1 hour, the clear solution decanted and the soaked goods stirred about in it until it takes on the desired color.

Note. This dyer's weed [färggräs] is called Scharte in Germany, Serette in France, and is Serratula tinctoria. It gives a good yellow dye, but of poorer appearance than wort produces, and tends toward green. Soakings and additions, for which we use common salt, alum, gypsum, vitriol, etc., produce different useful variations, which for the most part are resistant to sun and tvål, also for cotton and linen.

§ 372. Tin solution also does a quite good job with saw-wort, since if the goods is soaked during boiling with 1/16 tin solution and a similar amount of crystalli tartari for 2 hours, rinsed, and finally boiled out in the described manner, the color becomes much brighter than in § 371.

§ 373. If silk is soaked as in § 370 (a), and then boiled out with saw-wort, it becomes yellow-gold, but not particularly attractive.

To Dye Wool Yellow With Willow Leaf

§ 374 (a). Soaking takes place (per pound of goods) with 8 lod of Roman alum and 2 white tartar dissolved in 3 to 4 kanna of water over heat, but the goods is not put in before it cools off, and left there overnight, the solution wrung out, rinsed quickly in running water, and wrung out rather well. Cream of tartar in place of tartar makes the dye clearer.

(b) Boiling out is carried out in the previously de- scribed manner.

To Dye Linen Yellow With Willow Leaf

§ 375 (a). Soaking takes place with 8 lod alum dissolved with heating in 3 to 4 kanna of water in which the goods is left overnight, when it becomes cold, as was mentioned for silk.

(b) Boiling out is carried out as before.

Note 1. Dir. SCHEFFER told the wife of Commerce-Counselor ALSTRÖMER, the elder, confidentially, that [this is] the genuine yellow dye with Willow for which in the latter years there is great premium offerred in England; the color is not very bright, and somewhat changing in to green, but resists the sun well and boiling with tvål. Mr. Commerce-Counselor and Knight P. ALSTRÖMER has found through research that the color becomes fuller if the goods is soaked with 12 lod of alum, wrung out and dried before the dyeing is begun; he has also found that 1 lod of potash is needed per pound to develop the color sufficiently.

Note 2. Common aqua fortis gives both wool and silk in 3 to 4 minutes a beautiful light yellow, quite resistent dye; the goods should be immediately rinsed in pure water. Indeed the more phlogisticated the acid is the better; it becomes especially useful if indigo, litmus, orleana, or other similar substance is put in. If such phlogisticated acid of saltpeter is mixed with 6 to 8 times as much water, the goods can be boiled in it, and the longer this occurs the stronger the yellow, and that even should be replenished which fumes away. Aqua fortis cannot dye more than a certain amount of goods, but with addition of iron, and distillation, it can be restored. Soaking with alum is useful. Silk does not lose its luster. If the goods is first dyed with blood-lye and acid of vitriol (§ 379, Note 1), it becomes green in aqua fortis.

To Dye Silk With Lett Blossoms

§ 376. Pure water is boiled with Lett blossoms, while a little tin solution and a solution of crystalli tartari, or tin solution saturated with the last named salt, is dripped in until the decoction becomes moderately yellow. Into this [solution], which is kept hot but not boiling, is put the goods, and stirred until it gets to the desired yellowness, which is quite pale, provided that we use good water which will not precipitate tin solution.

Note. Lett blossoms are called Anthemis tinctoria; it contains a copious extractive yellow dye.

To Dye Cotton Or Linen Rust-yellow

§ 377 (a). First the goods is boiled for 1/2 hour with tval and wrung out.

(b) It is then put in cold vitriol solution overnight, hung up and almost dried.

(c) It is further stirred about in water, boiled with unslaked lime which is continued until it gets yellow, which requires about 15 minutes; it is taken out and wrung out.

(d) After this it is left in vitriol solution until it gets quite green, is hung up, dried, and then rinsed well.

The color resembles rust stains, and so is not especially attractive, but is very resistant to both sun and washing.

Note. In madder extract this gets dark or brown with many variations, with its own strength and length of dousing; consequently it is useful in calico- and linen-printing.

To Dye Wool With Yellowwood

§ 378 (a). The goods is soaked with 1/4 Roman alum, and 1/12 white tartar while boiling for 2 hours, and hung up overnight.

(b) The dye-extract is prepared by boiling with 1/2 yellowwood for 15 minutes, tied in a linen bag, after which the bag is taken out, the vessel removed from the fire, the goods is rinsed, put in and stirred for 1/2 hour, when it is rinsed [again] and shows a somewhat attractive color, but one which is altered markedly by only a day's sunshine.

<u>Note.</u> This dyestuff is taken from Morus [mulberry] tinctoria, whose wood is scraped; it is commonly called Brazilian Yellow [Gul bresilia], Gelbholtz in Germany, Fustik in England, and Fustet in France. Soaking with alum or tin solution makes the color much brighter, but more fleeting, than in some other methods.

Besides the foregoing, many other substances are found which are used for yellow [dyes], such as Sandelwood (Santalum album); turmeric (curcuma longa), the root; genista tinctoria, the blossoms; barberry bark (berberis vulgaris); wild chervil, the blossom (chaerophyllum sylvestre); netsle roots (utica dioica); and many more which still do not give other than imitation, but often quite beautiful dyes.

THIRD CHAPTER: On Blue Dye.

To Dye Blue on Linen with Indigo

§ 379 (a). One and one-half lod of alkali fixum is boiled with 3/4 lod unslaked lime in 1 quart of water, until it boils away 1/3, after which it is allowed to settle and the clear portion decanted off. To the residue is added 1/2 quart of water, which is boiled until 1/3 of it vaporizes, it is allowed to settle and finally the clear portion is poured off.

(b) In this quart of lye is boiled first 1/4 lod of finely ground indigo, until it is well mixed in, after which 1/2 lod of pulverized auripigment is put in and the boiling continued until the decoction becomes green and the froth blue with a copper-colored film.

(c) In this boiler the goods is stirred until it has acquired the desired degree of color.

<u>Note 1.</u> The author [Scheffer] goes into detail as to how wool is dyed, because it seldom is done, but HELLOT has already described both how a blue dyer's tub can be set up, using only indigo, and also with addition of woad [Veide}.

Indigo is obtained from a plant called Indigofera, which is cut within the blossoms about an inch above the ground, and put into water to rot, whereupon blue particles separate, which are decanted with the water and by reaction are caused to coagulate. It is obtained both from the East Indies and West Indies; the best floats on water, is dark blue and displays a copper color, at least in a fresh cut, or that which is scraped with the nails.

This expensive dye we have found it expedient in later years to use quite handily and with more economy than previously, the so-called Saxon Blue; but its advantages cause a great inconvenience thereby, so that its otherwise fast blue indigo color becomes entirely fleeting and inconstant. Mr. Mining Counselor PÖRNER has quite recently discovered how to make it stick to wool. On 1 part finely powdered indigo is poured 8 [parts] clear and well concentrated acid of vitriol; this is left undisturbed for 24 hours, is then mixed successively with 96 parts of water so that when the last is added it is all quite finely ground and dissolves. When we now want to dye in a cold vat, we mix it with boiling water; if we take 2 parts of this solution with 24 [parts] of water, 1 [part] of goods becomes dark blue and unchanged in air, although it undergoes no other treatment than to have been completely doused previously in water. This dye is still immediately usable for its darkness, but if it stays for 1 day in 24 parts water in which is dissolved 1/2 part common salt with boiling and then somewhat cooled, it becomes brighter without loss in permanence. If the goods is soaked with alum it gets a beautiful dark blue artificial color; likewise if it lies for some hours in a cold solution of potash, the darker it is the more permanent. This method also seems alright for silk, but it does not get quite permanent on vegetable goods.

We can also dye [goods] blue in this manner: the goods is soaked in blood lye and then put into a solution of iron vitriol with some excess of acid. In this way we get a good dyeing of wool and silk, but it requires a particular procedure if it is to go on evenly. Berlin Blue itself, preferably made without any alum, and ground well with a little acid of vitriol, can be worked into cotton goods if they are dipped alternately in water and the dye solution. It is understood that all acid is well rinsed out, otherwise the goods becomes brittle.

A cold blue dyeing vat is prepared with 1 lod finely ground and slimey indigo per kanna of water, into which is mixed a solution of 1 lod green vitriol and 2 [lod] fresh unslaked lime. These materials quickly begin to react and can be used after a few hours; just as soon as the solution gets green, a copper-colored film forms and a blue scum. Cotton goods is doused well with boiling water, wrung out, put into the vat, wrung out, rinsed and put in again, [and this procedure] is continued until the color becomes that which is desired. The last rinse water should be acidic with a little acid of vitriol. The vat is freshened with potash when it is needed.

The author´s [Scheffer´s] method for dyeing linen I have investigated; it succeeds very well, but is not suitable for wool, which is attacked by the strong lye.

By and large, the setting up and regulating of a vat of wold and indigo is governed by regulations adopted by the Master Dyers. Wold (Isatis tinctoria) is prepared in this manner: the leaves are picked, mascerated, formed into balls and dried to hardness, after which the balls are beaten to pieces, moistened, tossed together into piles to ferment, and when internal reaction ceases they are spread out again, moistened and tossed together; this is continued until fermentation is completed, when the wold is packed into tins. It is called Waid in Germany, and Pastel in France, and gives a yellow-brown dye with water alone; the blue must be dissolved by alkali and fermentation. To 18 tuns [casks] of water is usually taken 16 lispund [20 pounds weight] of wold, the same of indigo, white potash, 8 marks each of wheat bran and madder, which when all are well stirred up give after 24 hours a vat of very serviceable dye; but if the fermentation is not controlled in this careful way, but goes on too vigorously, the whole batch is sometimes lost and all the blue color likewise is consumed.

Logwood, brown brazil or bluewood (haematoxylon brasilianum) is called bois d'Inde by the French, produces a brown-red dye with water and wool which is boiled in it, but cotton goods becomes dark brown tending to violet without any additives, and with common salt, gypsum or potash, dark blue, colors which are changed somewhat by being well boiled with tvål, but never removed. We can also dye wool blue with bluewood [blauholz], without soaking, with only a little addition of copper vitriol; but it must be changed basically to an artificial black, and with various varieties of gray and purple, all of which still fade away in the sun. Probably there is a method for fixing [these dyes], because Mr. DE LA FOLLIE has found a way to make them serviceable, through phlogisticated material, for manufacturing blue sugar paper [sockerpappers].

Blueness, like other colors, has various degrees, of which we distinguish only a few by their own names, such as blue-black, dark-blue, medium-blue, light-blue, and heavenly-blue; but the French have distinguished more varieties, like blue-white [bleu-blanc], fresh-blue, pale-blue, mourant-blue, mignon-blue, celeste-blue, queen's-blue, turquoise-blue, royal-blue, bleu de guesde, bleu pers, bleu aldego, and bleu d'enfer.

Note 2. Mixtures of yellow and blue make green; cconsequently the goods usually undergoes 2 dyeings in order to make it green, for which speciaql dyeing is required in order to make it even and without spots. We can also make the goods green in a single operation with blauholtz [blue wood] and yellow mixed in the same vegetable decoction, but it is unattractive; various plants, such as bromgrass (bromus secalinus), omogne berry by thrashing (Rhamnus frangula), wild chervil (chaerophyllum sylvestre), red trefoil (trifolium pratense), reed (arundo phragmites), etc., also give green dyes, but they are faded mostly by sun and washing.

FOURTH CHAPTER: On Black Dyes.

§ 380. Black is obtained by yellow, red and blue brought to full strength and mixed (§ 340), but the result can often be achieved more quickly, especially with astringent substances and iron vitriol.

Note 1. By itself black is not a color, unless it consists of a lack of anything able to reflect light rays, of whatever kind they may be. Nevertheless, it is usually included as one of the dyer's 5 principall dyes, namely red, yellow, blue, black and brown-yellow, all of which can be produced from substances occurring in nature. The author [Scheffer] reasonably includes only three.

To Dye Wool Blajck

§ 381. To get a perfect black requires a foregoing good basic blue (§ 382, e - g), and the remaining procedure can be gathered from the following example.

(a) One hundred pounds of medium blue cloth is boiled for 2 hours with 8 [pounds] of tartar and 16 green vitriol in this manner, that the former [tartar] is put in the vat and when it boils the latter [is added], after which the goods should be put in. The reason for this order is that the tartar will prevent iron ochre from depositing itself, but if both salts go in together the vitriol is dissolved first and the action of tartar is lost.

Note 1. On a white base with astringents and vitriol we get nothing but a brown color, if the usual order is changed (§ 382, e - g), but on a blue base [we get] black; blue seems in it- self to bwt nothing but a diluted black, and black a concentrated blueness, so the bluer the base the better.

(b) Finally the cloth is rinsed until it no longer tastes vitriolic, and is boiled 2 hours in the solution, which previously through several hours boiling takes the strength out of bearberry [mjolonris] twigs, from which a beautiful black dye is obtained.

Note 2. Usually the blue goods is first soaked with boiling in a decoction of gallapple, along with being taken out and wrung out many times. Bearberry twigs (arbutus uva ursi) can be used with the same if not greater use, and costs very much less. Although blauholz is an inferior dyestuff, it gets used for a most perfect black that is similarly enhanced. One hundred pounds of blue based [goods] with 8 [pounds] tartar, 16 vitriol, 2 Spanish green, and 10 blauholz, are finally boiled with a decoction of bearberry twigs, to obtain an excellent black. We believe that vitriol makes the goods brittle, especially after boiling, but there is little danger if enough care is used; strong heat dephlogisticates the vitriol, from which a lot of ochre must precipitate (§ 86, Note), and to prevent this, tartar is used. We ought to avoid vitriol, which contains excess acid, because it weakens the dye (§§ 85; 164; 382, Note). Spanish green makes a black immediately with blauholz, if not too much water is used, when the solution tends to- wards blue.

Gray is obtained on a white bease with the same substances as for black, but in smaller proportions, or by means of more hasty boiling of the goods; it becomes most attractive if boiled with gall, and boiled out with vitriol in not too strong a heat.

(c) On a light blue base a good black is also obtained.

§ 382. Astringent substances and green vitriol produce black, according to circumstances and different proportions.

(a) If the woolen goods is boiled 2 hours with 1/4 alum, rinsed and dried, and then boiled for a similar time with 1/4 gallapple powder, we get a light yellow-brown color.

(b) If the yellow-brown goods in (a) is left 1/2 hour in water with green vitriol, 1 lod per pound, and dried, it becomes dark brown tending toward green.

(c) If it is soaked again in the same solution with 1 lod vitriol per pound and dried it becomes still darker.

(d) If it is soaked further in the same solution 1 hour with 2 lod vitriol per pound it gets still darker.

(e) If the goods is first boiled in 1/4 green vitriol for 2 hours, and then after rinsing, for a similar time with 1/4 gallapple, it becomes almost completely black.

(f) If operation (e) is carried out with bearberry twigs instead of gallapple, the color becomes almost the same but tends somewhat into violet.

(g) If 100 pounds of goods is boiled with 8 [pounds] tartar and 16 green vitriol for 2 hours, rinsed, and then boiled out for a similar time with bearberry twigs, it then also becomes almost completely black.

Note. This astringent, which is found in a multitude of plants, is an odd ingredient which seems to consist of acid, lime and fine oil; it dissolves in water and spiritus vini, and precipitates all metallic salts, which do not possess an excess of acid, whereby the metal is combined with the astringent and both precipitate, often with different color. Iron, dissolved in whatever acid you please, is precipitated black in this way, and when the precipitate is gummy, which makes the solution viscous, it is prevented from depositing, which according to Mr. LEWIS' research is best prepared as follows: 6 lod gall-apple and 2 blauholz are pulverized and boiled in 5 quarts of water until only 1 stop [3 pints] remains, it is allowed to cool, is strained into a vessel in which the ink is to be kept, after which 23 [lod], or sometimes more, gum is put in. One such servicable ink attaches well to paper therefore, but not on other goods. If some stronger acid is dropped into the ink, it becomes as colorless as water, but black again if the addition is neutralized with alkali. From this we can learn what damage a vitriol with excess acid can cause.

§ 383. Blue vitriol makes other variations with astringents.

(a) If the woolen goods is boiled with 1/4 blue vitriol for 2 hours, rinsed and boiled out with 1/4 gall for 1-1/4 hours, it becomes yellow-brown, darker than in § 382 (a).

(b) If 100 pounds [of goods] is boiled 2 hours with 16 [pounds] blue and 9 green vitriol, rinsed and boiled out with 1/4 gall for 1-1/4 hours, a still darker yellow-brown [color] is obtained.

(c) If operation (b) is carried out with a decoction of bearberry twigs instead of gall, the color becomes more beautiful and falls more into chestnut brown.

To Blacken Silk

§ 384 (a). The silk is first boiled with 1/4 gall for 2 hours, or with a decoction of 2 parts oak bark and 1 [part] bearberry twigs, which was previously boiled 2 hours in water. Then it is dyed black in the black decoction with 4 lod gummi arabicum and 8 lod green vitriol per pound; the decoction should be boiling hot, and the silk taken out 4-5 times and cooled, by which at length it becomes fully black in about 2-3 hours. This is done best in a black pan.

(b) To fix a black pan is done as follows. The kettle is filled with crushed oak bark and filled with water for 2 hours, after which the bark is taken out, put into another kettle and boiled with as much water as was needed to fill it before, and finally collected in it.

Then 2 lod of gummi arabicum per kanna of solution is put in, and after complete cooling 1-1/2 [lod] green vitriol, and 2 lod of small [pieces of] iron, after which it is stirred daily until the vitriol is saturated, and with this blackening is better than with 69 species.

Note 1. In France they use a very difficult process, and with manifold useless additions, which can be inferred from Mr. Macquer's description. On a blue base we can blacken silk in the same way as with wool, but it is not used. In Genna and Tours it is carried out as follows: the silk is soaked in a solution of 1/3 to 1/2 gallapple compared to the [weight of] goods, and it is blackened in a clear solution of 20 percent gall to goods, with 3-1/2 percent green vitriol, 12 iron filings, and 20 gummi [arabicum], which stands at least 6 to 7 days. Using this it is warmed up, but not more than the hand can be held in, white galled silk is dipped in and taken out again after 10 minutes to be aired; this is repeated often up to 30 times, while some vitriol and iron filings are added, until the silk reaches the desired blackness. The vitriol by degrees is brought to exceed about 8 times that which was first put in.

Note 2. Cotton and linen cannot be brought to take on a permanent black in the described manner, unless special preparation is added to this. Proper processing with animal substances is used or similar preparation, then the goods should be made rust colored (§ 377), and finally boiled out with some astringent, so that the color becomes both attractive and permanent. This method, which otherwise is used with full success on cotton goods, is otherwise instituted, but ought not to be divulged here because, as I learned myself, it can have some harm.

To blacken linen, which does not need to be washed, we can avail ourselves of soaking with 1/4 lod Spanish green per pound in a boiling hot decoction of 3 kanna of water; the goods should be well stirred about in the beginning and then left in it overnight. Boiling out occurs with 12 lod blauholz, which has previously been boiled 1 hour in 4 kanna of water, then the clear [liquid] is collected, in which the goods is doused and left overnight; finally the goods is dried in the shade.

This does not dye by itself but resists washing well, and tends by degrees into blue as it is continued. It is unusual what a great effect the Spanish green has, since at first we use here only 1/128 to the goods, and secondly the least part of it is dissolved in water, but nonetheless the goods immediately becomes black when it is put into the blauholz solution.

Various plant saps become black in free air; Anacardium is known for this particularly, the nut of which yields a sap which makes linen red-brown at first, but then quite black, which neither soap nor lye can boil out, and therefore this fruit is known all over India by the name of Marknut, because linen is marked by the sap. The sap of Lycopus europaeus penetrates linen well, makes a dark dye, but is easily boiled out with soapy water; from Actaeae spicatae berry we get a violet dye which deserves much closer investigation.

Note 3. According to the Author's [Scheffer's} plan we should now take up in order Fashionable Colors, or such as are occasionally in especial favor, but I have not found anything in my remaining papers [i.e., notes by students of Scheffer]. Variously colored clothing are of 2, 3, or 4 kinds of variously colored wool, which is rubbed together and spun; variations of these consequently could be endless. They are called kettle-dyes which are obtained by boiling out, and which also after mixing give manifold varieties, and which do not all have separate names in Sweden. When we dye red on a blue base we get couleur de Roi, couleur de Prinse, pensée, violet, purple minime, rose seche, etc.; from blue and crimson we get colombin, purple, amaranthe, pensée, violet, etc.; from blue and green, verd jaune, verd naisant, verd gai, verd d'herbe, verd de laurier, verd molequin, verd brun, verd de mer, verd de perroquet, verd celadon sans bleu, etc.; from blue and yellow- brown, gris verdatre or olive color; from blue and black, gris de more; from red and yellow, aurore, couleur de soucy, orange, jaune d'or, couleur de langouste, and fleurs de grenades, of which the last two require scarlet red; from red and yellow-brown, couleur de canelle, de tabac, de chataine, muse, poils d'ours, etc.; from red and black, ecarlats brunes, cramoisis bruns, gris vineuse of various kinds of brownish nature; from yellow and brown-yellow, feuilles mortes, poils d'ours, etc.; from yellow and black, maure doré; from yellow-brown and black, caffé, maron, pruneau, muse, epine; from blue, brown- yellow, and a little red, grise ardoises, gris lavandes; from blue, red, and black, gris de sauge, de ramier, d'ardoise, gris plombé; from blue, yellow, and brown-yellow, verds, merde d'oye, olive, etc.; from red, yellow, and brown-yellow, carnations des veillards, canelles brules, orange, etc.; from yellow, brown-yellow, and black, poil de boeuf, noisette brun, etc.

§ 385 (a). When we dye in a large kettle, certain extra amounts of ingredients should be calculated. For example, let 30 pounds of goods be that which is to be dyed with cochenille, but if we have no other kettle available but one for 50, then the ordinary amount of cochenille to be added is unchanged, but of everything else we use more. Now to find out how large this augmentation should be, we multiply that which shall, and that which ought to be dyed with each other (30 x 50 = 1500), and from this extract the square root and subtract this amount from that which ought to be dyed (50 - 38 = 12), the residue which now gets added to that which shall be dyed (30 + 12 = 42), and so get the result, according to which the excess substance should be so proportioned, just as it should be dyed. We can also more easily find this, if what shall and should be dyed are added (30 + 50 = 80), and the sum divided in two (80/2 = 40).

(b) If a yarn, which is oily or greasy, is to be dyed, it is first twisted very tightly on two sticks, to eliminate curling, put into lukewarm water, which is brought to boiling for one hour, and left to stand until it is all cooled. Then it should be washed with hard soap [tvål] and hot water, yet not scrubbed, but only drawn through the hands, rinsed well then, and finally boiled with wet bran.

Note 1. Wool is seldom used undyed, but must nonetheless be very pure and white if the dye is to coat it well and become attractive..

(c) During by-boiling and out-boiling the water, when softening occurs, should never be poured directly on the goods, but to one side, because otherwise it becomes spotty.

Note 2. It is quite an art to dye in quantity and without measure, and it requires both special skill and practice, to process at one time a considerable quantity of goods in such a way that the dye goes on evenly and without spottiness. Mr. Scheffer was a great master of this; he was not content to be able to dye something by the yard, but must practice it to all possible perfection with whole pieces of cloth in his laboratory. This knack cannot be described simply, but must be learned by much practice.

§ 386. Dyed goods is altered variously by acids and alkalis.

(a) Scarlet red is made bluish by alkali, but is restored with dilute acid.

Note 1. This and similar knowledge of reactions show very basicly how stains should be removed. That a color may resist weak acids, such as vinegar, citron juice, and wine, is a special advantage, since the cloth may be exposed to such substances, and also now and then to urine.

(b) Rose red, dyed with safflower, yellows with strong alkali, gets blue with black, and is restored by citric acid.

(c) Artificial or Fernbock red is made blue by alkali, yellowed and made colorless by acid.

(d) Genuine blue is changed by either alkali or acid, but artificial [blue] is made red by acid.

Note 2. The redness of the artificial [blue] is the color belonging to blauholz, which is stripped away by acid.

(e) Green is made blue by acid and restored by alkali volatile.

(f) Genuine black is not changed by alkali, but is made blue by acid.

Artificial [black] is reddened by acid and made brown by alkali.

Note 3. The acid dissolves it with iron combined with astringents, and then the base is stripped away.

(g) Brown is reddened by acid, and restored by alkali, but becomes yellow spotted.

(h) Genuine gray is darkened by acid of vitriol, diluted with double its amount of water, but artifical [gray] reddens.

(i) Acid of saltpeter destroys all dyes, but then they can be restored.

Note 4. This should particularly be noted concerning vegetable dyes, since Berlin blue [a non-vegetable dye] resists this.

Note 5. To investigate whether or not a dye is genuine requires a careful test, to expose the dyed goods in open sunshine for many weeks, and now and then to moisten same with water; woolen [dyes] which are subsequently found to be unchanged, are judged in this way to be genuine. But since this test takes time, and only works in the summer, other tests, which are possible in haste, have been proposed by Mr. DU FAY, but only for woolen [goods], goods for which we are anxious to be certain about because they cost more, wear longer, and are of more general use than vegetable goods, about which not much is known up to now how to judge genuine dyes from those which are weak; [genuine dyes for vegetable goods] are desirable, and on the other hand should not only be resistant to sunshine, but also towards washing with soap; silk on the other hand is in a certain way an abundant commodity, needing also the highest sheen, which artificial dyestuffs often produce in greater fullness than genuine [dyestuffs]; indeed certain variations cannot be restored with the last named up to now, and consequently it is not always possible yet to control these goods, and in certain respects less useful.

As it is now very common for a purchaser to determine by himself the goodness of a dye, the Government has in many instances made itself the guardian in this respect and set up certain controls, by which the purchase of anything can be made safe and deception be avoided, and this is most important in particular with regard to woolen goods.

In France the test which dyed wool should withstand is the following, and is called de-bouilli.

(I) One quintin of goods is boiled 5 minutes with 1 pound of water, in which is dissolved 1 lod of Roman alum; [the goods] is then rinsed well when cooled; the alum is put in cold water , and when this boils completely, the goods. Here beong [dyes] crimson, scarlet, gray of flax [gris de lin], violet, and blue, with all their variations. By this test genuine crimson is blued a little, but artificial becomes flesh-colored or quite white; similarly with all scarlet; artificial violet, which is dyed on genuine blue, loses only its red, but otherwise almost all color; genuine gray of flax disappears somewhat, but less than the artificial; genuine purple, slate-gray and blue are changed a little, but artificial loses the greater part [of its color].

(II) <u>One quintin of goods is boiled 5 minutes in 1 pound of water, in which is dissolved 1/2 lod of hard soap [tvål]</u>. Here belong [dyes] yellow, green, red-madder, cinnamon- and tobacco-brown. Yellow with Grains d´Avignon, turmeric, Brazilian yellow, saffron, and orleana, is lost almost completely, but not with ängskar, genista, wold, foenum graecum, and so-called yellow- or citron-wood; artficial green becomes either colorless or blue; artificial madder-red becomes full in proportion afterwards, as does Brazilian [yellow], but the genuine becomes more attractive by boiling; genuine cinnamon- and tobacco-brown do not change, but [change] very much if made with Brazilian [yellow], orleana, etc.

An alum test fails here because it does not react with Brazilian [yellow] or orleana, but takes out a part of ängskar and genista.

(III) <u>One quintin of goods is boiled 5 minutes in 1 pound of water, in which is dissolved 1 lod of tartar.</u> Here belong so-called red-dyes, which are made with nuts, nut-wood root, alder bark, sumac, sandalwood and soot; the last two make the goods stiff and are not fully genuine; consequently the test shows [this] by strong weakening [of the dye] if it is put in in excess.

(IV) <u>One quintin of goods is boiled 15 minutes in 1 pound of water in which is dissolved 2 lod of Roman alum and 2 of red tartar.</u> Genuine black becomes blue in this test, but artificial [black] becomes gray. Although the decoction appears colored it is not always therefore the test for artificial, since gallaplle and vitriol is used often for browning, and does not come out, but it should still show the right base. Gray with gall and vitriol is always genuine and is not made otherwise.

Note 6. The art of dyeing is still a long way from complete, since it is not enough to get an attractive result, but the dyes should also be genuine, and such as can be used on all kinds of goods. Mr. Dir. SCHEFFER, in 1748, published for the States of the Kingdom, Manufacturers' Account a report on dyeing in the Kingdom, in which he has assured for himself the discovery of: (1) how to put genuine yellow on both wool and silk, using different Swedish substances; (2) similarly with genuine black; (3) that one whole barrel of wold does not yield more than 3/4 mark of indigo, but costs as much as 7-8 marks of the latter, and consequently is 9-10 times more expensive than the value of indigo, for which he knew Swedish plants which not only aroused more vigorous fermentation but, as the dyer says, sets the indigo in operation, [more] than wold and madder, which latter darkens somewhat and also is less permanent, so that the dye is never lost, which the dyer is accustomed to call gå durch [goes through, or penetrates]; that he found the way for safflower red to resist the sun; that of foreign dyestuffs only the fol- lowing were required: namely cochenille, madder, safflower, indigo, and further, for printing cal- ico, some blauholz, and French berries (grains d'Avignon). Mr. Scheffer then displayed a whole piece of cloth which he had dyed a genuine mineral-green, and he dyed cotton yarn a beautiful genuine red; but the most of these and more skills were never disclosed by him, and are still today unknown to our dyers.

To discover if a plant contains some useful dye requires now and then only boiling with the goods in enough water, but seldom or never is any noteworthy dye obtained in this manner; but we get from this no conclusion about its usefulness, unless we try adding alum, common salt, salmiak, potash, etc. in the investigation, salts which often reveal a hidden dyestuff. If a decoction of plants is diluted by half, filtered, and then mixed with various salts, we quickly get a large amount of information about its properties, through the changes which reveal themselves. Mr. HELLOT explains the durability of genuine dyes in this way, that the main cause seems to depend on certain salts to purify and open the pores of the goods so that the dyestuff can properly penetrate; and during cooling not only through contraction of the pores, but also through the process of removal of the salts, [the dyestuff] remains completely fixed. The most useful salts for this are alum and tartar. The former [alum] which attaches during soaking, is decomposed by the dyestuff, and so leaves in the pores a fine white base for the coloring particles to attach to, but this acid again makes with alkali vegetabile in tartar a tartarus vitriolatus, forms a crust which is not easily dissolved and does not darken or come off through efflorescence; but the dyestuff, because it is well conserved, stays the same. Tartar is similarly a difficultly soluble salt, which at the same time remains clear. Both have free acid which is useful in soaking. This resists very well and no doubt has by chance a certain place; but much durability of dyes cannot be explained in this way, and must be attributed either to the nature of the dyestuff, or other circumstances. {For example], here we have Brazilian yellow, turmeric, etc., which are fleeting with tartar and alum. That linen has a less spongy texture than animal goods is fairly evident, but that this is the only difficulty attributable to attaching the dye to it has not yet been determined with full certainty.

White is not properly a separate color, but a mixture of all; a body which reflects all rays is called white. Woolen goods is seldom white when used, but should be [white] when bright colors are put upon it; linen and also silk gain much value through whiteness, and are given 5 degrees [of whiteness] such as: Chinese White, Indian White, desirable or milk-white, silver-white, and blue-white.

Note 7. Printing on cotton and linen is a special art, halfway between dyeing and painting. Here we cannot blame the dye when it is pressed on by forms, when it shows itself finally, after boiling out in madder solution, washing and bleaching, to be quite different. Except for yellow and blue, most printed dyes are nothing else but certain solutions which not only affect how the madder solution goes on, but also how it produces different desired colors. So iron, principally dissolved in vegetable acids and mixed with blue vitriol or Spanish green, produces a material to print with which becomes violet through boiling in madder solution, and in different ways can be altered in proportions, addition of cochenille, kermes, etc.; the aforementioned iron solution, with gallapple or other astringents, becomes black in the decoction; for red we use soda salt, alum, borax, sal gemmae, saltpeter, salmiak, mercurius sublimatus corrosivus, mercurius praecipitatus albus, Spanish green, sugar of lead, white-, yellow-, or red-arsenic, liver of sulfur, and other substances probably unnecessary for the most part; soaking also now and then produces a regular dye, which only goes to show how smoothly printing goes. For yellow we generally use French berry boiled in vinegar and mixed with blue vitriol; orleana; turmeric; blue dyer's vat installed in various ways. All print dyes are mixed with gum or white starch, and are then called paste [papp]. When the goods is boiled with madder the dye resists well overall, but that which has not been soaked loosens again during ordinary usage and bleaching, and becomes entirely white. Now and then paste is also used which covers certain places and prevents the dye solution from going on, which then becomes either white or is painted with a pencil.

Note 8. A painter's preparation of dyes, if it is brought to completion, can seldom be renewed with the help of chemistry. Genuine Carmine is obtained with a solution of cochenille in water, which previously has been tinged with a permanent yellow dye, and finally precipitated wwith Roman alum; Florentine-lake is obtained when cochenille is boiled with fire-resistant pure alkali, after which alum solution is poured into the clear decoction, when fine earth takes the dye to itself during precipitation; a beautiful bright red and permanent lake is obtained, according to Mr. MARGGRAF'S account, from good madder dye in alum solution, which is precipitated with alkali, and can be prepared in similar manifold ways with general tinted juices, which still ought to be permanent; the costly Ultramarine is a fine powder of azure-stone, which by certain processing with fatty substances gets its high gloss; Neapolitan yellow is prepared in fire from lead and antimonium crudum, and also from white lead, antimonium diaphoreticum, salmiak, and alum; washable dyes should be transparent, but for pastel-paints dry powder is required, which with skill is brought to a reasonably hard chalk.

Inks of many colors can be prepared in various ways from vegetable juices mixed with alum, but to a large part are subject to change with time; from Berlin Blue ground with acid of saltpeter, which is then saturated with alkali, we get a beautiful blue [ink]; from a decoction of light blossoms [letblomsters] with cochenille [we get] a bright red; also with cinnobar finely ground and tempered with gun-water; from carmine, etc.; from purified Spanish green, ground with distilled vinegar, comes green [ink]; from the stamen of leis germanica is obtained with alum and gum-water, after fermentation, a juice-dye called Verd d´Iris, which can also be prepared with blue bells (campanula); yellow [ink] can be made from gutta gum, etc.

About cinnobar, minium, realgar, orpiment, massicot, turpentum minerale, musiv-gold, Berlin Blue, Spanish green, and more, mention was made previously.

The preparation of various products requires knowledge of the proper solvent for the substance and the exact procedure, if it is to be obtained in all perfection, consistent with its purpose, which includes not only its luster and enhanced color, by means of its transparent coverage, but also the preservation of the work itself; indeed Dr. Glaser´s product will, according to many tests performed, preserve from the fire itself which otherwise spares so few materials.

* * *

ADDITIONS BY TORBERN BERGMAN

Copies with related tables and Titlepage in copper etching, cost 13 Dal. 16 ore Kopp:mt.

Additions.

[Added by Torbern Bergman.]

Page 14, § 13, at the end of the Notes, is to be added: "A dark acid of vitriol can, by addition of strong acid of saltpeter, indeed simply through [addition of] a little saltpeter, be made clear and colorless; but this expedient, which is often misused, contaminates the acid of vitriol, in the former case with acidum nitri, and in the latter [case] similarly with tartarus vitriolatus. Acid of saltpeter attracts the combustible [phlogiston] from the vitriolic [acid]."

Page 22 [§ 20 (d)], at the end of Note 3, is added: "Fumes from phlogisticated acid of saltpeter provide a convenient means to measure the fitness of the air for respiration, since the good [air] reddens with it and the volume decreases, and the better it is the more this phenomenon shows itself. Two parts good air mixed with 1 [part] of the fumes mentioned, which should make a sum of 3 for both, [instead] usually comes to 1-2/3."

Page 71 [§51, Note], under imperfecti duplices, should appear: "Arsenic-acid in excess with alk. veget. yields alkali vegetabile arsenicatum (alkali minerale and [alkali] volatile can be brought to crystallization with this acid without excess of acid.)"

Same page, under perfecti triplices in the first row, should be added concerning Sal Seignetti: "(requires still for good crystallization some excess of alkali.)"

Same page, after Sal microcosmicus in the second row, is added: "Borax tartarisatus is a 4-fold notable neutral salt, with or without excess of acid or alkali in proportion."

Page 119, at the end of the text of § 94, is added: "Silver solution can be used for cold silvering. Just salmiak and tartar are ground well together, then worked in together with [silver] solution into a dough, formed into a ball or other shape as you please, which is dried; when something is to be silvered it is cleaned off with aqua fortis, rubbed with the ball previously dipped in water, and finally dried quickly over strong heat."

Page 120 [§ 94, Note], row 5, in place of "acid of salt" put "acid of saltpeter".

Page 149, § 142, at the end of the Notes, is added: "Also bismuth, tin, and gold are dissolved, which is noted readily when we pour in too much on precipitation with alkali volatile."

Page 151 [§ 143], at the end of the text, just before the Notes, is added: "We can, with linen rags saturated with gold solution, dried and burned brittle black (not to gray ash) gild [metal objects] in the cold; first the metal is made quite bright with aqua fortis, after which im- mediately a gold rag is dipped in spirits of turpentine or water and then in the brittle material, and [the metal] is rubbed with this until the gilding is heavy enough, then finally it is dried quickly over strong heat, since otherwise the acid used can cause damage."

Page 168 [§ 168 (d), Note, 2nd par.], row 19, put: "about 15" where it reads: "at most 3".

Page 199, § 188, at the end of Note 2, is added: "Mr. P. J. HJELM has also recently informed me that he believes he has found the earth of heavy spar [barium] combined with acid of salt in various waters in Vestergothland."

Page 201, § 189, at the end of Note 3, is added: "Mr. HJELM has for some time since informed me that he found brunsten [manganese] combined with acid of salt in various waters around Lake Venern."

Page 364, § 311, at the end of the Note, is added: "This is still properly true concern- ing the ordinary regulus, but after complete purification from cobalt it is mixed easily with silver and zinc."

Page 428 [§ 375 (b)], Note 1, Row 6, put: "against the sun", instead of: "not well against the sun". At the end of the same Note is added: "I am consequently very uncertain whether Mr. SCHEFFER by Jollster meant Salix pentandra, since he could not classify as a genuine dye that which the sun bleaches out for the most part in a few weeks."

* *

Other books by the same author:

Fragments from the Stream of Time (1968).
Preparing the Way (1970).
Torbern Bergman, A Man Before His Time (1985).
Springville Summers (1987).
Ariadne´s Thread; Essays in History and Philosophy (1988)
Of Birds and Woodcutting (1991)
Search For The Mountain (In preparation 1991)

Monographs:

Vicinal Water (1987).
Ion Exchange Dating of Arid Zone Sediments (1988).

Co-author:

Vacquero To Dominie (with Alfonso Esquibel) (1978).
Foundations of Physical Chemistry (with M. Venugopalan) (1981)
Topics in Physical Chemistry (with M. Venugopalan) (1984).

Translations:

Bergman´s Dissertation on Elective Attractions (from Latin) (1968).
Juan José D´Elhuyar, Discoverer of Tungsten (from Spanish) (1981).

Contributing author:

Aridity and Man (1963).
Teaching the History of Chemistry (1971).
From Sun Dagger to Space Exploration (1986).

Chemists and Chemistry

1. M. Morselli: *Amedeo Avogadro* [1776–1856]. A Scientific Biography. 1984 ISBN 90-277-1624-2
2. F.W.J. McCosh: *Boussingault* [1802–1887]. Chemist and Agriculturist. 1984 ISBN 90-277-1682-X
3. D.A. Stansfield: *Thomas Beddoes, MD (1760–1808).* Chemists, Physician Democrat. 1984 ISBN 90-277-1686-2
4. Pieter Eduard Verkade (1891-1979): *A History of the Nomenclature of Organic Chemistry.* Edited by F.C. Alderweireldt, H.J.T. Bos, L. Maat, P.J. Slootmaekers and B.M. Wepster. 1985 ISBN 90-277-1643-9
5. A. Thackray, J.L. Sturchio, P.T. Carroll and R. Bud: *Chemistry in America, 1876–1976.* Historical Indicators. 1985 ISBN 90-277-1720-6; Pb 90-277-2662-0
6. G.B. Kauffman (ed.): *Frederick Soddy (1877–1956).* Early Pioneer in Radiochemistry. 1986 ISBN 90-277-1926-8
7. R.B. Seymour and Tai Cheng (eds.): *History of Polyolefins.* The World's Most Widely Used Polymers. 1986 ISBN 90-277-2128-9
8. J.T. Stock and M.V. Orna, OSU (eds.): *The History and Preservation of Chemical Instrumentation.* 1986 ISBN 90-277-2269-2
9. N.A. Peppas (ed.): *One Hundred Years of Chemical Engineering.* From Lewis M. Norton (MIT, 1888) to Present. 1989 ISBN 0-7923-0145-8
10. R.B. Seymour (ed.): *Pioneers in Polymer Science.* 1989 ISBN 0-7923-0300-8
11. H. Benninga: *A History of Lactic Acid Making.* A Chapter in the History of Biotechnology. 1990 ISBN 0-7923-0625-2
12. R. Mierzecki: *The Historical Development of Chemical Concepts.* 1991 ISBN 0-7923-0915-4
13. S. Ross: *Nineteenth-Century Attitudes: Men of Science.* 1991 ISBN 0-7923-1308-9
14. Torbern Bergman (1775) (ed.): *Chemical Lectures of H.T. Scheffer.* Translated by J.A. Schufle. 1992 ISBN 0-7923-1760-2

KLUWER ACADEMIC PUBLISHERS – DORDRECHT / BOSTON / LONDON

MIX
Papier aus verantwortungsvollen Quellen
Paper from responsible sources
FSC® C105338

If you have any concerns about our products,
you can contact us on
ProductSafety@springernature.com

In case Publisher is established outside the EU,
the EU authorized representative is:
Springer Nature Customer Service Center GmbH
Europaplatz 3, 69115 Heidelberg, Germany

Printed by Libri Plureos GmbH
in Hamburg, Germany